# Temperature Trends in the Lower Atmosphere

## *Steps for Understanding and Reconciling Differences*

Synthesis and Assessment Product 1.1
Report by the U.S. Climate Change Science Program
and the Subcommittee on Global Change Research

EDITED BY:
Thomas R. Karl, Susan J. Hassol,
Christopher D. Miller, and William L. Murray

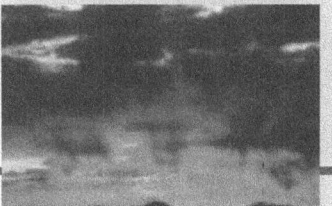

April 2006

Members of Congress:

We are pleased to transmit to you this report, *Temperature Trends in the Lower Atmosphere: Steps for Understanding and Reconciling Differences*, the first of a series of Synthesis and Assessment Products produced by the U.S. Climate Change Science Program (CCSP). This series of 21 reports is aimed at providing current evaluations of climate change science to inform public debate, policy, and operational decisions. These reports are also intended to help inform CCSP's consideration of future program priorities.

CCSP's guiding vision is to empower the Nation and the global community with the science-based knowledge to manage the risks and opportunities of change in the climate and related environmental systems. The Synthesis and Assessment Products are important steps toward that vision, helping translate CCSP's extensive observational and research base into informational tools that directly address key questions that are being asked of the research community.

This first Synthesis and Assessment Product addresses previously identified discrepancies between observations and simulations of surface and atmospheric temperature trends. It was developed with broad scientific input and reviewed by the National Research Council. Public comments were solicited and carefully reviewed at multiple stages in the process. It was prepared in accordance with the Information Quality Act and the Federal Advisory Committee Act. Further information on the process for preparing Synthesis and Assessment products and the CCSP itself can be found at www.climatescience.gov.

We commend the report's authors for both the thorough nature of their work and their adherence to an inclusive review process. This product sets a high standard for quality for subsequent Synthesis and Assessment Products.

Carlos M. Gutierrez
Secretary of Commerce
Chair, Committee on Climate Change
Science and Technology Integration

Samuel W. Bodman
Secretary of Energy
Vice Chair, Committee on Climate
Change Science and Technology
Integration

John H. Marburger III
Director, Office of Science and
Technology Policy
Executive Director, Committee
on Climate Change Science and
Technology Integration

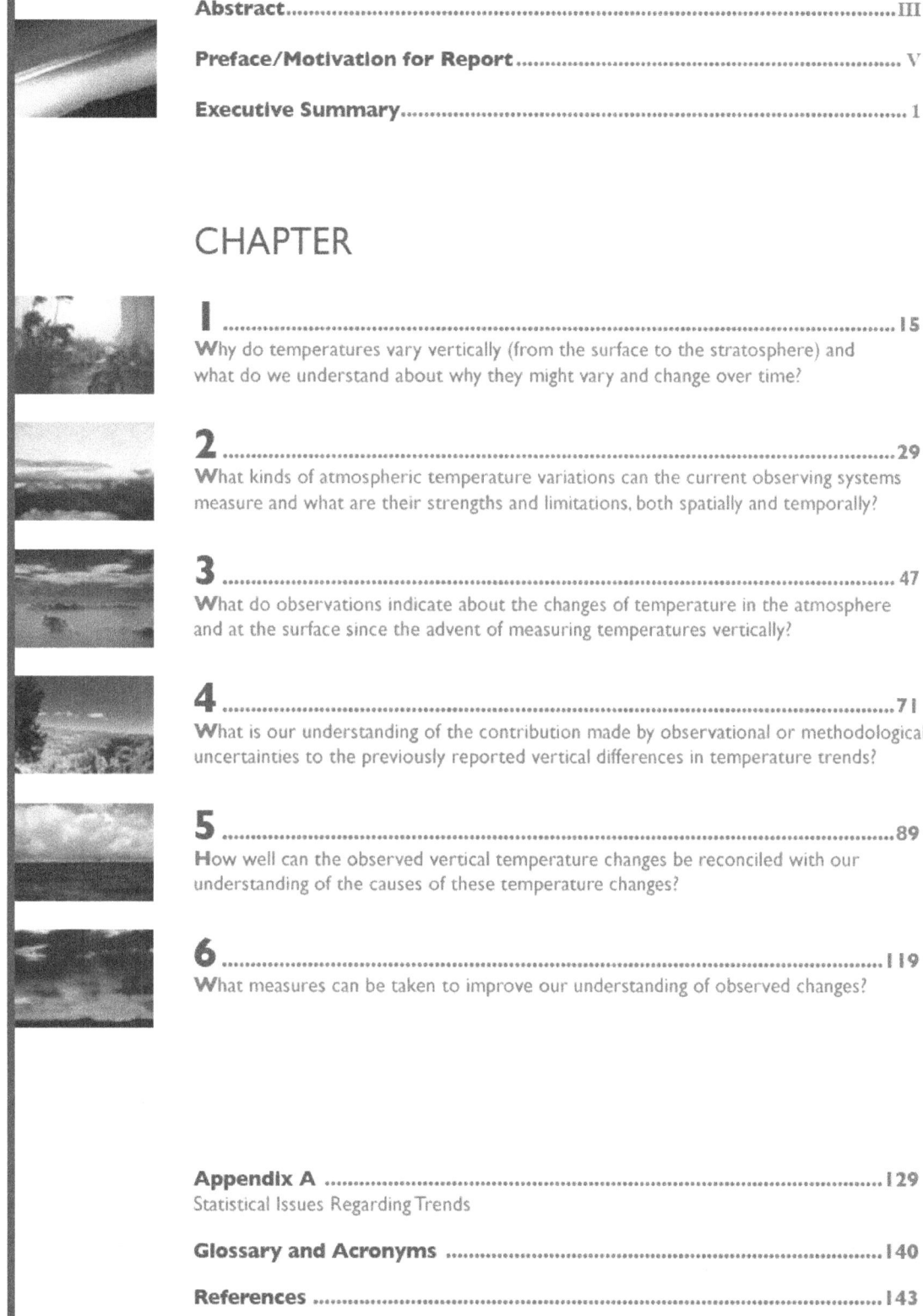

TABLE OF CONTENTS

## CHAPTER

Why do temperatures vary vertically (from the surface to the stratosphere) and what do we understand about why they might vary and change over time?

What kinds of atmospheric temperature variations can the current observing systems measure and what are their strengths and limitations, both spatially and temporally?

What do observations indicate about the changes of temperature in the atmosphere and at the surface since the advent of measuring temperatures vertically?

What is our understanding of the contribution made by observational or methodological uncertainties to the previously reported vertical differences in temperature trends?

How well can the observed vertical temperature changes be reconciled with our understanding of the causes of these temperature changes?

What measures can be taken to improve our understanding of observed changes?

# AUTHOR TEAM FOR THIS REPORT

Preface            Thomas R. Karl, NOAA; Christopher D. Miller, NOAA; William L. Murray, STG, Inc.

Executive Summary  **Convening Lead Author:** Tom M. L. Wigley, NSF NCAR
                   **Lead Authors:** V. Ramaswamy, NOAA; John R. Christy, Univ. of AL in Huntsville;
                   John R. Lanzante, NOAA; Carl A. Mears, Remote Sensing Systems; Benjamin D. Santer,
                   DOE LLNL; Chris K. Folland, U.K. Met. Office

Chapter 1          **Convening Lead Author:** V. Ramaswamy, NOAA
                   **Lead Authors:** James W. Hurrell, NSF NCAR; Gerald A. Meehl, NSF NCAR
                   **Contributing Authors:** Adam Phillips, NSF NCAR; Benjamin D. Santer, DOE LLNL;
                   M. Daniel Schwarzkopf, NOAA; Dian J. Seidel, NOAA; Steven C. Sherwood, Yale Univ.;
                   Peter W. Thorne, U.K. Met. Office

Chapter 2          **Convening Lead Author:** John R. Christy, Univ. of AL in Huntsville
                   **Lead Authors:** Dian J. Seidel, NOAA; Steven C. Sherwood, Yale Univ.
                   **Contributing Authors:** Ming Cai, FL State Univ.; Eugenia E. Kalnay, Univ. of MD;
                   Chris K. Folland, U.K. Met. Office; Carl A. Mears, Remote Sensing Systems;
                   Peter W. Thorne, U.K. Met. Office; John R. Lanzante, NOAA

Chapter 3          **Convening Lead Author:** John R. Lanzante, NOAA
                   **Lead Authors:** Thomas C. Peterson, NOAA; Frank J. Wentz, Remote Sensing Systems;
                   Konstantin Y. Vinnikov, Univ. of MD
                   **Contributing Authors:** Dian J. Seidel, NOAA; Carl A. Mears, Remote Sensing Systems;
                   John Christy, Univ. of AL in Huntsville; Chris E. Forest, MIT; Russell S. Vose, NOAA;
                   Peter W. Thorne, U. K. Met. Office; Norman C. Grody, NOAA

Chapter 4          **Convening Lead Author:** Carl A. Mears, Remote Sensing Systems
                   **Lead Authors:** Chris E. Forest, MIT; Roy W. Spencer, Univ. of AL in Huntsville;
                   Russell S. Vose, NOAA; Richard W. Reynolds, NOAA
                   **Contributing Authors:** Peter W. Thorne, U.K. Met. Office; John R. Christy,
                   Univ. of AL in Huntsville

Chapter 5          **Convening Lead Author:** Benjamin D. Santer, DOE LLNL
                   **Lead Author:** Joyce E. Penner, Univ. of MI; Peter W. Thorne, U.K. Met. Office
                   **Contributing Authors:** William D. Collins, NCAR; Keith W. Dixon, NOAA; Thomas L.
                   Delworth, NOAA; Charles Doutriaux, DOE LLNL; Chris K. Folland, U.K. Met. Office;
                   Chris E. Forest, MIT; James E. Hansen, NASA; John R. Lanzante, NOAA; Gerald A.
                   Meehl, NSF NCAR; V. Ramaswamy, NOAA; Dian J. Seidel, NOAA; Michael F. Wehner,
                   DOE LBNL; Tom M.L. Wigley, NSF NCAR

Chapter 6          **Convening Lead Author:** Chris K. Folland, U.K. Met. Office
                   **Lead Authors:** David E. Parker, U.K. Met. Office; Richard W. Reynolds, NOAA;
                   Steven C. Sherwood, Yale Univ.; Peter W. Thorne, U.K. Met. Office

Appendix A         Tom M.L. Wigley, NSF NCAR
                   **With contributions by:** Benjamin D. Santer, DOE LLNL; John R. Lanzante, NOAA

ABSTRACT

**P**reviously reported discrepancies between the amount of warming near the surface and higher in the atmosphere have been used to challenge the reliability of climate models and the reality of human-induced global warming. Specifically, surface data showed substantial global-average warming, while early versions of satellite and radiosonde data showed little or no warming above the surface. This significant discrepancy no longer exists because errors in the satellite and radiosonde data have been identified and corrected. New data sets have also been developed that do not show such discrepancies.

**T**his Synthesis and Assessment Product is an important revision to the conclusions of earlier reports from the U.S. National Research Council and the Intergovernmental Panel on Climate Change. For recent decades, all current atmospheric data sets now show global-average warming that is similar to the surface warming. While these data are consistent with the results from climate models at the global scale, discrepancies in the tropics remain to be resolved. Nevertheless, the most recent observational and model evidence has increased confidence in our understanding of observed climatic changes and their causes.

# Report Motivation and Guidance for Using this Synthesis/Assessment Report

**Authors:**

Thomas R. Karl, NOAA; Christopher D. Miller, NOAA; William L. Murray, STG, Inc.

A primary objective of the U. S. Climate Change Science Program (CCSP) is to provide the best possible scientific information to support public discussion and government and private sector decision-making on key climate-related issues. To help meet this objective, the CCSP has identified an initial set of 21 synthesis and assessment products that address its highest priority research, observation, and decision-support needs. This Synthesis/Assessment Report, the first of the 21 Reports, focuses on understanding the causes of the reported differences between independently produced data sets of atmospheric temperature trends from the surface through the troposphere to the lower stratosphere.

This topic is relevant to policy-makers because previous discrepancies between surface and tropospheric temperature observations challenged the correctness of climate model simulations and the reality of greenhouse gas-induced global warming. As described in the Executive Summary, considerable progress has been made in resolving many of these earlier discrepancies.

## Background

Measurements of global surface air temperature show substantial increases over the past several decades. In the early 1990s, data from the National Oceanic and Atmospheric Administration's (NOAA's) polar orbiting satellites were analyzed for multi-decadal trends. These initial analyses indicated that global-mean temperatures in the troposphere showed little or no increase, in contrast with surface air measurements from ships, land-based weather stations, and ocean buoys. This result led some to question the reality and/or the cause of reported global-mean surface temperature increases, on the basis that human influences, thought to be important contributors to observed change, were expected to increase temperatures both at the surface and in the troposphere, with the largest increases expected in the tropical troposphere. This led to an intensive effort by climate scientists to better understand

the causes of the apparent differences in the reported rates of temperature changes between the surface and the troposphere.

Scientists analyzing the data knew that there were complex and unresolved issues related to inadequacies of observing systems that could lead to misleading impressions or misinterpretation of the data. There were also uncertainties in our understanding of how the climate might respond to various forcings, as is often assessed through the use of climate models. In an attempt to resolve these issues, in 2000 the National Research Council (NRC) specifically addressed the issue of temperature trends in the troposphere and at the surface. In its Report, the NRC concluded that "the warming trend in global-mean surface temperature observations during the past 20 years is undoubtedly real and is substantially greater than the average rate of warming during the twentieth century. The disparity between surface and upper air trends in no way invalidates the conclusion that surface temperature has been rising." The NRC further found that corrections in the Microwave Sounding Unit (MSU) processing algorithms brought the satellite data record into slightly closer alignment with surface temperature trends. They concluded that the substantial disparity that remained probably reflected a less rapid warming of the troposphere than the surface in recent decades due to both natural and human-induced causes.

In 2001, the Intergovernmental Panel on Climate Change (IPCC) Third Assessment Report devoted additional attention to new analyses of the satellite, weather balloon, and surface data to evaluate the difference in temperature trends between the surface and the troposphere. Similar to the NRC, the IPCC concluded that it was very likely that the surface temperature increases were larger and differed significantly from temperature increases higher in the troposphere. They concluded, "during the past two decades, the surface, most of the troposphere, and the stratosphere have responded differently to climate forcings because different physical processes have dominated in each of the regions during that time." (IPCC, Climate Change 2001: The Scientific Basis, Chapter 2, p. 122-123; Cambridge University Press).

**Focus of this Synthesis/Assessment Report**

The efforts of the NRC and IPCC to address uncertainties about the temperature structure of the lower atmosphere (*i.e.,* from the surface through the lower stratosphere) have helped move us closer to a comprehensive understanding of observed trends of temperature. Although these documents provided a great deal of useful information, full resolution of the issue was hampered by the complexities of the climate system coupled with shortcomings of the available observing systems. To more fully address remaining fundamental questions, a broader examination has been undertaken here to answer the following questions:

1) Why do temperatures vary vertically (from the surface to the stratosphere) and what do we understand about why they might vary and change over time?

2) What kinds of atmospheric temperature variations can the current observing systems measure and what are their strengths and limitations, both spatially and temporally?

3) What do observations indicate about the changes of temperature in the atmosphere and at the surface since the advent of measuring temperatures vertically?

4) What is our understanding of the contribution made by observational or methodological uncertainties to the previously reported vertical differences in temperature trends?

5) How well can the observed vertical temperature changes be reconciled with our understanding of the causes of these changes?

6) What measures can be taken to improve the understanding of observed changes?

These questions provide the basis for the six main chapters in this Synthesis/Assessment Report (the chapter numbers correspond to the question numbers above). They highlight several of the fundamental uncertainties and differences between and within the individual components of the existing observational and modeling systems. The responses to the questions are written in a style consistent with major authoritative international scientific assessments (*e.g.,* IPCC assessments, and the Global Ozone Research and Monitoring Project of the World Meteorological Organiza-

tion [WMO]). The Executive Summary, which presents the key findings from the main body of the Report, is intended to be useful for those involved with policy-related global climate change issues. The Chapters supporting the Executive Summary are written at a more technical level suitable for non-climate specialists within the scientific community and well-informed lay audiences.

The Synthesis/Assessment Report is structured so as to compartmentalize, as much as possible, the answers to each of the six questions (above). However, given the interconnected nature of the questions, this is not entirely possible, or desirable. Occasionally topics extraneous to a particular chapter are mentioned in passing to make an important point or alert the reader to some issue(s) covered elsewhere in the report. However, as a general rule, in the interest of brevity this report does not always explicitly refer the reader to another chapter. The reader is advised to keep this in mind and refer to Table 1 (next page.) for guidance on locating the discussion of particular issues.

To help answer the questions posed, climate model simulations of temperature change based on time histories of important forcing factors have been compared with observed temperature changes. It is recognized that in a system containing internally generated variations, it is unrealistic to expect models to exactly replicate observed changes. If the ensemble of simulations replicates important aspects of the observed temperature changes (*e.g.,* global mean, tropical mean) this increases confidence in our understanding of the observed temperature record and reduces uncertainties about projected changes. If not, then this implies that the time histories of the important forcings are not adequately known, all of the important forcings are not included, the processes being simulated in the models have flaws, the observational record is incorrect, or some combination of these factors is present.

This CCSP Synthesis/Assessment Report assesses the uncertainties associated with the data used to determine changes of temperature, and whether such changes are consistent with our understanding of climate processes. This requires a detailed comparison of observations and climate models used to simulate observed changes, including an appreciation of why temperatures might respond differently at the surface compared to various layers higher in the atmosphere.

This CCSP Report also addresses the accuracy and consistency of the temperature records and outlines steps necessary to reconcile differences between individual data sets. Understanding exactly how and why there are differences in temperature trends reported by several analysis teams using different observation systems and analysis methods is a nec-

**Table I. Guide to readers to identify Chapter emphasis. The Executive Summary ties together all these aspects of the Synthesis/Assessment Report.**

| Report Section | Observations | Observational Uncertainty | Processes | Models | Comparing Model Simulations & Observations | Statistical Analysis |
|---|---|---|---|---|---|---|
| Chapter 1 | secondary | | primary | | | |
| Chapter 2 | | primary | | | | |
| Chapter 3 | primary | | | | | |
| Chapter 4 | | primary | | | | |
| Chapter 5 | | | secondary | primary | primary | secondary |
| Chapter 6 | | primary | | | secondary | |
| Appendix | secondary | | | | | primary |

essary step in resolving previously identified discrepancies between observations and model simulations.

## New observations and analyses since the IPCC and NRC Reports

Since the IPCC and NRC assessments, there have been intensive efforts to create new satellite and weather balloon data sets using a range of approaches. Having multiple tropospheric temperature data sets provides the opportunity for much greater understanding of observed changes and their uncertainty than was possible in the previous assessments. In addition, for the first time, a suite of models simulating observed climate since 1979 (when satellite data began) has provided a unique opportunity to inter-compare observed trends from various data sets with model simulations using various scenarios of historical climate forcings. Taken together, these advances lead to a greater understanding of the issues. The process of producing this Report has stimulated additional research and analysis on these topics, and helped to move the science forward.

This Report includes recent analyses of and corrections to data sets that have helped resolve inconsistencies among observational data sets and between observations and models. The science of upper air temperatures is a rapidly evolving field. During the preparation of this Report, new findings were published and are now included in the current draft. For example, a recent article demonstrated an error in the method used in the original satellite data set to correct for diurnal cycle errors due to satellite orbital drift.

When corrected, the data set yielded greater warming in the lower troposphere. Since it was possible for the error to be rectified fairly quickly, a new version of this data set was available for this Report. At the same time, another research team produced its first version of satellite-derived lower tropospheric temperatures, and yet another team updated its tropospheric temperature time series. All these results are included in this Report and are compared to a suite of recent climate model simulations. The authors certainly expect that new data and discoveries that follow the release of this Report will further improve our understanding.

Factors that guided the authors in the selection of the climate records considered extensively in this Report were: (a) publication heritage, (b) public availability, (c) use by the scientific community at-large, (d) updates on a monthly basis, and (e) period of record beginning in 1979 or earlier. The climate records considered in this Report are also global in scope.[1]

---

[1] Most analyses undertaken to date have considered temperature trends at the global scale or large-regional scale (*e.g.*, the tropics). Because this report was charged with assessing the current state of the science, it also necessarily focuses on these large scales. It is at these scales that the apparent discrepancies in temperature trends were first reported. We also currently have most capability in simulating climate at these scales. Until we can reconcile our understanding on the very large scales, little scientific value will be added by considering finer regional details. This does not imply that future analyses should not consider finer regional scales for a complete understanding of relative temperature trends at the surface and in the troposphere.

The three surface analyses that were used have many publications that describe their construction methods. These data sets are readily available and are widely used. Two of the three satellite data sets used, while relatively recent, are based on a heritage of published versions that have incorporated new adjustments as discoveries have been made. Each of these data sets allows ready access to the public and has been used in several research publications. A third, more recently developed data set has been updated during the preparation of this Report. Two data sets used were based on weather balloon data. One of these data sets publicly appeared in 2005, but the authors had made the preliminary versions and methodology available to scientists as early as 2002 and have built upon the extensive experience acquired from previous versions of these data sets. Another data set has a heritage dating back several decades and was recently updated.

The models selected for comparison with observations were those models available to the author team during the course of this assessment. They represent the state-of-the-science from every major global climate modeling center in the world. The model simulations selected include a large fraction of those that were run for the Fourth Assessment Report of the IPCC, due to be published in 2007. The simulations are freely available, and details regarding access to the model data can be obtained from the Program for Climate Model Diagnosis and Intercomparison (http://www.pcmdi.11n1.gov/ipcc/about_ipcc.php). The data used in this report are also openly available and a list of web sites where they can be obtained is included in Chapter 3.

### How to use this Synthesis/Assessment Report

This Report promises to be of significant value to decision-makers, and to the expert scientific and stakeholder communities. Readers of this Report will find that new observations, data sets, analyses, and climate model simulations enabled the Author Team to resolve many of the issues noted by the NRC and the IPCC in their earlier Reports. This

Synthesis/Assessment Report already has had an important impact on the content of the draft to the Fourth Assessment Report of the IPCC, due to be published in 2007.

This Synthesis/Assessment Report exposes the remaining differences among different observing systems and data sets related to recent changes in tropospheric and stratospheric temperature. Discrepancies between the data sets and the models have been reduced and our understanding of observed climate changes and their causes has increased. Given this, there is no longer sufficient evidence to conclude that there exists any notable discrepancy between our understanding of recent global average temperature changes and model simulations of these changes. This represents a change from conclusions of earlier reports (see above) and should constitute a valuable source of information to policymakers.

In addition, we expect the information generated here will be used both nationally and internationally, *e.g.*, by the Global Climate Observing System (GCOS) Atmospheric Observation Panel to help identify effective ways to reduce observational uncertainty. The findings regarding observations and comparisons between models and observations of lower stratospheric temperature trends may also be useful for future WMO/United Nations Environment Programme (UNEP) Ozone Assessments.

Some terms used in the Report may be unfamiliar to those without training in meteorology; a glossary and list of acronyms is included at the end of the Report. In addition, Table 2 on page X defines the terminology used in this Report for the layers of the atmosphere.

To integrate a wide variety of information, this Report also uses a lexicon of terms (See Fig. 1) to express the team's considered judgment about the likelihood of results. Confidence in results is highest at each end of the spectrum. Unless qualified by these expressions of likelihood, all statements are implied to be certain.

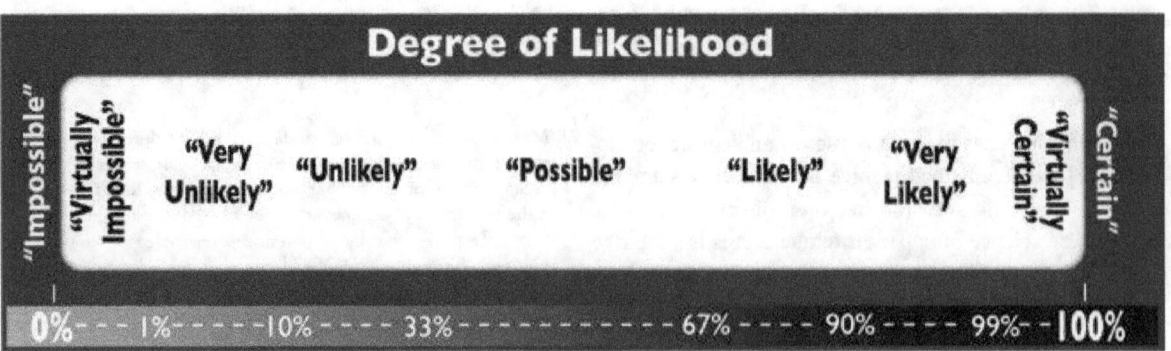

**Figure I.**

**The Synthesis/Assessment Product Team**

A full list of the Author Team (in addition to a list of lead authors provided at the beginning of each Chapter) is provided on page II of this Report. The Author Team Convening Lead Authors (CLAs), Lead Authors (LAs), and Chief Editor were constituted as a Federal Advisory Committee that was charged with advising the CCSP on the scientific and technical content of the Report. Contributing Authors (CAs) provided relevant input used in the development of the report, but CAs who were not also LAs or CLAs did not participate in the Federal Advisory Committee (FAC) committee deliberations upon which this Synthesis and Assessment Product was developed. The remainder of the Editorial Staff reviewed the scientific/technical input and managed the assembly, formatting and preparation of the Report. The focus of this Report follows the Prospectus guidelines developed by the Climate Change Science Program and posted on its website at http://www.climatescience.gov.

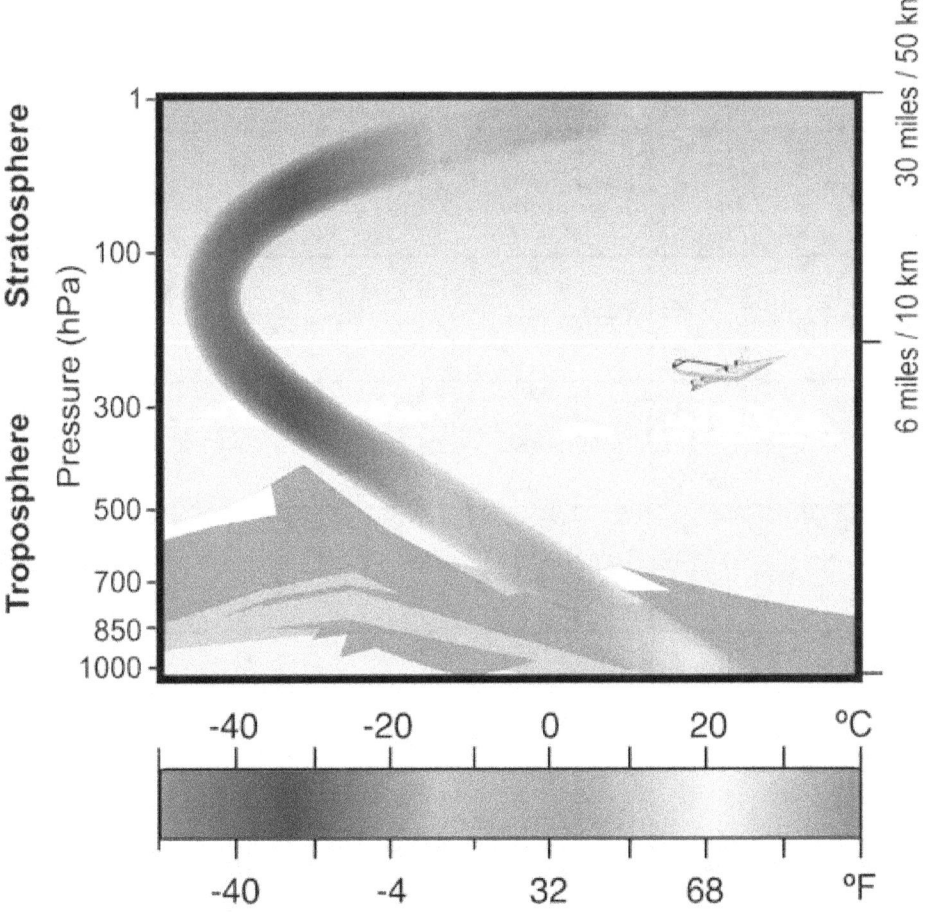

**Figure 2.** The illustration shows the layers of the atmosphere of primary interest to this Synthesis/Assessment Report. The multi-colored line on this diagram indicates the variations in temperature with altitude. The table on the following page defines the terminology used in this Report for the layers of the atmosphere.

**Table 2.** Abbreviated terms: Subscript "S," refers to the Surface. Subscripts "2" and "4" refer to MSU data from channels 2 and 4. Subscript "2LT" refers to a modification of channel 2 data to focus more directly on the °ower Troposphere and reduce the influence of stratospheric temperatures on channel 2 data. Subscripts "850–300" and "100–50" are specific atmospheric layers sampled by radiosondes. Subscript "*G" refers to a combination of channel 2 and channel 4 data derived by Fu and co-workers, applicable to global averages, and "*T" refers to applicable tropical averages. For the model-observation comparisons, the observation-based definitions were used as listed in the Table.

## Terms for Layers of the Atmosphere Used in this Report

| Common Term | Abbreviated Term for the temperature of that layer | Main region of Influence | Approximate altitude. (For satellite products: altitude range of bulk (90%) of layer measured) | Lower and upper pressure level boundaries |
|---|---|---|---|---|
| Surface | $T_S$ | Air: Just above surface <br> Water: Shallow depth | Surface Air: Land: 1.5 m above surface; Ocean: ship deck-height (5 – 25 m) above surface (NMATs). <br> Surface Water: 1 - 10 m depth in ocean (SSTs) | Surface (or ~1000 hPa at sea level) |
| Lower Troposphere | $T_{2LT}$ | Lower to Mid-Troposphere | Sfc – 8 km | Sfc – 350 hPa |
| Troposphere (radiosonde) | $T_{(850-300)}$ | Troposphere | 1.5 – 9 km | 850 – 300 hPa |
| Troposphere (satellite) | $T^*_G$ | Troposphere | Sfc – 13 km | Sfc – 150 hPa |
| Tropical Troposphere (satellite) | $T^*_T$ | Troposphere (tropics only) | Sfc – 16 km | Sfc – 100 hPa |
| Mid Troposphere to Lower Stratosphere | $T_2$ | Mid and Upper Troposphere to Lower Stratosphere[2] | Sfc – 18 km | Sfc – 75 hPa |
| Lower Stratosphere (satellite) | $T_4$ | Lower Stratosphere | 14 – 29 km | 150 – 15 hPa |
| Lower Stratosphere (radiosonde) | $T_{(100-50)}$ | Lower Stratosphere | 17 – 21 km | 100 – 50 hPa |

[2] Only about 10% of this layer extends into the lower stratosphere.

**Convening Lead Author:** Tom M. L. Wigley, NSF NCAR

**Lead Authors:** V. Ramaswamy, NOAA; J.R. Christy, Univ. of AL in Huntsville; J.R. Lanzante, NOAA; C.A. Mears, Remote Sensing Systems; B.D. Santer, DOE LLNL; C.K. Folland, U.K. Met Office

## Abstract

Previously reported discrepancies between the amount of warming near the surface and higher in the atmosphere have been used to challenge the reliability of climate models and the reality of human-induced global warming. Specifically, surface data showed substantial global-average warming, while early versions of satellite and radiosonde data showed little or no warming above the surface. This significant discrepancy no longer exists because errors in the satellite and radiosonde data have been identified and corrected. New data sets have also been developed that do not show such discrepancies.

This Synthesis and Assessment Product is an important revision to the conclusions of earlier reports from the U.S. National Research Council and the Intergovernmental Panel on Climate Change. For recent decades, all current atmospheric data sets now show global-average warming that is similar to the surface warming. While these data are consistent with the results from climate models at the global scale, discrepancies in the tropics remain to be resolved. Nevertheless, the most recent observational and model evidence has increased confidence in our understanding of observed climatic changes and their causes.

## NEW RESULTS AND FINDINGS

This Report is concerned with temperature changes in the atmosphere, differences in these changes at various levels in the atmosphere, and our understanding of the causes of these changes and differences. Considerable progress has been made since the production of reports by the NRC and the IPCC in 2000 and 2001. Data sets for the surface and from satellites and radiosondes (temperature sensors on weather balloons) have been extended and improved, and new satellite and radiosonde data sets have been developed[1]. Many new model simulations of the climate of the 20th century have been carried out using improved climate models[2] and better estimates of past forcing changes, and numerous new and updated comparisons between model and observed data have been performed. The present Report reviews this progress. A summary and explanation of the main results is presented first. Then, to address the issues in more detail, six questions that provide the basis for the six main chapters in this Synthesis and Assessment Report are posed and answered in Sections 1 through 5 below.

### The important new results presented in this Report include:
### Global Average Temperature Results

- For observations since the late 1950s, the start of the study period for this Report, the most recent versions of all available data sets show that both the surface and troposphere have warmed, while the stratosphere has cooled[3]. These changes are in accord with our understanding of the effects of radiative forcing agents[4] and with the results from model simulations.

---

[1]  For details of new observed data see Table 3.1 in Chapter 3.

[2]  For details of new models and model simulations see Chapter 5 and http://www-pcmdi.llnl.gov/ipcc/model.documentation.

[3]  We use the words "warming" and "cooling" here to refer to temperature increases or decreases, as is common usage. Technically, these words refer to changes in heat content, which may occur through changes in either the moisture content and/or the temperature of the atmosphere. When we say that the atmosphere has warmed (or cooled) over a given period, this means that there has been an overall positive (or negative) temperature change based on a linear trend analysis. For more on the use of linear trends, including a discussion of their strengths and weaknesses, see Appendix A.

[4]  The main natural forcing agents are changes in solar output and the effects of explosive volcanic eruptions. The main human-induced ("anthropogenic") factors are: the emissions of greenhouse gases (e.g., carbon dioxide [$CO_2$], methane [$CH_4$], nitrous oxide [$N_2O$]); aerosols (tiny droplets or particles such as smoke) and the gases that lead to aerosol formation (most importantly,

- Since the late 1950s, all radiosonde data sets show that the low and mid troposphere have warmed at a rate slightly faster than the rate of warming at the surface. These changes are in accord with our understanding of the effects of radiative forcing agents on the climate system and with the results from model simulations.

- For observations during the satellite era (1979 onwards), the most recent versions of all available data sets show that both the low and mid troposphere have warmed. The majority of these data sets show warming at the surface that is greater than in the troposphere. Some of these data sets, however, show the opposite - tropospheric warming that is greater than that at the surface. Thus, due to the considerable disagreements between tropospheric data sets, it is not clear whether the troposphere has warmed more than or less than the surface.

- The most recent climate model simulations give a range of results for changes in global-average temperature. Some models show more warming in the troposphere than at the surface, while a slightly smaller number of simulations show the opposite behavior. There is no fundamental inconsistency among these model results and observations at the global scale.

- Studies to detect climate change and attribute its causes using patterns of observed temperature change in space and time show clear evidence of human influences on the climate system (due to changes in greenhouse gases, aerosols, and stratospheric ozone).

- The observed patterns of change over the past 50 years cannot be explained by natural processes alone[5], nor by the effects of short-lived atmospheric constituents (such as aerosols and tropospheric ozone) alone.

**Tropical Temperature Results (20°S to 20°N)**
- Although the majority of observational data sets show more warming at the surface than in the troposphere, some observational data sets show the opposite behavior. Almost all model simulations show more warming in the troposphere than at the surface. This difference between models and observations may arise from errors that are common to all models, from errors in the observational data sets, or from a combination of these factors. The second explanation is favored, but the issue is still open.

sulfur dioxide); and changes in land cover and land use (see Chapter 1, Table 1.1). Since these perturbations act to drive or "force" changes in climate, they are referred to as "forcings". Tropospheric ozone [$O_3$], which is not emitted directly, is also an important greenhouse gas. Tropospheric ozone changes occur through the emissions of gases like carbon monoxide, nitrogen oxides and volatile organic compounds, which, by themselves, are not important directly as greenhouse gases.

5    "Natural processes" here refers to the effects of natural external forcing agents such as volcanic eruptions and solar variability, and/or internally generated variability.

## EXPLANATION OF FINDINGS

These results for the globe and for the tropics characterize important changes in our understanding of the details of temperature changes at the surface and higher in the troposphere. In 2000 and 2001, the NRC and the IPCC both concluded that global-average surface temperature increases were larger and differed significantly from temperature increases in the troposphere. The new and improved observed data sets and new model simulations that have been developed require modifications of these conclusions.

The issue of changes at the surface relative to those in the troposphere is important because larger surface warming (at least in the tropics) would be inconsistent with our physical understanding of the climate system, and with the results from climate models. The concept here is referred to as "vertical amplification" (or, for brevity, simply "amplification"): greater changes in the troposphere would mean that changes there are "amplified" relative to those at the surface.

For global averages, observed changes from 1958 through 2004 exhibit amplification: i.e., they show greater warming trends in the troposphere compared with the surface. Since 1979, however, the situation is different: most data sets show slightly greater warming at the surface.

Whether or not these results are in accord with expectations based on climate models is a complex issue, one that we have been able to address more comprehensively now using new model results. Over the period since 1979, for global-average temperatures, the range of recent model simulations is almost evenly divided among those that show a greater global-average warming trend at the surface and others that show a greater warming trend aloft. The range of model results for global average temperature reflects the influence of the mid- to high-latitudes where amplification results vary considerably between models. Given the range of model results and the overlap between them and the available observations, there is no conflict between observed changes and the results from climate models.

In the tropics, the agreement between models and observations depends on the time scale considered. For month-to-month and year-to-year variations, models and observations both show amplification (*i.e.,* the month-to-month and year-to-year variations are larger aloft than at the surface). This is a consequence of relatively simple physics, the effects of the release of latent heat as air rises and condenses in clouds. The magnitude of this amplification is very similar in models and observations. On decadal and longer time scales, however, while almost all model simulations show greater warming aloft (reflecting the same physical processes that operate on the monthly and annual time scales), most observations show greater warming at the surface.

These results could arise either because "real world" amplification effects on short and long time scales are controlled by different physical mechanisms, and models fail to capture such behavior; or because non-climatic influences remaining in some or all of the observed tropospheric data sets lead to biased long-term trends; or a combination of these factors. The new evidence in this Report favors the second explanation.

# I. HOW DO WE EXPECT VERTICAL TEMPERATURE PROFILES TO CHANGE?

**Why do temperatures vary vertically (from the surface to the stratosphere) and what do we understand about why they might vary and change over time?**

When all forcings are considered, we expect the troposphere to have warmed and the stratosphere to have cooled since the late 1950s.

This question is addressed in both Chapter 1 and Chapter 5 of this Report.

In response to this question, Chapter 1 notes:

(1) TEMPERATURES VARY VERTICALLY

- The global temperature profile of the Earth's atmosphere reflects a balance between radiative, convective and dynamical heating and cooling processes in the surface-atmosphere system. Radiation from the Sun is the source of energy for the Earth's climate. Physical properties of the atmosphere and dynamical processes mix heat vertically and horizontally, yielding the highest temperatures, on average, at the surface, with marked seasonal and spatial variations. In the atmosphere above the surface, the distribution of moisture and the lower air pressure at progressively higher altitudes result in decreasing temperatures with height up to the tropopause (marking the top of the troposphere, *i.e.,* the lower 8 to 16 km of the atmosphere, depending on latitude). Above this, the physical properties of the air produce a warming with height through the stratosphere (extending from the tropopause to ~50 km).

(2) TEMPERATURE TRENDS AT THE SURFACE CAN BE EXPECTED TO BE DIFFERENT FROM TEMPERATURE TRENDS HIGHER IN THE ATMOSPHERE BECAUSE:

- The physical properties of the surface vary substantially according to location and this produces strong horizontal variations in near-surface temperature. Above the surface, on monthly and longer time scales, these contrasts are quickly smoothed out by atmospheric motions so the patterns of change in the troposphere must differ from those at the surface. Temperature trend variations with height must, therefore, vary according to location.
- Changes in atmospheric circulation or modes of atmospheric variability (*e.g.,* the El Niño-Southern Oscillation [ENSO]) can produce different temperature trends at the surface and aloft.
- Under some circumstances, temperatures may increase with height near the surface or higher in the troposphere, producing a "temperature inversion." Such inversions are more common at night over continents, over sea ice and snow in winter, and in the trade wind regions. Since the air in inversion layers is resistant to vertical mixing, temperature trends can differ between inversion layers and adjacent layers.
- Forcing factors, either natural or human-induced, can result in differing temperature trends at different levels in the atmosphere, and these vertical variations may change over time.

---

As noted above, temperatures in the atmosphere vary naturally as a result of internal factors and natural and human-induced perturbations ("forcings"). These factors are expected to have different effects on temperatures near the surface, in the troposphere, and in the stratosphere, as summarized in Table 1. When all forcings are considered, we expect the troposphere to have warmed and the stratosphere to have cooled since the late 1950s (and over the whole 20th century). The relative changes in the troposphere and stratosphere provide information about the causes of observed changes.

Within the troposphere, the relative changes in temperature at different levels are controlled by different processes according to latitude. In the tropics, the primary control is the thermodynamics of moist air *(i.e.,* the effects of evaporation at the surface and the release of latent heat through condensation that occurs in clouds as moist air rises due to convection), and the way these effects are distributed and modified by the atmospheric circulation. Thermodynamic principles require that temperature changes in the tropics will be larger in the troposphere than near the surface ("amplification"), largely independent of the type of forcing. In mid to

**Table 1: Summary of the most important global-scale climate forcing factors and their likely individual effects on global-, annual-average temperatures; based on Figure 1.3 (which gives temperature information) and Table 1.1 (which gives information on radiative forcing) in Chapter 1, and literature cited in Chapter 1. The stated effects are those that would be expected if the change specified in column 1 were to occur. The top two rows are the primary natural forcing factors, while the other rows summarize the main human-induced forcing factors. The relative importance of these different factors varies spatially and over time. For example, volcanic effects last only a few years in the stratosphere, and slightly longer in the troposphere; while the effects of well-mixed greenhouse gases last for decades to centuries.**

| | Theoretically expected change in annual-global-average temperature | | |
|---|---|---|---|
| **Forcing Factor** | **Surface** | **Low to Mid Troposphere** | **Stratosphere** |
| Increased solar output | Warming | Warming | Warming |
| Volcanic eruptions | Cooling | Cooling | Warming |
| Increased concentrations of well-mixed greenhouse gases ($CO_2$, $CH_4$, $N_2O$, halocarbons) | Warming | Warming | Cooling |
| Increased tropospheric ozone ($O_3$) | Warming | Warming | Slight cooling |
| Decreased stratospheric ozone | Negligible except at high latitudes | Slight cooling | Cooling |
| Increased loading of tropospheric sulfate ($SO_4$) aerosol – sum of direct plus indirect effects | Cooling | Cooling | Negligible |
| Increased loading of carbonaceous aerosol (black carbon [BC] and organic matter [OM]) in the troposphere – sum of direct plus indirect effects | Regional cooling or warming – possible global-average cooling | Warming | Uncertain |
| Land use and land cover changes | Regional cooling or warming – probably slight global-average cooling | Uncertain | Negligible |

high latitudes, the processes controlling how temperature changes in the vertical are more complex, and it is possible for the surface to warm more than the troposphere. These issues are addressed further in Chapter 1 and Chapter 5.

## 2. STRENGTHS AND LIMITATIONS OF THE OBSERVATIONAL DATA

**What kinds of atmospheric temperature variations can the current observing systems detect and what are their strengths and limitations, both spatially and temporally?**

This question is addressed in Chapter 2 of this Report. Chapter 2 draws the following main conclusions:

(1) The observing systems available for this Report are able to detect small surface and upper air temperature variations from year to year as well as trends[6] in climate since the late 1950s (and over the last century for surface observations), once the raw data are successfully adjusted for changes over time in observing systems and practices, and micro-climate exposure. Measurements from all systems require such adjustments. This Report relies solely on adjusted data sets.

---

6  Many of the results in this Report (and here in the Executive Summary) are quantified in terms of linear trends, *i.e.*, by the value of the slope of a straight line that is fitted to the data. A simple straight line is not always the best way to describe temperature data, so a linear trend value may be deceptive if the trend number is given in isolation, removed from the original data. Nevertheless, used appropriately, linear trends provide the simplest and most convenient way to describe the overall change over time in a data set, and are widely used. For a more detailed discussion, see Appendix A.

(2) Independently performed adjustments to the land surface temperature record have been sufficiently successful that trends given by different data sets are reasonably similar on large (*e.g.,* continental) scales, despite the fact that spatial sampling is uneven and some errors undoubtedly remain. This conclusion holds to a lesser extent for the ocean surface record, which suffers from more serious sampling problems and changes in observing practice.

(3) Adjustments for changing instrumentation are most challenging for upper-air data sets. While these show promise for trend analysis, and it is very likely that current upper-air climate records give reliable indications of directions of change (*e.g.,* warming of the troposphere, cooling of the stratosphere), some questions remain regarding the accuracy of the data after adjustments have been made to produce homogeneous time series from the raw measurements.

- Upper-air data sets have been subjected to less scrutiny than surface data sets.

- Adjustments are complicated, can be large compared to the linear trend signal, involve expert judgments, and cannot be stringently evaluated because of lack of traceable standards.

- Unlike surface trends, reported upper-air trends vary considerably between research teams beginning with the same raw data owing to their different decisions on how to remove non-climatic factors.

_____

Many different methods are used to measure temperature changes at the Earth's surface and at various levels in the atmosphere. Near-surface temperatures have been measured for the longest period, over a century, and are measured directly by thermometers. Over land, these data come from fixed meteorological stations. Over the ocean, measurements are of both air temperature and sea-surface (top 10 meters) temperature taken by ships or from buoys.

The next-longest records are upper-air data measured by radiosondes (temperature sensors

All data sets require careful examination for instrument biases and reliability, and adjustments are made to remove changes that might have arisen for non-climatic reasons.

carried aloft by weather balloons). These have been collected routinely since 1958. There are still substantial gaps in radiosonde coverage.

Satellite data have been collected for the upper air since 1979 with almost complete global coverage. The most important satellite records come from Microwave Sounding Units (MSU) on polar orbiting satellites. The microwave data from MSU instruments require calculations and adjustments in order to be interpreted as temperatures. Furthermore, these satellite data do not represent the temperature at a particular level, but, rather, the average temperature over thick atmospheric layers (see Figure 2.2 in Chapter 2). As such, they cannot reveal the detailed vertical structure of temperature changes, nor do they completely isolate the troposphere from the stratosphere. Channel 2 data (mid troposphere to lower stratosphere, $T_2$) have a latitudinally dependent contribution from the stratosphere, while Channel 4 data (lower stratosphere, $T_4$) have a latitudinally dependent contribution from the troposphere, factors that complicate their interpretation. However, retrieval techniques can be used both to approximately isolate specific layers and to check for vertical consistency of trend patterns.

All measurement systems have inherent uncertainties associated with: the instruments employed; changes in instrumentation; and the way local measurements are combined to produce area averages. All data sets require careful examination for instrument biases and reliability, and adjustments are made to remove changes that might have arisen for non-climatic reasons. We refer to these as "adjusted" data sets. The term "homogenization" is also used to describe this adjustment procedure.

Reanalyses[7] and other multi-system products that synthesize observational data with model results to ensure spatial and inter-variable consistency have the potential for addressing issues of surface and atmospheric temperature trends by making better use of available information and allowing analysis of a more comprehensive,

_____

7 Reanalyses are mathematically blended products based upon as many observing systems as practical. Observations are assimilated into a global weather forecasting model to produce globally comprehensive data sets that are most consistent with both the available data and the assimilation model.

internally consistent, and spatially and temporally complete set of climate variables. At present, however, these products contain biases, especially in the stratosphere, that affect trends and that cannot be readily removed because of the complexity of the data products.

## 3. WHAT TEMPERATURE CHANGES HAVE BEEN OBSERVED?

**What do observations indicate about the changes of temperature in the atmosphere and at the surface since the advent of measuring temperatures vertically?**

**What is our understanding of the contribution made by observational or methodological uncertainties to the previously reported vertical differences in temperature trends?**

These questions are addressed in Chapters 3 and 4 of this Report. The following conclusions are drawn in these chapters. Supporting information is given in Figure 1 and Figure 2.

**(1) Surface temperatures:** For global-average changes, as well as in the tropics (20°S to 20°N), all data sets show warming at the surface since 1958, with a greater rate of increase since 1979. Differences between the data sets are small.

- Global-average temperature increased at a rate of about 0.12°C per decade since 1958, and about 0.16°C per decade since 1979. In the tropics, temperature increased at about 0.11°C per decade since 1958, and about 0.13°C per decade since 1979.

- Systematic local biases in surface temperature trends may exist due to changes in station exposure and instrumentation over land[8], or changes in measurement techniques by ships and buoys in the ocean. It is likely that these biases are largely random and

therefore cancel out over large regions such as the globe or tropics, the regions that are of primary interest to this Report.

**(2) Tropospheric temperatures:** All data sets show that the global- and tropical-average troposphere has warmed from 1958 to the present, with the warming in the troposphere being slightly more than at the surface. For changes from 1979, due to the considerable disagreements between tropospheric data sets, it is not clear whether the troposphere has warmed more than or less than the surface.

- Global-average tropospheric temperature increased at a rate of about 0.14°C per decade since 1958 according to the two radiosonde data sets. For the period from 1979, temperature increased by 0.10°C to 0.20°C per decade according to the two radiosonde and three satellite data sets. In the tropics, temperature increased at about 0.13°C per decade since 1958, and between 0.02°C and 0.19°C per decade since 1979.

- Errors in observed temperature trend differences between the surface and the troposphere are more likely to come from errors in tropospheric data than from errors in surface data.

- It is very likely that estimates of trends in tropospheric temperatures are affected by errors that remain in the adjusted radiosonde data sets. Such errors arise because the methods used to produce these data sets are only able to detect and remove the more obvious causes, and involve many subjective decisions. The full consequences of these errors for large-area averages, however, have not yet been fully resolved. Nevertheless, it is likely that a net spurious cooling corrupts the area-averaged adjusted radiosonde data in the tropical troposphere, causing these data to indicate less warming than has actually occurred there.

- For tropospheric satellite data, a primary cause of trend differences between different versions is differences in how the data from different satellites are merged together. Corrections required to account for drifting measurement times are also important.

Errors in observed temperature trend differences between the surface and the troposphere are more likely to come from errors in tropospheric data than from errors in surface data.

---

[8]    Some have expressed concern that land temperature data might be biased due to urbanization effects. Recent studies specifically designed to identify systematic problems using a range of approaches have found no detectable urban influence in large-area averages in the data sets that have been adjusted to remove non-climatic influences *(i.e.,* "homogenized").

- Comparisons between satellite and radiosonde temperatures for the mid troposphere to lower stratosphere layer (MSU channel 2; $T_2$) are very likely to be corrupted by excessive stratospheric cooling in the radiosonde data.

**(3) Lower stratospheric temperatures:** All data sets show that the stratosphere has cooled considerably from 1958 and from 1979 to the present, although there are differences in the linear trend values from different data sets.

- The largest differences between data sets are in the stratosphere, particularly between the radiosonde and satellite-based data sets. It is very likely that the discrepancy between satellite and radiosonde trends arises primarily from uncorrected errors in the radiosonde data.

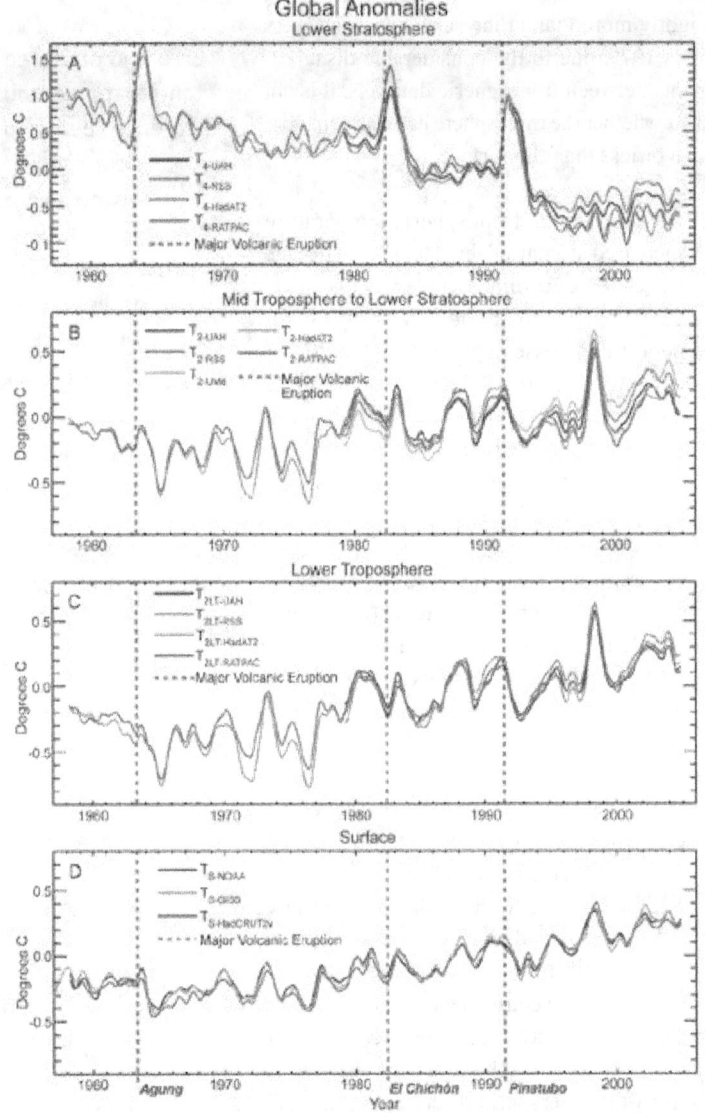

**Figure 1:** Observed surface and upper air global-average temperature records. From top to bottom: A, lower stratosphere (denoted $T_4$) records from two satellite analyses (UAH and RSS) together with equivalently weighted radiosonde records based on HadAT2 and RATPAC data; B, mid-troposphere to lower stratosphere ($T_2$) records from three satellite analyses (UAH, RSS and UMd) together with equivalently weighted radiosonde records based on HadAT2 and RATPAC; C, lower troposphere ($T_{2LT}$) records from UAH and RSS (satellite), and from HadAT2 and RATPAC (equivalently weighted radiosonde); D, surface ($T_S$). All time series are based on monthly-average data smoothed with a 7-month running average, expressed as departures from the Jan. 1979 to Dec. 1997 average. Note that the $T_2$ data (panel B) contain a small contribution (about 10%) from the lower stratosphere. Information here is from Figures 3.1, 3.2 and 3.3 in Chapter 3.

Figure 1 shows the various temperature time series examined in this Report.

For the lower stratosphere, the cooling trend since the late 1950s (which is as expected due to the effects of greenhouse-gas concentration increases and stratospheric ozone depletion) is punctuated by short-term warming events associated with the explosive volcanic eruptions of Mt. Agung (1963), El Chichón (1982) and Mt. Pinatubo (1991).

Both the troposphere and the surface show warming since the late 1950s. For the surface, most of the temperature increase since 1958 occurs starting around 1976, a time coincident with a previously identified climate shift. For the balloon-based tropospheric data, a major part of the temperature increase since 1958 also occurs around 1976, in the form of a relatively rapid rise in temperature. The shift in 1976 is important because it occurs just before the start of the satellite era.

The dominant shorter time scale fluctuations are those associated with the El Niño-Southern Oscillation phenomenon (ENSO). The major ENSO warming event in 1998 is obvious in all records. Cooling following the eruptions of Mt. Agung and Mt. Pinatubo is also evident, but the cooling effect of El Chichón is masked by an ENSO warming that occurred at the same time. The changes following volcanic eruptions (*i.e.*, surface and tropospheric cooling and stratospheric warming) are consistent with our physical understanding and with model simulations.

Global-average temperature changes over the periods 1958 through 2004 and 1979 through 2004 are shown in Figure 2 in degrees Celsius and degrees Fahrenheit.

# 4. ARE MODEL SIMULATIONS CONSISTENT WITH THE OBSERVED TEMPERATURE CHANGES?

Computer-based climate models encapsulate our understanding of the climate system and the driving forces that lead to changes in climate. Such models are the only tools we have for simulating the likely patterns of response of the climate system to different forcing mechanisms. The crucial test of our understanding is to compare model simulations with observed changes to address the question:

**How well can the observed vertical temperature changes be reconciled with our understanding of the causes of these changes?**

In addressing this question, Chapter 5 draws the following conclusions ...

FINGERPRINT PATTERN STUDIES
(1) Results from many different pattern-based "fingerprint"[9] studies (see Box 5.5 in Chapter 5) provide consistent evidence for human influences on the three-dimensional structure of atmospheric temperature changes over the second half of the 20th century.

• Fingerprint studies have identified greenhouse gas and sulfate aerosol signals in observed surface temperature records, a stratospheric ozone depletion signal in stratospheric temperatures, and the combined effects of these forcing agents in the vertical structure of atmospheric temperature changes.

(2) Natural factors (external forcing agents like volcanic eruptions and solar variability and/or internally generated variability) have influenced surface and atmospheric temperatures, but cannot fully explain their changes over the past 50 years.

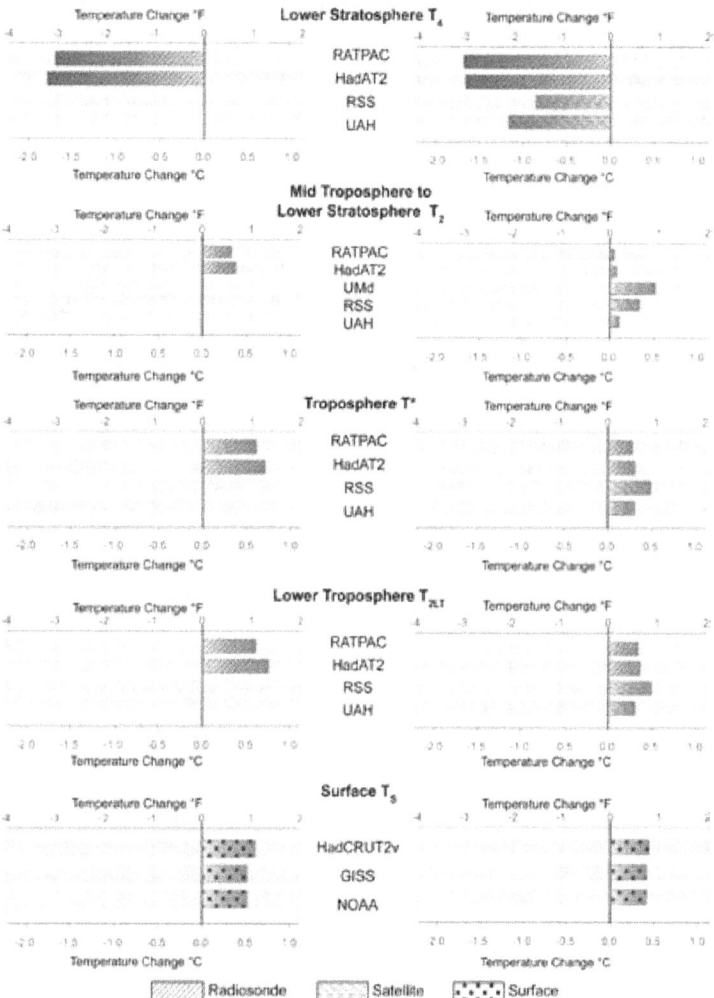

**Total Global-Average Temperature Changes**

Figure 2: Total global-average temperature changes for the surface and different atmospheric layers, from different data sets and over two periods, 1958 to 2004 and 1979 to 2004. The values shown are the total change over the stated period in both degrees Celsius (°C; lower scales) and degrees Fahrenheit (°F; upper scales). All changes are statistically significant at the 5% level except RSS $T_4$ and RATPAC, HadAT2 and UAH $T_2$. Total change in °C is the linear trend in °C per decade (see Tables 3.2 and 3.3 in Chapter 3) times the number of decades in the time period considered. Total change in °F is this number times 1.8 to convert to °F. For example, the Table 3.2 trend for NOAA surface temperatures over January 1958 through December 2004 is 0.11°C/decade. The total change is therefore 0.11 times 4.7 decades to give a total change of 0.53°C, Multiplying this by 1.8 gives a total change in degrees Fahrenheit of 0.93°F. Warming is shown in red, and cooling in blue.

---

[9]  Fingerprint studies use rigorous statistical methods to compare the patterns of observed temperature changes with model expectations and determine whether or not similarities could have occurred by chance. Linear trend comparisons are less powerful than fingerprint analyses for studying cause-effect relationships, but can highlight important differences and similarities between models and observations.

When models are run with natural and human-induced forcings, simulated global-average temperature trends for individual atmospheric layers are consistent with observations.

LINEAR TREND COMPARISONS

(3) When models are run with natural and human-induced forcings, simulated global-average temperature trends for individual atmospheric layers are consistent with observations.

(4) Comparing trend differences between the surface and the troposphere exposes potentially important discrepancies between model results and observations in the tropics.

• In the tropics, most observational data sets show more warming at the surface than in the troposphere, while almost all model simulations have larger warming aloft than at the surface.

AMPLIFICATION OF SURFACE WARMING IN THE TROPICAL TROPOSPHERE

(5) Amplification means that temperatures show larger changes aloft than at the surface. In the tropics, on monthly and inter-annual time scales, both models and observations show amplification of temperature variability in the troposphere relative to the surface. This amplification is of similar magnitude in models and observations. For multi-decadal trends, models show the same amplification that is seen on shorter time scales. The majority of the most recent observed data sets, however, do not show this amplification.

• This inconsistency between model results and observations could arise either because "real world" amplification effects on short and long time scales are controlled by different physical mechanisms, and models fail to capture such behavior; or because non-climatic influences remaining in some or all of the observed tropospheric datasets lead to biased long-term trends; or a combination of these factors. The new evidence in this Report - model-to-model consistency of amplification results, the large uncertainties in observed tropospheric temperature trends, and independent physical evidence supporting substantial tropospheric warming (such as the increasing height of the tropopause) - favors the second explanation. However, the large observational uncertainties that currently exist make it difficult to determine whether or not models still have significant errors. Resolution of this issue requires reducing these uncertainties.

OTHER FINDINGS

(6) Because of differences between different observed data sets and differences between models, it is important to account for both model and observational uncertainty in comparisons between modeled and observed temperature changes.

• Large "construction" uncertainties in observed estimates of global-scale atmospheric temperature change can critically influence the outcome of consistency tests between models and observations.

(7) Inclusion of previously ignored, spatially variable forcings in the most recent climate models does not fundamentally alter conclusions about the amplification of warming in the troposphere relative to the surface.

• Changes in sulfate aerosols and tropospheric ozone, which have spatially variable forcings, have been incorporated routinely in climate model experiments for a number of years. It has been suggested that the spatially heterogeneous forcing effects of black carbon aerosols and land use/land cover changes may have had significant effects on regional temperatures that might modify previous conclusions regarding vertical temperature changes. These forcings have been included for the first time in about half of the global model simulations considered here. Within statistical uncertainties, model simulations that include these forcings show the same amplification of warming in the troposphere relative to the surface at very large spatial scales (global and tropical averages) as simulations in which these forcings are neglected.

---

Chapter 5 analyzes state-of-the-art model simulations from 19 institutions from around the world, run using combinations of the most important natural and human-induced forcings. The Chapter compares the results of these simulations with a number of different observational data sets for the surface and different atmospheric layers, resulting in a large number of possible model/observed data comparisons.

Figures 3 and 4 summarize the new model results used in this Report, together with the corresponding observations. Figure 3 gives results for global-average temperature, while Figure 4 gives results for the tropics (20°S to 20°N). Model and observed results are compared in these Figures using linear trends over the period January 1979 through December 1999[10] for the surface, for individual layers, and (right-hand panels) for surface changes relative to the troposphere. Rectangles are used to illustrate the ranges of both model trends (red rectangles) and observed trends (blue rectangles). Individual observed-data trends are also shown.

Since statistical uncertainties (see Appendix A) are not shown in these Figures, the rectangles do not represent the full ranges of uncertainty. However, they allow a useful first-order assessment of similarities and differences between observations and model results. Overlapping rectangles in the Figures indicate consistency, while rectangles that either do not overlap or show minimal overlap point to potential inconsistencies between observations and model results.

For global averages (Fig. 3), models and observations generally show overlapping rectangles. A potentially serious inconsistency, however, has been identified in the tropics. Figure 4G shows that the lower troposphere warms more rapidly than the surface in almost all model simulations, while, in the majority of observed data sets, the surface has warmed more rapidly than the lower troposphere. In fact, the nature of this discrepancy is not fully captured in Fig. 4G as the models that show best agreement with the observations are those that have the lowest (and probably unrealistic) amounts of warming (see Chapter 5, Fig. 5.6C). On the other hand, as noted above, the rectangles do not express the full range of uncertainty, as they do not account for the large statistical uncertainties in the individual model trends or the large constructional and statistical uncertainties in the observed data trends.

The potential discrepancy identified here is a different way of expressing the amplification discrepancy described in Section 4, item (5)

above. It may arise from errors that are common to all models, from errors in the observational data sets, or from a combination of these factors. The second explanation is favored, but the issue is still open.

A potentially serious inconsistency has been identified in the tropics. The favored explanation for this is residual error in the observations, but the issue is still open.

---

[10] This is the longest period common to all model simulations.

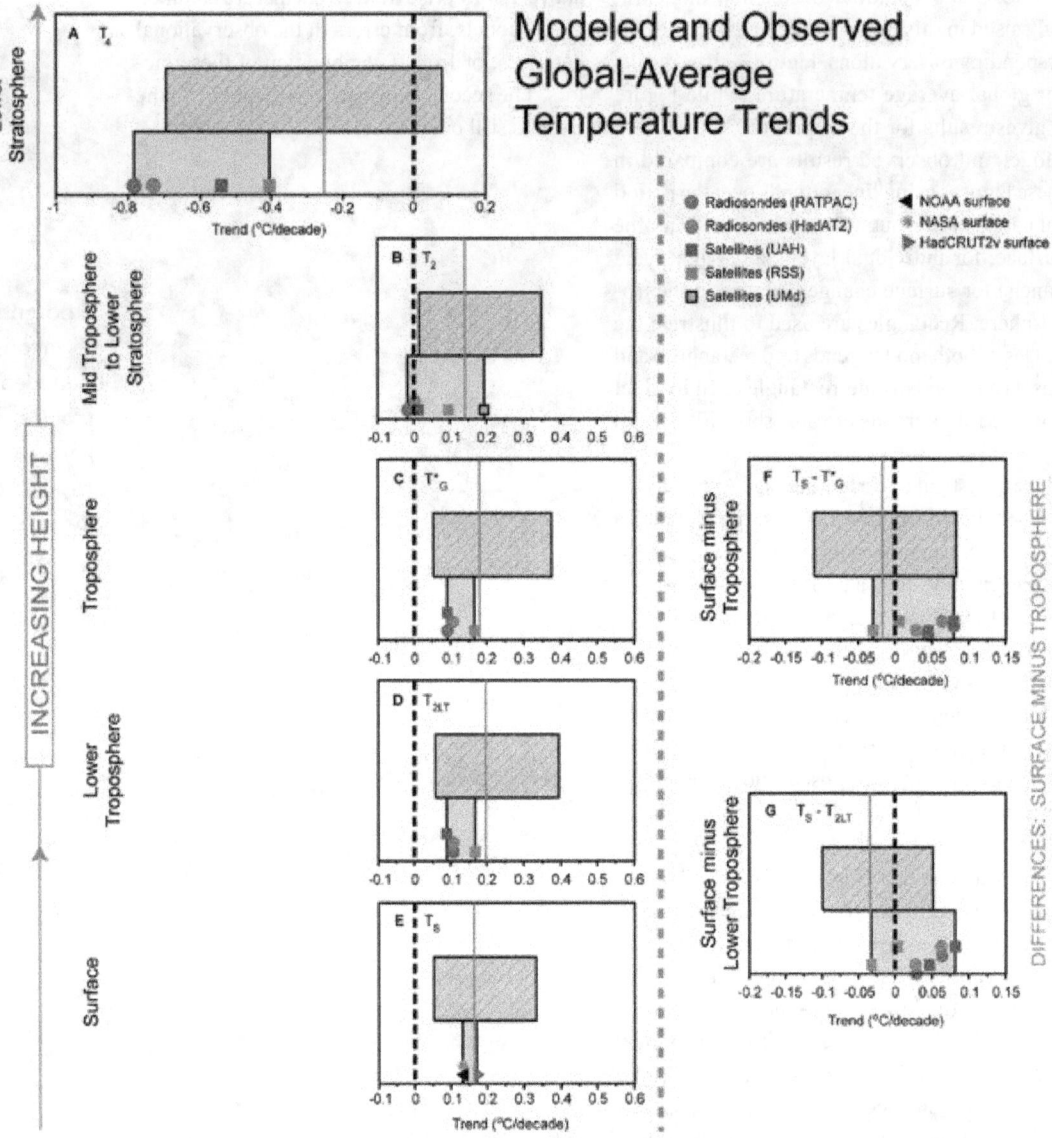

## Modeled and Observed Global-Average Temperature Trends

**Figure 3:** Comparison of observed and model-simulated global-average temperature trends (left-hand panels) and trend differences (right-hand panels) over January 1979 through December 1999, based on Table 5.4A and Figure 5.3 in Chapter 5. The upper red rectangles in each box show the range of model trends from 49 model simulations. The lower blue rectangles show the range of observed trends, with the individual trends from different data sets indicated by the symbols. From bottom to top, the left-hand panels show trends for the surface ($T_S$), the lower troposphere ($T_{2LT}$), the troposphere ($T^*$), the mid troposphere to lower stratosphere ($T_2$), and the lower stratosphere ($T_4$). The right-hand panels show differences in trends between the surface and either the troposphere or the lower troposphere, with a positive value indicating a stronger warming at the surface. The red vertical lines show the average of all model results. The vertical black dashed lines show the zero value. For the observed trend differences, there are eight values corresponding to combinations of the four upper-air data sets (as indicated by the symbols) and either the HadCRUT2v surface data or the NASA/NOAA surface data (which have almost identical trends).

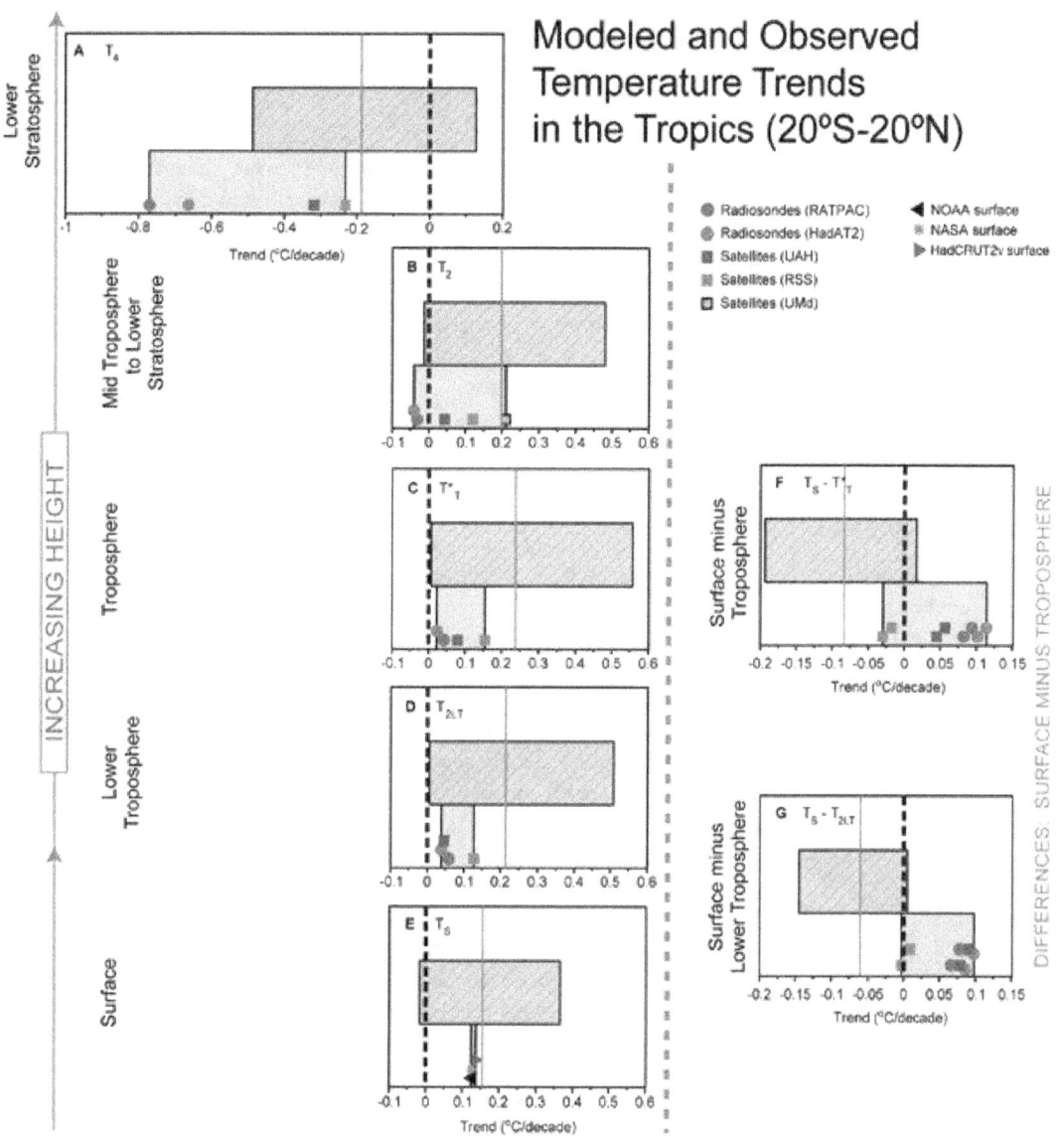

**Figure 4:** As Figure 3, but for the tropics (20°S to 20°N), based on Table 5.4B and Figure 5.4 in Chapter 5. Note that, in the tropics, the tropospheric radiosonde data (green and purple filled circles in panels C and D) may have a cooling bias and that it is unlikely that this bias has been completely removed from the adjusted data used here. Note also that the (small) overlap in panel G is deceptive because the models in this overlap area have unrealistically small amounts of warming. On the other hand, the rectangles do not express the full range of uncertainty, as they do not account for uncertainties in the individual model or observed data trends.

## 5. RECOMMENDATIONS

### What measures can be taken to improve the understanding of observed changes?

In answer to this question, drawing on the material presented in the first five chapters of this Report, a set of primary recommendations has been developed and is described in detail in Chapter 6. The items described in Chapter 6 expand and build upon existing ideas, emphasizing those that are considered to be of highest utility. The seven inter-related recommendations are:

(1) The independent development of data sets and analyses by several scientists or teams will help to quantify structural uncertainty. In order to encourage further independent scrutiny, data sets and their full metadata (i.e., information about instrumentation used, observing practices, the environmental context of observations, and data-processing procedures) should be made openly available. Comprehensive analyses should be carried out to ascertain the causes of remaining differences between data sets and to refine uncertainty estimates.

(2) Efforts should be made to archive and make openly available for independent analysis surface, balloon-based, and satellite data and metadata that have not previously been exploited. Emphasis should be placed on the tropics and on the recovery of satellite data before 1979 (which may allow better characterization of the climate shift in the mid-1970s).

(3) Efforts should be made to develop new or reprocess existing data to create climate quality data sets[11] for a range of variables other than temperature (e.g. atmospheric water vapor content, ocean heat content, the height of the tropopause, winds and clouds, radiative fluxes, and cryospheric changes). These data sets should subsequently be compared with each other and with temperature data to determine whether they are consistent with our physical understanding. It is important to create several independent estimates for each variable in order to assess the magnitude of construction uncertainties.

(4) Efforts should be made to create several homogeneous atmospheric reanalyses. Particular care needs to be taken to identify and homogenize critical input climate data. Identification of critical data requires, in turn, observing system experiments where the impacts and relative importance of different observation types from land, radiosonde, and space-based observations are assessed.

(5) Models that appear to include the same forcings often differ in both the way the forcings are quantified and how these forcings are applied to the model. Hence, efforts are required to separate more formally uncertainties arising from model structure from the effects of forcing uncertainties. This requires running multiple models with standardized forcings, and running the same models individually under a range of plausible scenarios for each forcing.

(6) The GCOS (Global Climate Observing System) climate monitoring principles should be fully adopted. In particular, when any type of instrument for measuring climate is changed or re-sited, there should be a period of overlap between old and new instruments or configurations that is sufficient to allow analysts to adjust for the change with small uncertainties that do not prejudice the analysis of climate trends. The minimum period is a full annual cycle of the climate. Thus, replacement satellite launches should be planned to take place at least a year prior to the expected time of failure of a key instrument.

(7) A small subset (about 5%) of the operational radiosonde network should be developed and implemented as reference sites for all kinds of climate data from the surface to the stratosphere.

---

[11]  Climate quality data sets are those where the best possible efforts have been made to identify and remove non-climatic effects that might produce spurious changes over time.

CHAPTER 1

**W**hy do temperatures vary vertically (from the surface to the stratosphere) and what do we understand about why they might vary and change over time?

**Convening Lead Author:** V. Ramaswamy, NOAA
**Lead Authors:** J.W. Hurrell, NSF NCAR; G.A. Meehl, NSF NCAR
**Contributing Authors:** A. Phillips, NCAR, Boulder;
B.D. Santer, DOE LLNL; M.D. Schwarzkopf, NOAA;
D.J. Seidel, NOAA; S.C. Sherwood, Yale Univ.;
P.W. Thorne, U.K. Met. Office

## SUMMARY

### Temperatures Vary Vertically

The global temperature profile of the Earth's atmosphere reflects a balance between the radiative, convective and dynamical heating/cooling of the surface-atmosphere system. Radiation from the Sun is the source of energy for the Earth's climate, with most of it absorbed at the surface. Combined with the physical properties of the atmosphere and dynamical processes, the heat is mixed vertically and horizontally, yielding the highest temperatures, on average, at the surface, with marked seasonal and spatial variations. In the atmosphere, the distribution of moisture and the lower air pressure at progressively higher altitudes result in decreasing temperatures with height up to the tropopause, with the rate of decrease depending on geographical factors and meteorological conditions. The tropopause marks the top of the troposphere, i.e., the lower 8 to 16 km of the atmosphere (see Preface, Fig. 2), and varies with latitude and longitude. Above this altitude, the physical properties of the air produce a warming with height through the stratosphere (extending from the tropopause to ~50 km).

*Temperature trends at the surface can be expected to be different from temperature trends higher in the atmosphere.*

Temperature trends at the surface can be expected to be different from temperature trends higher in the atmosphere because:

- Surface types (sea, snow, ice, and different vegetative covers of land) differ considerably in their physical properties. Near the surface, these differing conditions can produce strong horizontal variations in temperature. Above the surface layer, these contrasts are quickly smoothed out by the atmospheric motions, contributing to varying temperature trends with height at different locations.
- Changes in atmospheric circulation or modes of atmospheric variability (e.g., El Niño-Southern Oscillation [ENSO]) can produce different temperature trends at the surface and aloft.
- Under some circumstances, temperatures may increase with height near the surface or higher in the troposphere, producing a "temperature inversion." Such inversions are more common at night; over continents, sea ice and snow during winter; and in the trade wind regions. Since the air in inversion layers is resistant to vertical mixing, temperatures trends can differ between inversion layers and adjacent layers.
- Forcing factors, either natural (e.g., volcanoes and solar) or human-induced (e.g., greenhouse gas, aerosols, ozone, and land use) can result in differing temperature trends at different altitudes, and these vertical variations may change over time. This can arise due to spatial and temporal changes in the concentrations or properties of the forcing agents.

This Chapter describes the temperature profile of the layers of the atmosphere from the surface through the stratosphere and discusses the basic reasons for this profile. We also use results from global climate model simulations to show how changes in natural and human-induced factors can produce different temperature trends in the various layers of the atmosphere. This discussion provides the background for the presentation of the observed changes (Chapters 2-4), and for the understanding of their causes (Chapter 5). We also describe temperature changes in the stratosphere in recent decades and the influences of these changes on the troposphere. Finally, making use of surface and satellite observations, we examine the physical processes that can result in different temperature trends at the surface and in the troposphere.

## 1.1 THE THERMAL STRUCTURE OF THE ATMOSPHERE

Radiation input from the Sun is the source of energy for the Earth's climate system (Hartmann, 1994). Most of the solar radiation absorbed is at the surface, the rest being absorbed by the atmosphere. The global-and-annual-mean solar radiation absorbed and outgoing longwave emission from Earth balance each other to yield a steady-state climate for the planet as a whole. Both radiation components have a maximum in the tropics and decrease towards the poles. There is an excess of net radiative (solar+longwave) heating of the climate system in the tropics, with a deficit in the high latitudes (poleward of ~40⁰). Dynamical motions arising as a consequence of this equator-to-pole gradient, combined with convective processes and the influence due to the rotation of the Earth, result in a heat transfer from the low to middle and high latitudes, thereby setting up the climatological horizontal and vertical thermal structure (Hartmann, 1994; Salby, 1996).

Surface temperatures are at their warmest in the tropics, where the largest amount of solar radiation is received during the course of the year, and decrease towards the polar regions where the annual-mean solar radiation received is at a minimum (Oort and Peixoto, 1992). The temperature contrast between summer and winter is greatest at the poles and least at the Equator. Since land areas heat up and cool more rapidly than oceans, and because of the preponderance of land in the Northern Hemisphere, there is a larger contrast in temperature between summer and winter in the Northern Hemisphere.

The rate at which the temperature changes with height is termed the "lapse rate." The lapse rate can vary with location and season, and its value depends strongly on the atmospheric humidity.

Figure 1.1 illustrates the climatological vertical temperature profiles for December, January, February and June, July, August mean conditions, as obtained from the National Centers for Environmental Prediction (NCEP) reanalyses (Kalnay *et al.,* 1996; updated). It is convenient to think first in terms of climatological conditions upon which spatial and temporal variations/trends are superimposed. The solid line in the plot estimates the tropopause, which separates the troposphere below from the stratosphere above (see Preface, Fig. 2). The tropopause is at its highest level in the tropics (~20°N-20°S). It descends sharply in altitude from ~16 km at the equator to ~12 km at ~30-40° latitude, and to as low as about 9 km at the poles.

Temperatures generally decrease with height from the surface although there are important exceptions. The rate at which the temperature changes with height is termed the "lapse rate." The lapse rate can vary with location and season, and its value depends strongly on the atmospheric humidity, *e.g.,* the lapse rate varies from ~4°C/km near the surface in the moist tropical regions (near the equator) to much larger values (~8-9°C/km) in the drier subtropics (~20-30°). Important departures from nominal lapse rate values can occur near the surface and in the upper troposphere. In the equatorial tropics, the tropopause region (~16 km) is marked by a smaller value of the lapse rate than in the lower troposphere.

The thermal structure of the lowest 2-3 km, known as the "planetary boundary layer," can be complicated, even involving inversions (in which temperature increases rather than decreases with height) occurring at some latitudes due to land-sea contrasts, topographic influences, radiative effects and meteorological conditions. Inversions are particularly common during winter over some middle and high latitude land regions and are a climatological feature in the tropical trade wind regions. The presence of inversions acts to decouple surface temperatures from tropospheric temperatures on daily or even weekly timescales.

Above the tropopause is the stratosphere, which extends to ~50 km and in which the temperature increases with height. In the vicinity of the tropical tropopause, (*i.e.,* the upper troposphere and lower stratosphere regions, ~15-18 km), the temperature varies little with height. The extratropical (poleward of 30°) lower stratosphere (at ~8-12 km) also exhibits a similar feature (Holton, 1979). The lapse rate change with altitude in the upper troposphere/lower stratosphere region is less sharp in the extratropical latitudes than in the tropics.

The global temperature profile of the atmosphere reflects a balance between the radiative, convective and dynamical heating/cooling of the surface-atmosphere system. From a global, annual-average point of view, the thermal profile of the stratosphere is the consequence of a balance between radiative heating and cooling rates due to greenhouse gases, principally carbon dioxide ($CO_2$), ozone ($O_3$) and water vapor ($H_2O$) (Andrews *et al.,* 1987). The vertical profile in the troposphere is the result of a balance of radiative processes involving greenhouse gases, aerosols, and clouds (Stephens and Webster, 1981; Goody and Yung, 1989), along with the strong role of moist convection and dynamical motions (Holton, 1979; Held, 1982; Kiehl, 1992). An important difference between the troposphere and stratosphere is that the stratosphere is characterized by weak vertical motions, while in the troposphere, the vertical mo-

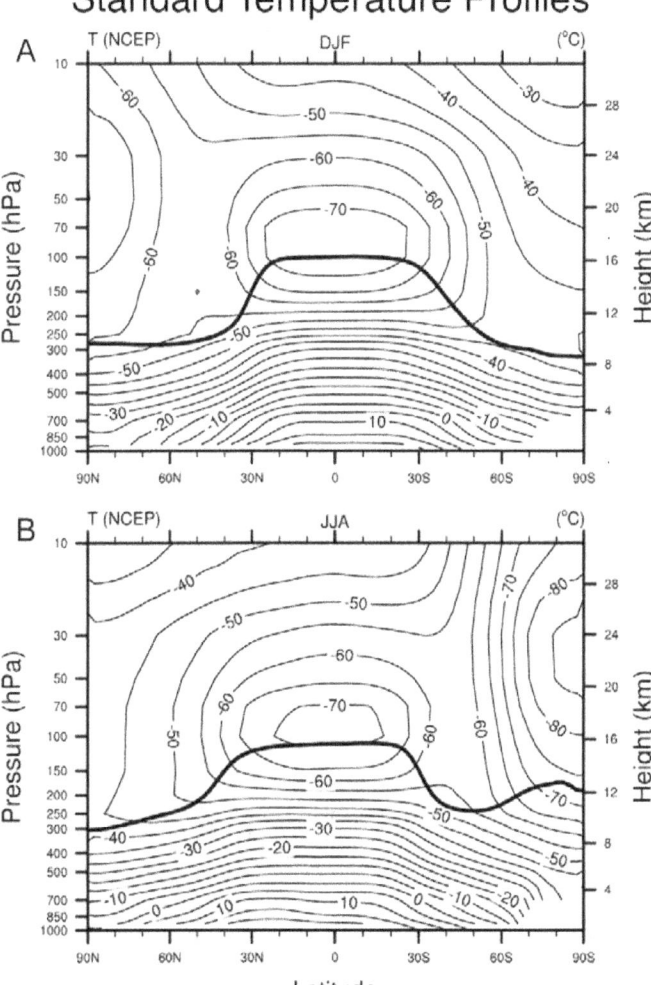

## Standard Temperature Profiles

**Figure 1.1.** Global climatological vertical temperature profiles from surface to troposphere and extending into the stratosphere for December-January-February (DJF) and June-July-August (JJA) mean conditions (1979-2003), as obtained from the National Centers for Environmental Prediction (NCEP) reanalyses (Kalnay *et al.,* 1996; updated). The solid line denotes the tropopause which separates the stratosphere from the surface-troposphere system. The tropopause pressure level is defined by the standard lapse rate criterion: it is identified by the lowest level (above 450 hPa) where the temperature lapse rate becomes less than 2°C/km. The tropopause pressure archived in the reanalyses is estimated by deriving the lapse rate at each model level and estimating (by interpolation in height) the pressure where the threshold value of 2°C/km is reached. This algorithm produces tropopause estimates which vary smoothly in space and time.

> The presence of inversions acts to decouple surface temperatures from tropospheric temperatures on daily or even weekly timescales.

tions are stronger. Most significantly, the moist convective processes that are a characteristic feature of the troposphere include the transfer of large amounts of heat due to evaporation or condensation of water.

A useful conceptual picture of the thermal structure can be had by considering the radiative-convective balance that is approximately applicable to the tropics taken as a whole. If radiative processes alone were considered, that would cause the surface to be significantly warmer than it is actually (Goody and Yung, 1989). This would occur because the atmosphere is relatively transparent to the Sun's radiation. This would lead to a strong heating of the surface accompanied by a net radiative cooling of the atmosphere (Manabe and Wetherald, 1967). However, the resulting convective motions remove the excess heating from the surface in the form of sensible and latent heat, the latter involving the evaporation of moisture from the surface (Ramanathan and Coakley, 1978). As air parcels rise and expand, they cool due to decompression, leading to a decrease of temperature with height. The lapse rate for a dry atmosphere, when there are no moist processes and the air is rising quickly enough to be unaffected by other heating/cooling sources, is close to 10°C/km. However, because of moist convection, there is condensation of moisture, formation of clouds and release of latent heat as the air parcels rise, causing the lapse-rate to be much less, as low as 4°C/km in very humid atmospheres (Houghton, 1977).

Actual thermal profiles are more complex than above owing to the additional roles of the large-scale circulation and convection-cloud physical interactions. For example, a more detailed picture in subtropical regions consists of a surface mixed layer (up to about 500 m) and a trade wind boundary layer (up to about 2 km) above which is the free troposphere. Each of the boundary layers is topped by an inversion which tends to isolate the region from the layer above (Sarachik, 1985). This indicates the limitations in assuming nominal lapse rate values from the surface to the tropopause everywhere. In the tropical upper troposphere, moisture- and cloud-related features related to convection (*e.g.,* upper tropospheric relative humidity, cirrus cloud microphysics, and mesoscale cir-

culations in anvil clouds) are important factors in shaping the thermal profile (Ramaswamy and Ramanathan, 1989; Donner *et al.,* 2001; Sherwood and Dessler, 2003).

In the more general sense, the interactions between radiation, moist convection, and dynamical motions (ranging from large- to meso- and small-scales) govern the quantitative rate at which temperature decreases with height at any location. Large-scale dynamical mechanisms tend to result in more spatially uniform temperatures (on monthly-mean and longer time scales) above the boundary layer, and over horizontal scales (Rossby radius; Hartmann, 1994) that vary from planetary scale near the equator to a couple of thousand kilometers at middle latitudes and to a few hundred kilometers near the poles. The major circulation patterns in the atmosphere such as the Hadley and Walker circulations (Holton, 1979; Hartmann, 1994) play a key role in the atmospheric energy balance of the tropics and subtropics (~30 degrees in latitude), and this crucially affects the thermal structure in those regions (Trenberth and Stepaniak, 2003). The low latitudes are characterized by a vertical coherence in the vertical temperature structure, with variations of opposite signs in temperature below and above the tropopause associated with upward motion and subsidence, respectively (Trenberth and Smith, 2006). In the extra-tropics (poleward of 30° latitude), the lapse rate and tropopause height are affected by instabilities associated with large-scale eddies of the familiar weather systems ("baroclinic instability"; Holton, 1979; Held, 1982). In the polar regions (~60-90°), planetary-scale waves forced by the influences of mountains and that of land-sea contrasts upon the flow of air play a significant role in the determination of the wintertime temperatures at the poles.

The sense of the radiative-convective-dynamical balance above, together with the requirement of radiative balance at the top-of-the-atmosphere (namely, equilibrium conditions wherein the net solar energy absorbed by the Earth's climate system must be balanced by the infrared radiation emitted by the Earth), can help illustrate the significance of long-lived infrared absorbing gases in the global atmosphere. The presence of such greenhouse gases (*e.g.,* carbon dioxide, methane, nitrous

As air parcels rise and expand, they cool due to decompression, leading to a decrease of temperature with height.

oxide, halocarbons) increases the radiative heating of the surface and troposphere. As specific humidity is strongly related to temperature, it is expected to rise with surface warming (IPCC, 1990), The increased moisture content of the atmosphere amplifies the initial radiative heating due to the greenhouse gas increases (Manabe and Wetherald, 1967; Ramanathan, 1981). The re-establishment of a new thermal equilibrium in the climate system involves the communication of the added heat input to the troposphere and surface, leading to surface warming (Goody and Yung, 1989; IPCC, 1990; Lindzen and Emanuel, 2002). From the preceding discussions, the lapse rate can be expected to decrease with the resultant increase in humidity, and also to depend on the resultant changes in atmospheric circulation. In general, the lapse rate can be expected to decrease with warming such that temperature changes aloft exceed those at the surface. As a consequence, the characteristic infrared emission level of the planet is shifted to a higher altitude in the atmosphere.

## 1.2 NATURAL AND ANTHROPOGENIC FORCING OF CLIMATE CHANGE

Potentially significant variations and trends are superimposed on the basic climatological thermal profile, as revealed by observational data in the subsequent chapters. While the knowledge of the climatological mean structure discussed in the previous section involves considerations of radiative, convective, and dynamical processes, understanding the features and causes of the magnitude of changes involves a study of the perturbations in these processes, which then frame the response of the climate system to the forcing. While the understanding of climate variability is primarily based on observations of substantial changes (*e.g.,* sea-surface temperature changes during El Niño), the vertical temperature changes being investigated in this report are changes on the order of a few tenths of degrees on the global-mean scale (local changes could be much greater), as discussed in the subsequent chapters.

"Unforced" variations, *i.e.,* changes arising from internally generated variations in the atmosphere-ocean-land-ice/snow climate system, can influence surface and atmospheric temperatures substantially, *e.g.,* due to changes in equatorial sea-surface temperatures associated with ENSO. Climate models indicate that global-mean unforced variations on multidecadal timescales are likely to be smaller than, say, the 20th century global-mean increase in surface temperature (Stouffer *et al.,* 2000). However, for specific regions and/or seasons, this may not be valid and the unforced variability could be substantial. Chapter 5 provides more detail on models and their limitations (see particularly Box 5.1 and 5.2).

Because of the influence of radiative processes on the thermal structure, any agent external to the climate system that perturbs the planet's radiative heating distribution can cause climate changes, and thus is potentially important in accounting for the observed temperature changes (Santer *et al.,* 1996). The radiative (solar plus longwave) heat balance of the planet can be perturbed (forced) by:

- natural factors such as changes in the Sun's irradiance, and episodic, explosive volcanic events (leading to a build-up of particulates in the stratosphere);

- human-induced factors such as changes in the concentrations of radiatively active gases (carbon dioxide, methane, etc.) and aerosols.

Important external forcing agents of relevance for the surface and atmospheric temperature changes since pre-industrial time (1750) are summarized in Table 1.1. As illustrative examples, global-mean forcing estimates for the period 1750 to ~2000 (late 1990s) are listed. For more details, see Ramaswamy *et al.* (2001) and NRC (2005).

From Table 1.1, the important anthropogenic contributions to the global-mean forcing since pre-industrial times to ~2000 are due to well-mixed greenhouse gases and ozone, aerosols, and land-use (albedo). The natural factors are comprised of solar irradiance variations and stratospheric aerosols in the aftermath of explo-

The lapse rate can be expected to decrease with warming such that temperature changes aloft exceed those at the surface.

**Table 1.1. Agents potentially causing an external radiative forcing of climate change between 1750 and 2000. Estimates of the global-mean forcing and uncertainty (in percent or range about the mean) for the period 1750 to ~2000 are listed (adapted from Ramaswamy *et al.*, 2001). See reference and notes below for explanations.**

| Forcing Agent | Natural (N) or Anthropogenic (A) | Solar Perturbation | Long-wave Perturbation | Surface Radiation Effect | Tropospheric Radiation Effect | Stratospheric Radiation Effect | Geographic Distribution (global G or localized, L) | Global-mean Forcing (with uncertainty or range) Estimates [W/m²] | Level of Confidence |
|---|---|---|---|---|---|---|---|---|---|
| Well-mixed greenhouse gases | A | (small) | Y | Y | Y | Y | G | 2.43 [10%] | High |
| Tropospheric ozone | A | Y | Y | Y | Y | (small) | L | 0.35 [43%] | Medium |
| Stratospheric ozone | A | Y | Y | (small) | Y | Y | L | -0.15 [67%] | Medium |
| Sulfate aerosols (direct) | A | Y | - | Y | (small) | - | L | -0.40 [2X] | Low |
| Black carbon aerosols (direct) | A | Y | (small) | Y | Y | - | L | 0.20 [2X] | Very low |
| Organic carbon aerosols (direct) | A | Y | - | Y | (small) | - | L | -0.10 [3X] | Very low |
| Biomass burning aerosols (direct) | A | Y | - | Y | Y | - | L | -0.20 [3X] | Very low |
| Indirect aerosol effect | A | Y | Y | Y | Y | (small) | L | 0 to -2.0 | Very low |
| Land-use | A | Y | (small) | Y | - | - | L | -0.20 [100%] | Very low |
| Aircraft contrails | A | (small) | (small) | (small) | (small) | - | L | 0 to 0.1 | Very low |
| Sun | N | Y | - | Y | (small) | Y | G | 0.30 [67%] | ▲Very low |
| Volcanic aerosols | N | Y | Y | Y | (small) | Y | ● | ♦ | ■ |

## Table 1.1 Notes:

Natural (N) and Anthropogenic (A) sources of the forcing agents. Direct aerosol forcing is to be contrasted with the indirect effects; only the first aerosol indirect effect is estimated here. "Black" and "organic" carbon aerosol estimates relate to fossil-fuel burning. Note that dust aerosol is ignored here as there are considerable uncertainties about the anthropogenic aspect of this species and its forcing. The forcing estimate for land-use is due to albedo change only. Y denotes a significant perturbation, "small" indicates considerably less important but not a negligible perturbation, while no entry denotes a negligible perturbation. Forcings other than well-mixed greenhouse gases and solar are spatially localized, with the degree of localization having a considerable variation amongst the different agents, depending on their respective source locations. In addition, for short-lived species such as ozone and aerosols, the long-range meteorological transport plays an important role in their global distributions. Number before "X" denotes the multiplicative factor for the uncertainty range about the mean estimate. Level of confidence represents our subjective judgement about the scientific reliability of the estimate based on assumptions invoked, knowledge of the physical/chemical mechanisms and uncertainties associated with the estimate. For more details, see Ramaswamy *et al.* (2001).

● Typically, the forcing becomes near-global a few months after an intense tropical eruption.

■ Volcanic aerosols are formed in the stratosphere in the aftermath of explosive volcanic eruptions. These aerosols are distinct from the other aerosol entries which are in the troposphere and arise from human-influenced activity. In the case of the naturally-occurring volcanic aerosols, the level of confidence in the forcing from the most recent intense eruption, that of Mt. Pinatubo in 1991, is reasonably good because of reliable observing systems in place; for prior explosive eruptions, observations were absent or sparse which affects the reliability of the quantitative estimates for the previous volcanic events.

♦ As volcanic aerosol forcing from episodic events is a transitory phenomenon, no attempt is made to construct a forcing estimate with respect to a particular reference year.

▲ Although solar irradiance variations before 1980 have a very low level of confidence, direct observations of the Sun's output from satellite platforms since 1980 are considered to be accurate (Lean *et al.*, 2005). Thus, the forcing due to solar irradiance variations from 1980 to present are known to a much greater degree of confidence than from pre-industrial to the present time.

sive, episodic volcanic eruptions. In Chapter 5, the responses of climate models to the temporal evolution of these important forcing agents are analyzed.

The quantitative estimates of the forcing due to the well-mixed greenhouse gases (comprised of carbon dioxide, methane, nitrous oxide and halocarbons) are known to a higher degree of scientific confidence than for other agents. The forcing agents differ in terms of whether their radiative effects are felt primarily in the stratosphere or troposphere or both, and whether the perturbations occur in the solar or longwave spectrum or both. Among aerosols, black carbon is distinct because it strongly absorbs solar radiation (see also Box 5.3). In contrast to sulfate aerosols, which cause a perturbation of solar radiation mainly at the surface (causing a cooling effect), black carbon acts to warm the atmosphere while cooling the surface (Chung *et al.,* 2002; Menon *et al.,* 2002). This could have implications for convective activity and precipitation (Ramanathan *et al.,* 2005), and the lapse rate (Chung *et al.,* 2002; Erlick and Ramaswamy, 2003). The response to radiative forcings is in general not localized. Atmospheric motions and processes can lead to perturbations in climate variables at locations far away from the location of the forcing. The vertical partitioning of the radiative perturbation determines how the surface heat and moisture budgets respond, how the convective interactions are affected, and hence how the surface temperature and the atmospheric thermal profile are altered. "Indirect" aerosol effects arising from aerosol-cloud interactions can lead to potentially significant changes in cloud characteristics such as cloud lifetimes, frequencies of occurrence, microphysical properties, and albedo (reflectivity) (*e.g.,* Lohmann *et al.,* 2000; Sherwood, 2002; Lohmann and Feichter, 2005). As clouds are important components in both solar and longwave radiative processes and hence significantly influence the planetary radiation budget (Ramanathan *et al.,* 1989; Wielicki *et al.,* 2002), any effect caused by aerosols in perturbing cloud properties is bound to exert a significant effect on the surface-troposphere radiation balance and thermal profile.

Estimates of forcing from anthropogenic land-use changes have consisted of quantification of the effect of snow-covered surface albedos in the context of deforestation (Ramaswamy *et al.,* 2001). However, there remain considerable uncertainties in these quantitative estimates. There are other possible ways in which land-use change can affect the heat, momentum and moisture budgets at the surface (*e.g.,* changes in transpiration from vegetation) (see also Box 5.4), and thus exert a forcing of the climate (Pielke *et al.,* 2004; NRC, 2005). In addition to the forcings shown in Table 1.1, NRC (2005) has evoked a category of "nonradiative" forcings such as aerosols, land-cover, and biogeochemical changes which may impact the climate system first through nonradiative mechanisms, *e.g.,* modifying the hydrologic cycle and vegetation dynamics. Eventually, radiative impacts could occur, though no metrics for quantifying these nonradiative forcings have been accepted as yet (NRC, 2005).

Even for the increases in well-mixed greenhouse gases, despite their globally near-uniform mixing ratios, the resulting forcing of the climate system is at a maximum in the tropics, due to the temperature contrast between the surface and troposphere there and therefore the increased infrared radiative energy trapping. Owing to the dependence of infrared radiative transfer on clouds and water vapor, which have substantial spatial structure in the low latitudes, the greenhouse gas forcing is non-uniform in the tropics, being greater in the relatively drier tropical domains. For short-lived gases, the concentrations themselves are not globally uniform so there tends to be a distinct spatial character to the resulting forcing, *e.g.,* stratospheric ozone, whose forcing is confined essentially to the mid-to-high latitudes, and tropospheric ozone whose forcing is confined to tropical to midlatitudes. For aerosols, which are even more short-lived than ozone, the forcing has an even more localized structure (see also Box 5.3). However, although tropospheric ozone and aerosol forcing are maximized near the source regions, the contribution to the global forcing from remote regions is not negligible. The natural factors, namely solar irradiance changes and stratospheric aerosols from tropical volcanic eruptions, exert a forcing that is global in scope.

The forcing agents differ in terms of whether their radiative effects are felt primarily in the stratosphere or troposphere or both.

In terms of the transient changes in the climate system, it is also important to consider the temporal evolution of the forcings. For well-mixed greenhouse gases, the evolution over the past century, and in particular the past four decades, is very well quantified because of reliable and robust observations. However, for the other forcing agents, there are uncertainties concerning their evolution that can affect the inferences about the resulting surface and atmospheric temperature trends. Stratospheric ozone changes, which have primarily occurred since ~1980, are slightly better known than tropospheric ozone and aerosols. For solar irradiance and land-surface changes, the knowledge of the forcing evolution over the past century is poorly known. Only in the past five years have climate models included time varying estimates of a subset of the forcings that affect the climate system. In particular, current models typically include GHGs, ozone, sulfate aerosol direct effects, solar influences, and volcanoes. Some very recent model simulations also include time-varying effects of black carbon and land use change. Other forcings either lack sufficient physical understanding or adequate global time- and space-dependent datasets to be included in the models at this time. As we gain more knowledge of these other forcings and are better able to quantify their space- and time-evolving characteristics, they will be added to the models used by groups around the world. Experience with these models so far has shown that the addition of more forcings generally tends to improve the realism and details of the simulations of the time evolution of the observed climate system (*e.g.,* Meehl *et al.,* 2004).

Whether the climate system is responding to internally generated variations in the atmosphere itself, to atmosphere-ocean-land-surface coupling, or to externally applied forcings by natural and/or anthropogenic factors, there are feedbacks that arise which can play a significant role in determining the changes in the vertical and horizontal thermal structure. These include changes in the hydrologic cycle involving water vapor, clouds (including aerosol-cloud interactions), sea-ice, and snow, which by virtue of their strong interactions with solar and longwave radiation, amplify the effects of the initial perturbation (Stocker *et al.,* 2001; NRC, 2003) in the heat balance, and thus influence

the response of the climate system. Convection and water vapor feedback, and cloud feedback in particular, are areas of active observational studies; they are also being pursued actively in climate modeling investigations to increase our understanding and reduce uncertainties associated with those processes.

## 1.3 STRATOSPHERIC FORCING AND RELATED EFFECTS

Observed changes in the stratosphere in recent decades have been large and several recent studies have investigated the causes. WMO (2003) and Shine *et al.* (2003) conclude that the observed vertical profile of cooling in the global, annual-mean stratosphere (from the tropopause up into the upper stratosphere) can, to a substantial extent, be accounted for in terms of the known changes that have taken place in well-mixed greenhouse gases, ozone, and water vapor (Figure 1.2). Even at the zonal, annual-mean scale, the lower stratosphere temperature trend is discernibly influenced by the changes in the stratospheric gases (Ramaswamy and Schwarzkopf, 2002; Langematz *et al.,* 2003). In the tropics, there is considerable uncertainty about the magnitude of the lower stratospheric cooling (Ramaswamy *et al.,* 2001a). In the high northern latitudes, the lower stratosphere becomes highly variable both in the observations and model simulations, especially during winter, such that causal attribution is difficult to establish. In contrast, the summer lower stratospheric temperature changes in the Arctic and the springtime cooling in the Antarctic can be attributed in large part to the changes in the greenhouse gases (WMO, 2003; Schwarzkopf and Ramaswamy, 2002).

Owing to the cooling of the lower stratosphere, there is a decreased infrared emission from the stratosphere into the troposphere (Ramanathan and Dickinson, 1979; WMO, 1999), leading to a radiative heat deficit in the upper troposphere, and a tendency for the upper troposphere to cool. In addition, the depletion of ozone in the lower stratosphere can result in ozone decreases in the upper troposphere due to reduced transport from the stratosphere (Mahlman *et al.,* 1994). This too affects the heat balance in the upper troposphere. Further, lapse rate near the tropopause can be affected by changes in radia-

The addition of more forcings in the models tends to improve the realism and details of simulations of the observed climate system.

tively active trace constituents such as methane (WMO, 1986; Pyle *et al.*, 2005).

The height of the tropopause (the boundary between the troposphere and stratosphere) is determined by a number of physical processes (*e.g., Holton et al.*, 1995) that make up the integrated heat content of the troposphere and the stratosphere. Changes in the heat balance within the troposphere and/or stratosphere can consequently affect the tropopause height. For example, when a volcanic eruption puts a large aerosol loading into the stratosphere where the particles absorb solar and longwave radiation and produce stratospheric heating and tropospheric cooling, the tropopause height shifts downward. Conversely, a warming of the troposphere moves the tropopause height upward (*e.g., Santer et al.*, 2003). Changes in tropopause height and their potential causes will be discussed further in Chapter 5.

The episodic presence of volcanic aerosols affects the equator-to-pole heating gradient, both in the stratosphere and troposphere. Temperature gradients in the stratosphere or troposphere can affect the state of the polar vortex in the northern latitudes, the coupling between the stratosphere and troposphere, and the propagation of temperature perturbations into the troposphere and to the surface. This has been shown to be plausible in the case of perturbations due to volcanic aerosols in observational and modeling studies, leading to a likely causal explanation of the observed warming pattern seen in northern Europe and some other high latitude regions in the first winter following a tropical explosive volcanic eruption (Robock, 2000, and references therein; Jones *et al.*, 2003; Robock and Oppenheimer, 2003;

Shindell *et al.*, 2001; Stenchikov *et al.*, 2002). Ozone and well-mixed greenhouse gas changes in recent decades can also affect stratosphere-troposphere coupling (Thompson and Solomon, 2002; Gillett and Thompson, 2003), propagating radiatively-induced temperature perturbations from the stratosphere to the surface in the high latitudes during winter.

## 1.4 SIMULATED RESPONSES IN VERTICAL TEMPERATURE PROFILE TO DIFFERENT EXTERNAL FORCINGS

Three-dimensional computer models of the coupled global atmosphere-ocean-land surface climate system have been used to systematically analyze the expected effects of different forcings on the vertical structure of the temperature response and compare these with the observed changes (*e.g., Santer et al.*, 1996; 2003; Hansen *et al.*, 2002). A climate model can be run with time-varying specification of just one forcing over the 20th century to study its effect on the

> Ozone and well-mixed greenhouse gas changes in recent decades can affect stratosphere-troposphere coupling.

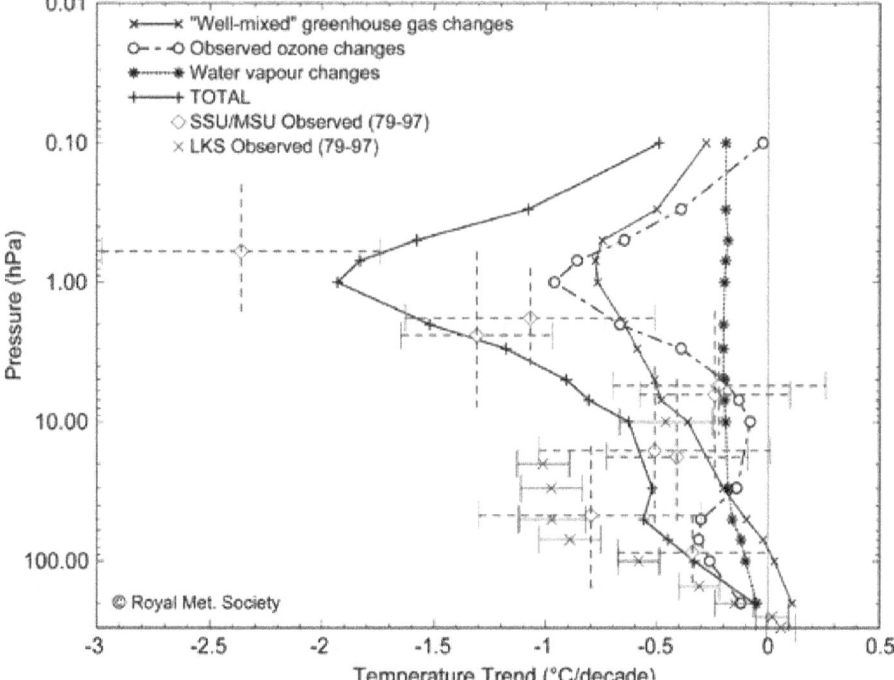

**Figure 1.2.** Global- and annual-mean temperature change over the 1979-1997 period in the stratosphere. Observations: LKS (radiosonde), SSU and MSU (satellite) data. Vertical bars on satellite data indicate the approximate span in altitude from where the signals originate, while the horizontal bars are a measure of the uncertainty in the trend. Computed: effects due to increases in well-mixed gases, water vapor, and ozone depletion, and the total effect (Shine *et al.*, 2003).

Models are able to reproduce the basic time evolution of globally averaged surface air temperature over the 20th century, and show that the warming in the late 20th century is mostly due to human induced increases of greenhouse gases.

vertical temperature profile. Then, by running more single forcings, a picture emerges concerning the relative effect of each forcing. The model can then be run with a combination of forcings to determine the degree to which the simulation resembles the observations made in the 20th century. Note that a linear additivity of responses, while approximately valid for certain combinations of forcings, need not hold in general (Ramaswamy and Chen, 1997; Hansen *et al.*, 1997; Santer *et al.*, 2003; Shine *et al.*, 2003a). To first order, models are able to reproduce the basic time evolution of globally averaged surface air temperature over the 20th century, with the warming in the first half of the century mostly due to natural forcings and internally generated variations, and the warming in the late 20th century mostly due to human-induced increases of GHGs (Stott *et al.*, 2000; Mitchell *et al.*, 2001; Meehl *et al.*, 2003; 2004; Broccoli *et al.*, 2003). Such modeling studies used various observed estimates of the forcings, but uncertainties remain regarding details of such factors as solar variability (Frohlich and Lean, 2004), historical volcanic forcing (Bradley, 1988), and tropospheric aerosols (Charlson *et al.*, 1992; Anderson *et al.*, 2003).

Analyses of model responses to external forcings also require consideration of the internal variability of the climate system for a proper causal interpretation of the observed surface temperature record (*e.g.*, Trenberth and Hurrell, 1994). The relationship between external forcing and internal decadal variability of the climate system (*i.e.*, can the former influence the latter, or are they totally independent?) is another intriguing research problem that is being actively studied (*e.g.*, Lindzen and Giannitsis, 2002; Wigley *et al.*, 2005).

In addition to the analyses of surface temperatures outlined above, climate models can also show the expected effects of different forcings on temperatures in the vertical. For example using simplified ocean representations for equilibrium 2XCO$_2$, Hansen *et al.* (2002) show that changes of various anthropogenic and natural forcings produce different patterns of temperature change horizontally and vertically. Hansen *et al.* (2002) also show considerable sensitivity of the simulated vertical temperature response to the choice of ocean representation,

particularly for the GHG-only and solar-only cases. For both of these cases, the "Ocean A" configuration (SSTs prescribed according to observations) lacks a clear warming maximum in the upper tropical troposphere, thus illustrating that there could be some uncertainty in our model-based estimates of the upper tropospheric temperature response to GHG forcing (see Chapter 5).

An illustration of the effects of different forcings on the trends in atmospheric temperatures at different levels from a climate model with time-varying forcings over the latter part of the 20th century is shown in Figure 1.3. Here, the temperature changes are calculated over the time period of 1958-1999, and are averages of four-member ensembles. As in Hansen *et al.*, this model, the NCAR/DOE Parallel Climate Model (PCM, *e.g.*, Meehl *et al.*, 2004) shows warming in the troposphere and cooling in the stratosphere for an increase of GHGs, warming through most of the stratosphere and a slight cooling in the troposphere for volcanic aerosols, warming in a substantial portion of the atmosphere for an increase in solar forcing, warming in the troposphere from increased tropospheric ozone and cooling in the stratosphere due to the decrease of stratospheric ozone, and cooling in the troposphere and slight warming in the stratosphere from sulfate aerosols. The multiple-forcings run shows the net effects of the combination of these forcings as a warming in the troposphere and a cooling in the stratosphere. Note that these simulations may not provide a full accounting of all factors that could affect the temperature structure, *e.g.*, black carbon aerosols, land use change (Ramaswamy *et al.*, 2001; Hansen *et al.*, 1997; 2002; Pielke, 2001; NRC, 2005; Ramanathan *et al.*, 2005).

The magnitude of the temperature response for any given model is related to its climate sensitivity. This is usually defined either as the equilibrium warming due to a doubling of CO$_2$ with an atmospheric model coupled to a simple slab ocean, or the transient climate response (warming at time of CO$_2$ doubling in a 1% per year CO$_2$ increase experiment in a global coupled model). The climate sensitivity varies among models due to a variety of factors (Cubasch *et al.*, 2001; NRC, 2004).

The important conclusion here is that representations of the major relevant forcings are important to simulate 20th century temperature trends since different forcings affect temperature differently at various levels in the atmosphere.

## 1.5 PHYSICAL FACTORS, AND TEMPERATURE TRENDS AT THE SURFACE AND IN THE TROPOSPHERE

Tropospheric and surface temperatures, although linked, are separate physical entities (Trenberth *et al.,* 1992; Hansen *et al.,* 1995; Hurrell and Trenberth, 1996; Mears *et al.,* 2003). Insight into this point comes from an examination of the correlation between anomalies in the monthly-mean surface and tropospheric temperatures over 1979-2003 (Figure 1.4). The correlation coefficients between monthly surface and tropospheric temperature anomalies (as represented by temperatures derived from MSU satellite data) reveal very distinctive patterns, with values ranging from less than zero (implying poor vertical coherence of the surface and tropospheric temperature anomalies) to over 0.9. The highest correlation coefficients (>0.75) are found across the middle and high latitudes of Europe, Asia, and North America, indicating a strong association between the surface and tropospheric monthly temperature variations. Correlations are generally much less (~0.5) over the tropical continents and the North Atlantic and North Pacific Oceans. Correlations less than 0.3 occur over the tropical and southern oceans and are

lowest (<0.15) in the tropical western Pacific. Relatively high correlation coefficients (>0.6) are found over the tropical eastern Pacific where the ENSO signal is large and the sea-surface temperature fluctuations influence the atmosphere significantly.

Differences between the surface and tropospheric temperature records are found where there is some degree of decoupling between the layers of the atmosphere. For instance, as discussed earlier, over portions of the subtropics and tropics, variations in surface temperature are disconnected from those aloft by a persistent trade-wind inversion. Shallow temperature

*Tropospheric and surface temperatures, although linked, are separate physical entities.*

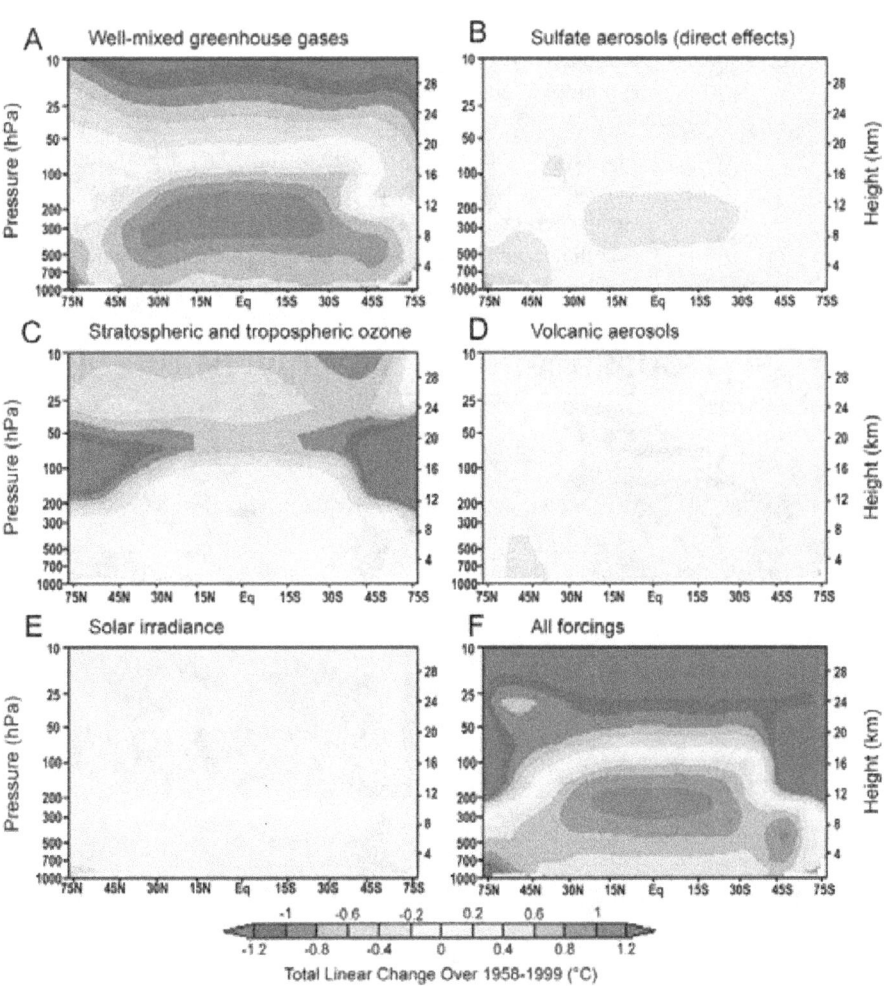

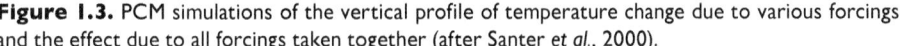

**Figure 1.3.** PCM simulations of the vertical profile of temperature change due to various forcings, and the effect due to all forcings taken together (after Santer *et al.,* 2000).

inversions are also commonly found over land in winter, especially in high latitudes on subseasonal timescales, so that there are occasional large differences between monthly surface and tropospheric temperature anomalies.

More important than correlations for trends, however, is the variability of the two temperature records, assessed by computing the standard deviation of the measurement samples of each record (Figure 1.5). The figure exhibits pronounced regional differences in variability between the surface and tropospheric records. Standard deviations also help in accounting for the differences in correlation coefficients, because they yield information on the size and persistence of the climate signal relative to the noise in the data. For instance, large variations in eastern tropical Pacific sea surface temperature associated with ENSO dominate over measurement uncertainties, as do large month-to-month swings in surface temperatures over extratropical continents.

The largest variability in both surface and tropospheric temperature is over the Northern Hemisphere continents. The standard deviation over the oceans in the surface data set is much smaller than over land except where the ENSO

phenomenon is prominent. The standard deviations of tropospheric temperature, in contrast, exhibit less zonal variability. Consequently, the standard deviations of the monthly tropospheric temperatures are larger than those of the surface data by more than a factor of two over the North Pacific and North Atlantic. Over land, tropospheric temperatures exhibit slightly less variability than surface temperatures. These differences in variability are indicative of differences in physical processes over the oceans versus the continents. Of particular importance are the roles of the land surface and ocean as the lower boundary for the atmosphere and their very different abilities to store heat, as well as the role of the atmospheric winds that help to reduce regional differences in tropospheric temperature through the movement of heat from one region to another.

Over land, heat penetration into the surface involves only the upper few meters, and the ability of the land to store heat is low. Therefore, land surface temperatures vary considerably from summer to winter and as cold air masses replace warm air masses and vice versa. The result is that differences in magnitude between surface and lower-atmospheric temperature anomalies are relatively small over the continents: very warm or cold air aloft is usually associated with very warm or cold air at the surface. In contrast, the ability of the ocean to store heat is much greater than that of land, and mixing in the ocean to typical depths of 50 meters or more considerably moderates the sea surface temperature response to cold or warm air. Over the northern oceans, for example, a very cold air mass (reflected by a large negative temperature anomaly in the tropospheric record) will most likely be associated with a relatively small negative temperature anomaly at the sea surface. This is one key to understanding the differences in trends between the two records.

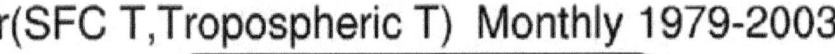

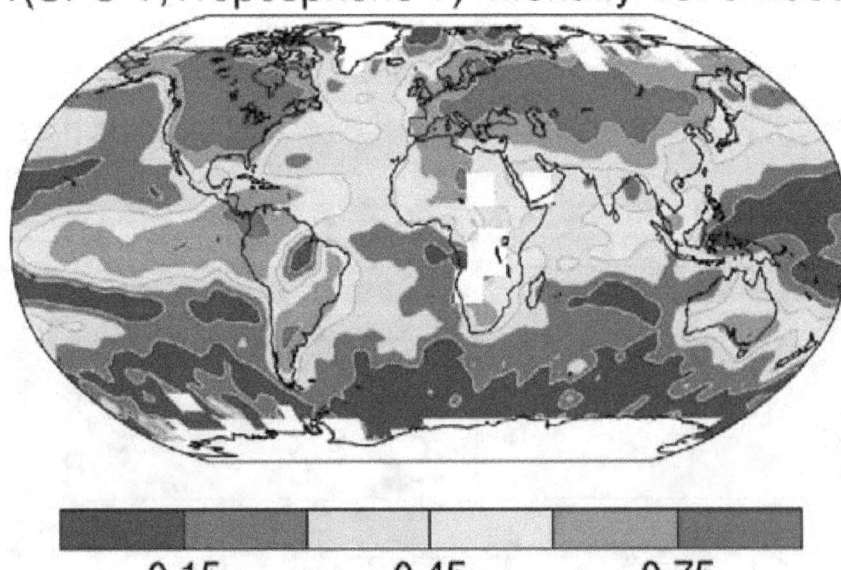

**Figure I.4.** Gridpoint correlation coefficients (r) between monthly surface and tropospheric temperature anomalies over 1979-2003. The tropospheric temperatures are derived from the MSU T$_2$ satellite data (Christy *et al.*, 2003).

Although the mechanisms for observed long-term changes in the atmospheric circulation are not fully understood, such changes are reflected by trends in indices of patterns (or modes) of natural climate variability such as ENSO, the North Atlantic Oscillation (NAO; also known as the Northern annular mode, or NAM), and the Southern annular mode (SAM). The exact magnitudes of the index trends depend on the period of time examined. Over the past several decades, for instance, changes in atmospheric circulation (reflected by a strong upward trend in indices of the NAO) have contributed to a Cold Ocean Warm Land (COWL) surface temperature pattern over the Northern Hemisphere (Hurrell, 1996; Thompson and Wallace, 2000). In the lower atmosphere, winds are much stronger than at the surface, and these stronger winds tend to moderate east-west variability in the tropospheric temperatures (Figure 1.5). Thus, the recent warm anomalies over the continents are roughly cancelled by the cold anomalies over the oceans in the tropospheric dataset. This is not the case for the surface temperature record, which is dominated by the warmth over the continents. The result is that the changes in the Northern Hemisphere atmospheric circulation over the past few decades have produced a significant difference in surface and tropospheric temperature trends (Hurrell and Trenberth, 1996). Similarly, Thompson and Solomon (2002) showed that recent tropospheric temperature trends at high southern latitudes were related to changes in the SAM.

Physical differences between the two measures of temperature are also evident in their dissimilar responses to volcanic eruptions and ENSO (*e.g.,* Santer *et al.* 2000). These phenomena have a greater effect on tropospheric than surface temperature, especially over the oceans (Jones, 1994). Hegerl and Wallace (2002) show that, in the tropics and subtropics, a distinctive signature of ENSO is apparent in the interannual variations in lapse rate, but that observed, longer-term

changes in the statistics of ENSO account only for a small fraction of the observed trend in lapse rate. Changes in concentrations of stratospheric ozone could also be important, as the troposphere is cooled more by observed ozone depletion than is the surface (Hansen *et al.,* 1995; Ramaswamy *et al.,* 1996). Another contributing factor could be that at the surface, the daily minimum temperature has increased at a faster rate than the daily maximum, resulting in a decrease in the diurnal temperature range over many parts of the world (*e.g.,* Easterling *et al.,* 1997; Dai *et al.,* 1999). Because of nighttime temperature inversions, the increase in the daily minimum temperatures likely involves only

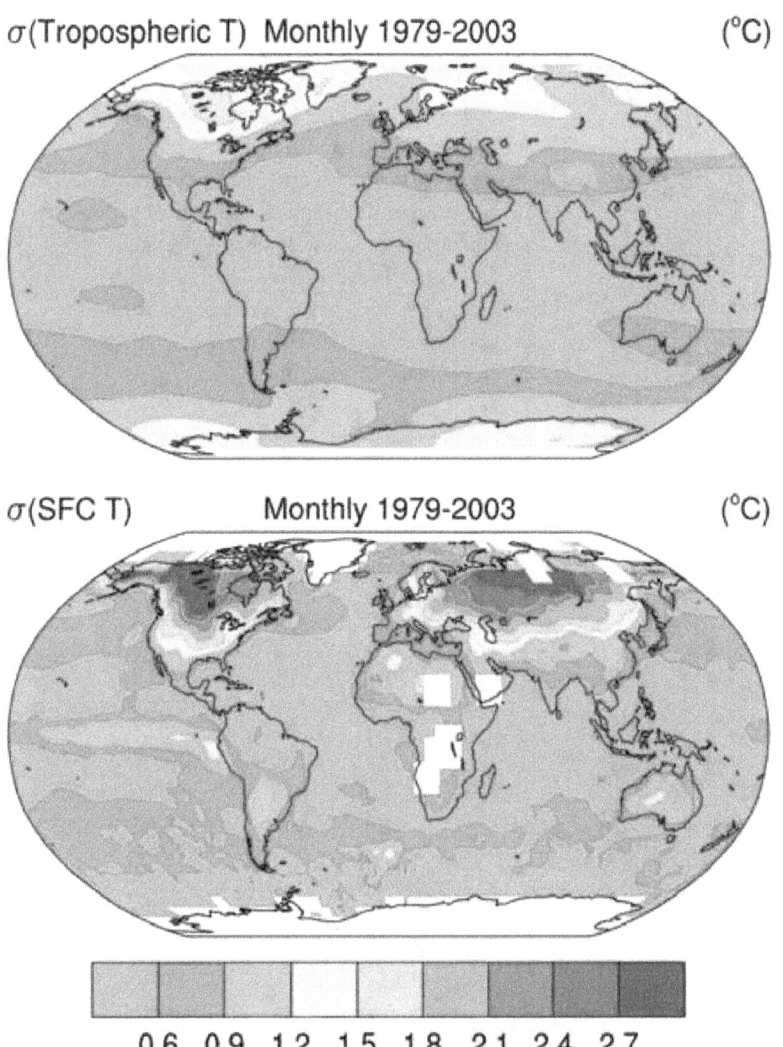

**Figure 1.5.** Standard deviations (σ) of monthly mean temperature anomalies from the surface and tropospheric temperature records over 1979-2003. The tropospheric temperatures are derived from the MSU T$_2$ satellite data (Christy *et al.,* 2003).

a shallow layer of the atmosphere that would not be evident in upper-air temperature records. However, during the satellite era, maximum and minimum temperatures have been rising at nearly the same rate, so that there has been almost no change in the diurnal temperature range (Vose *et al.*, 2005).

At the surface, the daily minimum temperature increased at a faster rate than the daily maximum since 1958. But since 1979, maximum and minimum temperatures have been rising at nearly the same rate.

These physical processes provide indications of why trends in surface temperatures are expected to be different from trends in the troposphere, especially in the presence of strong interannual variability, even if both sets of measurements were perfect. Of course they are not, as described in more detail in Chapter 2, which deals with the strengths and limitations of current observing systems. An important issue implicit in Figure 1.5 is that of spatial sampling, and the accompanying caveats about interpretation of the truly global coverage provided by satellites versus the incomplete space and time coverage offered by radiosondes. These are discussed in depth in Chapters 2 and 3.

# CHAPTER 2

**W**hat kinds of atmospheric temperature variations can the current observing systems measure and what are their strengths and limitations, both spatially and temporally?

*Convening Lead Author:* John R. Christy, Univ. of AL in Huntsville
*Lead Authors:* D.J. Seidel, NOAA; S.C. Sherwood, Yale Univ.
*Contributing Authors:* M. Cai, FL State Univ.; E. Kalnay, Univ. of MD; C.K. Folland, U.K. Met. Office; C.A. Mears, Remote Sensing Systems; P.W. Thorne, U.K. Met. Office; J.R. Lanzante, NOAA

## KEY FINDINGS

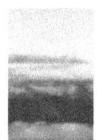

- The observing systems available for this report are able to detect small surface and upper air temperature variations from year to year, for example, those caused by El Niño or volcanic eruptions.
- The data from these systems also have the potential to provide accurate trends in climate over the last few decades (and over the last century for surface observations), once the raw data are successfully adjusted for changes over time in observing systems, practices, and micro-climate exposure (*e.g.*, urban heat island effect) to produce usable climate records. Measurements from all systems require such adjustments and this report relies solely on adjusted data sets.
- Adjustments to the land surface temperature record have been sufficiently successful that trends are reasonably similar on large (*e.g.*, continental) scales, despite the fact that spatial sampling is uneven and some errors undoubtedly remain. This conclusion holds to a lesser extent for the ocean surface record, which suffers from more serious sampling problems and changes in observing practice.
- Adjustments for changing instrumentation are most challenging for upper-air data sets. While these show promise for trend analysis, and it is very likely that current upper-air climate records give reliable indications of directions of change (*e.g.*, warming of the troposphere, cooling of the stratosphere), some questions remain regarding the accuracy of the measurements.
- Upper-air data sets have been subjected to less scrutiny than surface data sets.
- Adjustments are complicated, sometimes as large as the trend itself, involve expert judgments, and cannot be stringently evaluated because of lack of traceable standards.
  - Unlike surface trends, reported upper-air trends vary considerably between research teams beginning with the same raw data owing to their different decisions on how to remove non-climatic factors.
  - The diurnal cycle, which must be factored into some adjustments for satellite data, is well observed only by surface observing systems.
  - No available observing system has reference stations or multi-sensor instrumentation that would provide stable calibration over time.
  - Most observing systems have not retained complete metadata describing changes in observing practices which could be used to identify and characterize non-climatic influences.
- Relevant satellite data sets measure broad vertical layers and cannot reveal the detailed vertical structure of temperature changes, nor can they completely isolate the troposphere from the stratosphere. However, retrieval techniques can be used both to approximately isolate these layers and to check for vertical consistency of trend patterns. Consistency between satellite and radiosonde data can be tested by proportionately averaging radiosonde profiles.

- Reanalyses and other multi-system products have the potential for addressing issues of surface and atmospheric temperature trends by making better use of available information and allowing analysis of a more comprehensive, internally consistent, and spatially and temporally complete set of climate variables. At present, however, they contain biases, especially in the stratosphere, that affect trends and that cannot be readily removed because of the complexity of the data products.
- There are as yet under-exploited data archives with potential to contribute to our understanding of past changes, and new observing systems that may improve estimates of future changes if designed for long-term measurement stability and operated for sufficient periods.

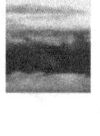

### Chapter 2: Recommendations

- Current and future observing systems should adhere to the principles for climate observations adopted internationally under the Framework Convention on Climate Change and documented in NRC 2000b and the Strategic Plan for the U.S. Climate Change Science Program (CCSP 2003) to significantly mitigate the limitations listed above.

- The ability to fully and accurately observe the diurnal cycle should be an important consideration in the design and implementation of new observing systems.

- When undertaking efforts to retrieve data it is important also to collect detailed metadata which could be used to reduce ambiguity in the timing, sign, and magnitude of non-climatic influences in the data.

- New climate-quality reanalysis efforts should be strongly encouraged and specifically designed to minimize small, time-dependent biases arising from imperfections in both data and forecast models.

- Some largely overlooked satellite data sets should be reexamined to try to extend, fortify, or corroborate existing microwave-based temperature records for climate research, e.g., microwave data from the Nimbus E Microwave Spectrometer (NEMS) (1972) and Scanning Microwave Spectrometer (SCAMS) (1975), infrared from the High-resolution Infrared Radiation Sounder (HIRS) suite, and radio occultation from GPS.

## I. MAIN OBSERVING SYSTEMS AND SYNTHESIS DATA PRODUCTS

Temperature is measured in three main ways; (1) *in situ*, where the sensor is immersed in the substance of interest; (2) by radiative emission, where a remote sensor detects the intensity or brightness of the radiation emanating from the substance; and (3) radiative transmission, where radiation is modified as it passes through the substance in a manner determined by the

substance's temperature. All observations contain some level of random measurement error, which is reduced by averaging; bias, which is not reduced by averaging; and sampling errors (see Appendix A).

### a) Surface and Near-surface Air Temperatures

Over land, "near-surface" air temperatures are those commonly measured about 1.5 to 2.0 meters above the ground level at official weather stations, at sites run for a variety of scientific purposes, and by volunteer ("coopera-tive") observers (*e.g.*, Jones and Moberg, 2003). These stations often experience relocations, changes in instrumentation and/or exposure (including changes in nearby thermally emit-ting structures), effects of land-use changes (*e.g.*, urbanization), and changing observing practices, all of which can introduce biases into their long-term records. These changes are often undocumented.

"Near-surface" air temperatures over the ocean ("Marine Air Temperatures" or MATs) are measured by ships and buoys at various heights from 2 to more than 25 meters, with poorer temporal and spatial coverage than over land (*e.g.*, Rayner *et al.*, 2003). To avoid the con-tamination of daytime solar heating of the ships' surfaces that may affect the MAT, it is generally preferred to limit these to night MAT (NMAT) readings only. Observations of the water tem-perature near the ocean surface or "Sea Surface Temperatures" (SSTs) are widely used and are closely tied to MATs; ships and buoys measure SSTs within a few meters of the surface. The scale of the spatial and temporal coherence of SST and MAT anomalies is greater than that of near-surface air temperatures over land; thus a lower rate of oceanic sampling, in theory, can provide an accuracy similar to the more densely monitored land area.

Incomplete geographic sampling, changing measurement methods, and land-use changes all introduce errors into surface temperature compilations (Jones *et al.*, 2001). The spatial coverage, indicated in Figure 2.1, is far from uniform over either land or ocean areas. The southern oceans, polar regions, and interiors of Brazil and Africa are not well sampled by in-situ networks. However, creating global surface temperature analyses involves not only merging land and ocean data but also considering how best to represent areas where there are few or no observations. The most conservative approach is to use only those grid boxes with data, thus avoiding any error associated with interpola-tion. Unfortunately, the areas without data are not evenly or randomly distributed around the world, leading to considerable uncertainties in the analysis, though it is possible to make an estimate of these uncertainties. Using the conservative approach, the tropical land surface

areas would be under-represented, as would the southern ocean. Therefore, techniques have been developed to interpolate data to some extent into surrounding data-void regions. A single group may produce several different such data sets for different purposes. The choice may depend on whether the interest is a particular local region, the entire globe, or use of the data set with climate models (Chapter 5). Estimates of global and hemispheric scale averages of near-surface temperatures generally begin around 1860 over both land and ocean.

Data sets of near-surface land and ocean tem-peratures have traditionally been derived from in-situ thermometers. With the advent of satel-lites, some data sets now combine both *in-situ*

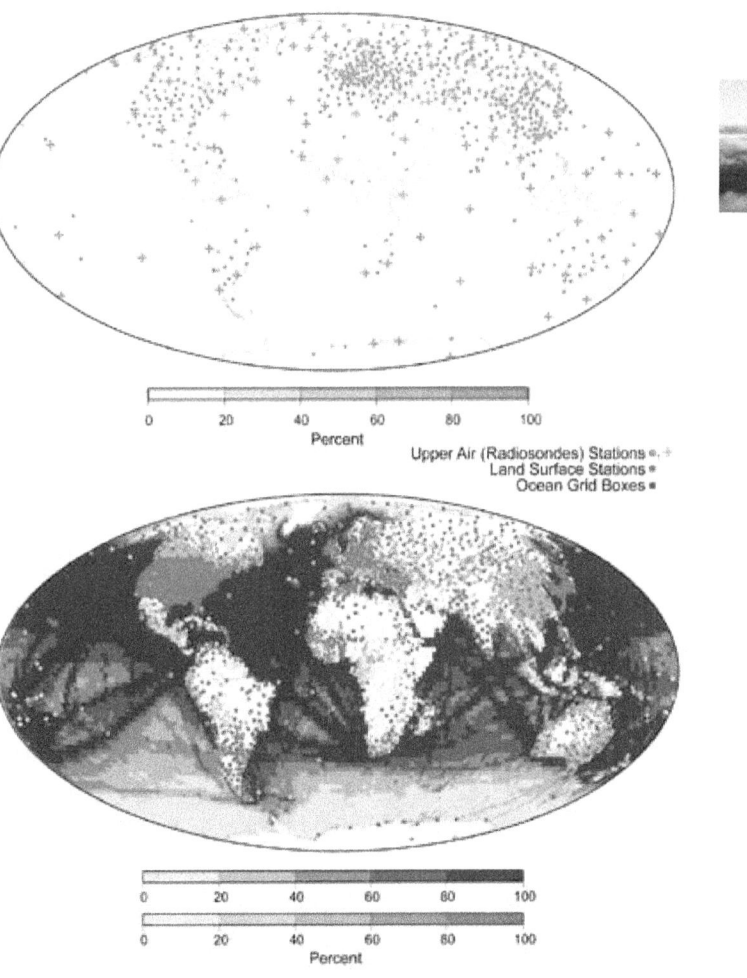

**Figure 2.1** Top: Location of radiosonde stations used in the HadAT2 upper air data set with those also in the RATPAC as crosses. Bottom: Distribution of land stations (green) and SST observations (blue) reporting temperatures used in the surface temperature data sets over the period 1979-2004. Darker colors represent locations for which data were reported with greater frequency. See chapter 3 for definitions of data sets.

and remotely sensed data (Reynolds *et al.,* 2002; Smith and Reynolds, 2005), or use exclusively remotely sensed data (Kilpatrick *et al.,* 2001) to produce more geographically complete distributions of surface temperature. Because the satellite sensors measure infrared or microwave emission from the Earth's surface (a "skin" typically tens of microns thick that may have a temperature different from either the air above or material at greater depths), calculations are required to convert the skin temperature into the more traditional near-surface air or SST observation (in this context SSTs are called "bulk sea surface temperatures," Chelton, 2005.) Typically, *in situ* observations are taken as "truth" and satellite estimates (which may be affected by water vapor, clouds, volcanic aerosols, *etc.*) are adjusted to agree with them (Reynolds, 1993.) With continued research, data sets with surface temperatures over land, ice, and ocean from infrared and microwave sensors should provide expanded coverage of surface temperature variations (*e.g.,* Aires *et al.,* 2004).

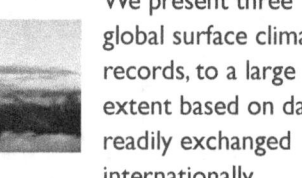

We present three global surface climate records, to a large extent based on data readily exchanged internationally.

Sampling errors in ship and buoy SST data typically contribute more to large-scale averages than random measurement errors as shown in Smith and Reynolds (2004), especially as the temperature record extends backward in time. Biases depend on observing method. Most ship observations since the 1950s were made from insulated buckets, hull contact sensors, and engine intake temperatures at depths of one to several meters. Historical correction of ship data prior to 1942 is discussed by Folland and Parker (1995) and Folland (2005) and bias and random errors from ships are summarized by Kent and Taylor (2006) and Kent and Challenor (2006). They report that engine intake temperatures are typically biased 0.1-0.2°C warmer than insulated buckets. This is primarily due to engine room heating of the water temperatures although there is also evaporative cooling of the water in the insulated buckets. Hull contact sensors are the most accurate though much less common. The bias correction of the ship SST data (Kent and Kaplan, 2006) requires information on the type of measurement (*e.g.,* insulated bucket, *etc.*) which becomes more difficult to determine prior to 1990s due to incomplete documentation. Kent and Kaplan (2006) also found that insulated bucket temperatures may be too cold by 0.12 to 0.16°C. When the bucket bias is used, engine intake temperatures in the mid-to-late 1970s and the 1980s were found to be smaller than that suggested by previous studies, ranging from 0.09 to 0.18°C. In addition, their study indicates that engine intake SSTs may have a cold bias of -0.13°C in the early 1990s. The reliability of these biases is subject to revision due to sample sizes that for these comparisons tend to be small with large random errors. Buoy observations became more plentiful following the start of the Tropical Ocean Global Atmosphere (TOGA) Program (McPhaden, 1995) in 1985. These observations are typically made by an immersed temperature sensor or a hull contact sensor, and are more accurate because they do not have the bias errors of ship injection or insulated bucket temperatures.

The global surface air temperature data sets used in this report are to a large extent based on data readily exchanged internationally, *e.g.,* through CLIMAT reports and the WMO publication Monthly Climatic Data for the World. Commercial and other considerations prevent a fuller exchange, though the United States may be better represented than many other areas. In this report, we present three global surface climate records, created from available data by NASA Goddard Institute for Space Studies, NOAA National Climatic Data Center, and the cooperative project of the U.K. Hadley Centre and the Climate Research Unit of the University of East Anglia (HadCRUT2v). These are identified as $T_{S-NASA}$, $T_{S-NOAA}$ and $T_{S-HadCRUT2v}$ respectively.

## b) Atmospheric "Upper Air" Temperatures

### I. RADIOSONDES

Radiosonde or balloon-based observations of atmospheric temperature are in-situ measurements as the thermometer (often a thermistor or a capacitance-based sensor), suspended from a balloon, is physically carried through the atmospheric column. Readings are radio-transmitted back to a data recorder. Balloons are released once or twice a day (00 and/or 12 Coordinated Universal Time or UTC) at about 1,000 stations around the globe, many of which began operations in the late 1950s or 1960s. These sites are unevenly distributed, with only the extra-tropical Northern Hemisphere land areas and the Western Pacific Ocean/Indonesia/Australia region being well-sampled in space and time. Useful temperature data can be collected from near the surface through the lower and middle stratosphere (though not all balloons survive to these heights). Radiosonde data in the first hundred meters or so above the surface are sometimes erroneous if the sensors have not been allowed to reach equilibrium with the atmosphere before launch, and may not be representative of regional conditions, due to microclimatic and terrain effects.

Although most radiosonde data are transmitted to meteorological centers around the world and archived, in practice many soundings do not reach this system and are collected later. No definitive archive of radiosonde data exists, but several archives in the U.S. and abroad contain nearly complete collections, though several different schemes have been employed for quality control. To monitor climate, it is desirable to have a long, continuous record of measurements from many well-distributed fixed sites. There are about 700 radiosonde stations that have operated in the same location for at least three decades; many of these are clustered in a few areas, further reducing the effective coverage (Figure 2.1). Thus, a dilemma exists for estimating long-term changes: whether to use a smaller number of stations having long segments of continuous records, or a larger number of stations with shorter records that do not always overlap well. Various analysis groups have approached this dilemma differently (see Chapters 3 and 4).

Typically, radiosonde-based data sets are developed for specific atmospheric pressure surfaces known as "mandatory reporting levels" (Figure 2.2). Such data at discrete vertical levels provide unique information for assessing changes in the structure of the atmosphere. Two such data sets are featured in this report, the Hadley Centre Atmospheric Temperatures from the U.K. (HadAT2), and Radiosonde Atmospheric Temperatures Products for Assessing Climate (RATPAC) from NOAA. A product such as $T_{850-300}$, for example, will be identified as $T_{850-300-HadAT2}$ and $T_{850-300-RATPAC}$ for HadAT2 and RATPAC respectively.[1]

Throughout the radiosonde era there have been numerous changes in stations, types of instrumentation, and data processing methods that can create data discontinuities. Because radiosondes are expendable instruments, instruments are more easily changed than for the more permanent surface sites. The largest discontinuities appear to be related to solar heating of the temperature sensor and changes in design and/or data adjustments intended to deal with this problem. These discontinuities have greatest impact at stratospheric levels (the stratosphere's lower boundary is ~16 km in the tropics, dropping to < 10 km in the high latitudes, Figure 2.2), where direct sunlight can cause radiosonde-measured temperatures to rise several °C above ambient temperatures. For example, when Australia and U.S. stations changed instrumentation to Vaisala RS-80, processed stratospheric temperatures shifted downward by 1 to 3°C (Parker *et al.,* 1997, Christy *et al.,* 2003). Many other sources of system-dependent bias exist (which often affect the day

---

[1] A third radiosonde data set was generated by comparing radiosonde observations against the first-guess field of the ERA-40 simulation forecast model (Haimberger, 2004). Adjustments were applied when the relative difference between the radiosonde temperatures and the forecast temperatures changed by a significant amount. The data were not yet in final form for consideration in this report, although the tropospheric values appear to have general agreement with HadAT2 and RATPAC.

and night releases differently, Sherwood *et al.,* 2005), including icing of the sensors in regions of super-cooled water, software errors in some radiosonde systems, poor calibration formulae, and operator errors. Documentation of these many changes is limited, especially in the earlier decades.

### 2. PASSIVE SATELLITE INSTRUMENTATION

Unlike radiosondes, passive satellite observations of microwave and infrared brightness temperatures sample relatively thick atmospheric layers (and may include surface emissions), depicted as weighting functions in Figure 2.2. These measurements may be thought of as bulk atmospheric temperatures, as a single value describes the entire layer. Although this bulk measurement is less informative than the detailed information from a radiosonde, horizontal coverage is far superior, and consistency can be checked by comparing the appropriate vertical average from a radiosonde station against nearby satellite observations. Furthermore, because there are far fewer instrument systems than in radiosonde data sets, it is potentially easier to isolate and adjust problems in the data.

The space and time sampling of the satellites varies according to the orbit of the spacecraft, though the longer satellite data sets are based on polar orbiters. These spacecraft circle the globe from near pole to pole while maintaining a nominally constant orientation relative to the sun (sun-synchronous). In this configuration, the spacecraft completes about 14 roughly north-south orbits per day as the Earth spins eastward beneath it, crosses the equator at a constant local time, and provides essentially global coverage. Microwave measurements utilized in this report begin in late 1978 with the Television Infrared Observation Satellite (TIROS-N) spacecraft using a 4-channel radiometer (Microwave Sounding Unit or "MSU") which was upgraded in 1998 to a 16-channel system (advanced MSU or "AMSU") with better calibration, more stable station-keeping (*i.e.,* the timing and positioning of the satellite in its orbit - see discussion of "Diurnal Sampling" below), and higher spatial and temporal sampling resolution.

Laboratory estimates of precision (random error) for a single MSU measurement are 0.25°C. Thus with 30,000 observations per day, this error is inconsequential for global averages. Of far more importance are the time varying biases that arise once the spacecraft is in orbit: diurnal drifting, orbital decay, inter-satellite biases, and calibration changes due to heating and cooling of the instrument in space (see section 3 below.)

While bulk-layer measurements offer the robustness of a large-volume sample, variations within the observed layer

> Unlike radiosondes, passive satellite observations sample relatively thick atmospheric layers.

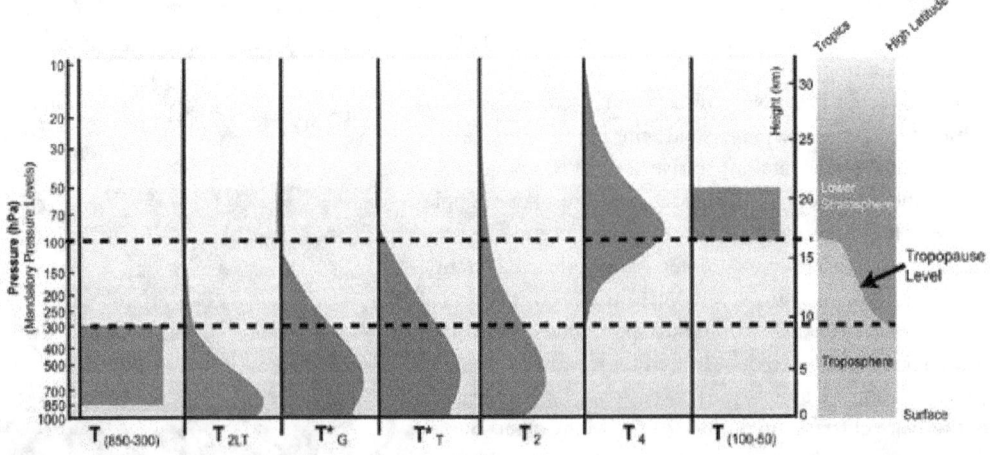

**Figure 2.2** Terminology and vertical profiles for the temperature products referred to in this report. Radiosonde-based layer temperatures ($T_{850-300}$, $T_{100-50}$) are height-weighted averages of the temperature in those layers. Satellite-based temperatures ($T_{2LT}$, $T_2$, and $T_4$) are mass-weighted averages with varying influence in the vertical as depicted by the curved profiles, *i.e.,* the larger the value at a specific level, the more that level contributes to the overall satellite temperature average.

**Notes:** (1) because radiosondes measure the temperature at discrete (mandatory) levels, their information may be used to create a temperature value that mimics a satellite temperature (Text Box 2.1), (2) layer temperatures vary from equator to pole so the pressure and altitude relationship here is based on the atmospheric structure over the conterminous U.S., (3) about 10% (5%) of the value of $T_{2LT}$ ($T_2$) is determined by the surface character and temperature, (4) $T^*_T$ and $T^*_G$ are simple retrievals, being linear combinations of 2 channels, $T_2$ and $T_4$.

are masked. This is especially true for the layer centered on the mid-troposphere ($T_2$) for which the temperatures of both lower stratospheric and tropospheric levels, which generally show opposite variations, are merged (Figure 2.2). Three MSU/AMSU-based climate records are presented in this report, prepared by Remote Sensing Systems (RSS) of Santa Rosa, California, The University of Alabama in Huntsville (UAH), and The University of Maryland (UMd).

Some polar orbiters also carry the Stratospheric Sounding Unit (SSU), an infrared sensor for monitoring deep layer temperatures above about 15 km. SSU data have been important in documenting temperature variations at higher elevations than observed by MSU instruments on the same spacecraft (Ramaswamy *et al.*, 2001). Generally, the issues that complicate the creation of long-term MSU time series also affect the SSU, with the added difficulty that infrared channels are more sensitive to variations in atmospheric composition (*e.g.*, volcanic aerosols, water vapor, *etc.*).

Future observing systems using passive-satellite methods include those planned for the National Polar-orbiting Operational Environmental Satellite System (NPOESS) series: the microwave sensors Conical scanning Microwave Imager/Sounder (CMIS) (which will succeed the Special Sensor Microwave/Imager [SSM/I]), Special Sensor Microwave Imager/Sounder (SSMI/S) and Advanced Technology Microwave Sounder (ATMS) (which will succeed the AMSU), and the infrared sensor Cross-track Infrared Sounder (CrIS) (following the High-resolution Infrared Radiation Sounder [HIRS]). Each of these will follow measuring strategies that are both similar (polar orbit) and dissimilar (*e.g.,* CMIS's conical scanner vs. AMSU's cross-track scanner) but add new spectral and more detailed resolution.

3. "ACTIVE" SATELLITE INSTRUMENTATION
A relatively recent addition to temperature monitoring is the use of Global Positioning System (GPS) radio signals, whose time of transmission through the atmosphere is altered by an amount proportional to air density and thus temperature at levels where humidity can be ignored (Kursinski *et al.*, 1997). A key

advantage of this technique for climate study is that it is self-calibrating. Current systems are accurate in the upper troposphere and lower to middle stratosphere where moisture is insignificant, but at lower levels, humidity becomes a confounding influence on density. Future versions of this system may overcome this limitation by using shorter wavelengths to measure humidity and temperature independently. Because of the relatively short GPS record and limited spatial coverage to date, its value for long-term climate monitoring cannot yet be definitively demonstrated.

### c) Operational Reanalyses

Operational reanalyses (hereafter simply "reanalyses") will be discussed here in Chapter 2, but trends derived from them are presented only sparingly in the following chapters because of evidence that they are not always reliable, even during the recent period. All authors expressed concern regarding reanalyses trends, a concern that ranged from unanimous agreement that stratospheric trends were likely spurious to mixed levels of confidence regarding tropospheric trends (see chapter 3). Surface temperature trends are a separate issue as reanalyses values are indirectly estimated rather than observed (see below). However, reanalyses products hold significant potential for addressing many aspects of climate variability and change.

Reanalyses are not separate observing systems, but are mathematically blended products based upon as many observing systems as practical. Observations are assimilated into a global weather forecasting model to produce analyses that are most consistent with both the available data (given their imperfections) and the assimilation model, which represents in a theoretical manner how the atmosphere behaves. The model, which is constrained by known but parameterized atmospheric physics, generates a result that could be more accurate and physically self-consistent than can be obtained from any one observing system. Some data are rejected or adjusted based on detected inconsistencies. Importantly, the operational procedure optimizes only the accuracy of each near-instantaneous ("synoptic") analysis. Time-varying biases of a few hundredths or tenths of a degree, which contribute little to

Reanalyses will be discussed here, but trends derived from them are presented only sparingly in the following chapters because of evidence that they are not always reliable.

short time scale weather error, present a major problem for climate trends, and these are not minimized (*e.g.,* Sherwood, 2000). The two main reanalyses available at this time are the National Centers for Environmental Prediction (NCEP)/National Center for Atmospheric Research (NCAR) reanalysis of data since 1948 (Kalnay *et al.,* 1996) and the European Centre for Medium-Range Weather Forecasts Re-Analysis-40 (ECMWF ERA-40) beginning in 1957 (Uppala *et al.,* 2005).

Because many observational systems are employed, a change in any one will affect the time series of the final product unless flagged and corrected as ERA-40 attempts to do. Reanalyses would be more accurate than lower-level data products for climate variations only if the above shortcomings were outweighed by the benefits of using a state-of-the-art model to treat unsampled variability. Factors that would make this scenario likely include a relatively skillful forecast model and assimilation system, large sampling errors (which are reduced by reanalysis), and small systematic discrepancies between different instruments. However, current models tend to have significant intrinsic biases that can particularly affect reanalyses when sampling is sparse.

Reanalysis problems that influence temperature trend calculations arise from changes over time in (a) radiosonde and satellite data coverage, (b) radiosonde biases (or in the corrections applied to compensate for these biases), (c) the effectiveness of the bias corrections applied to satellite data and (d) the propagation of errors due to an imprecise formulation of physical processes in the models. For example, since few data exist for the Southern Hemisphere before 1979, temperatures were determined mainly by model forecasts; a cold model bias (in ERA-40, for example) then produces a spurious warming trend when real data become available. Indirect effects may also arise from changes in the biases of other fields, such as humidity and clouds, which affect the model temperature (Andrae *et al.,* 2004; Simmons *et al.,* 2004, Bengtsson *et al.* 2004.).

Different reanalyses do not employ the same data. NCEP/NCAR does not include surface temperature observations over land but the analysis still produces estimated near-surface temperatures based on the other data (Kalnay and Cai, 2003). On the other hand ERA-40 does incorporate these by blending the observations with a background forecast for complete geographic coverage (Simmons et al. 2004). Thus, the 2-meter air temperatures of both reanalyses may not track closely with surface observations over time (Kalnay and Cai, 2003). SSTs in both reanalyses are simply those of the climate records used as input.

Simultaneous assimilation of radiosonde and satellite data for upper-air temperatures in reanalyses is particularly challenging because the considerably different instrument characteristics and products make it difficult to achieve the consistency possible in theory. Despite data adjustments, artifacts still remain in both radiosonde and satellite analyses; these produce the largest differences in the lower stratosphere in current reanalysis data sets (*e.g.,* Pawson and Fiorino, 1999; Santer *et al.,* 1999; Randel and Wu, 2006). Some of these differences can now be explained, so that future reanalyses will very likely improve on those currently available. However any calculation of deep-layer temperatures from reanalyses which require stratospheric information are considered in this report to be suspect (see Figure 2.2, $T^*_T$, $T_2$, $T_4$, and $T_{100-50}$).

### d.) Simple Statistical Retrieval Techniques

A problem in interpreting MSU (*i.e.,* broad-layer) temperature trends is that many channels receive contributions from both the troposphere and stratosphere, yet temperatures tend to change oppositely in these two layers with respect to both natural variability and predicted climate change. In particular, MSU Channel 2 ($T_2$) receives 10-15% of its emissions from the stratosphere (Spencer and Christy, 1992), which is a significant percentage because stratospheric cooling in recent decades far exceeds tropospheric warming. In principle, subtracting an appropriate fraction of MSU 4 from MSU 2, as advocated by Fu *et al.* (2004), will produce a value more representative of the troposphere. The statistical retrieval has the form: Tropospheric Retrieval = $(1+y)\cdot(T_2) - (y)\cdot(T_4)$ where $y$ is determined by regression. However this will not work exactly, because the stratospheric

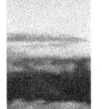

Despite data adjustments, artifacts still remain in both radiosonde and satellite analyses.

contribution to a signal in either channel will depend on the nature of the signal (*e.g.,* its time scale or cause). The seriousness of this problem for trends is not large (Gillett *et al.,* 2004; Johanson and Fu, 2006; Kiehl *et al.,* 2005) but some problems have been demonstrated (Tett and Thorne, 2004; Spencer *et al.,* 2006).

Fu *et al.* (2004) used a radiosonde data set to estimate values for y (for the globe, tropical region, and Northern and Southern Hemispheres) that most closely reproduced the monthly variability of mean temperature from 850 to 300 hPa, spanning most of the troposphere. From physical arguments, however, it is clear that the true physical contributions to the retrieval come from a broader range of altitudes, which, in the tropics, approximately span the full troposphere (Fu and Johanson, 2005). In the following chapters, two simple statistical retrievals will be utilized in comparison studies with the products of the observing systems. The tropospheric retrieval generated from global mean values of $T_2$ and $T_4$, is identified as $T^*_G$ where y = 0.143 (Johanson and Fu, 2006), and when applied to tropical mean values is identified as $T^*_T$ where y = 0.100 (Fu and Johanson, 2005).

A summary of the sources of biases and uncertainties for the data sets and other products described above is given at the end of this chapter. There are several data sets yet to be generated (or not yet at a stage sufficient for climate analysis) from other sources that have the potential to address the issue of vertical temperature distribution. A generic listing of these data sets with a characterization of their properties is given in Table 2.1.

## 2. ANALYSIS OF CLIMATE RECORDS

Two factors can interfere with the accurate assessment of climate variations over multi-year periods and relatively large regions. First, much larger variability (weather or "atmospheric noise") on shorter time or smaller space scales can, if inadequately sampled by the observing network, bias estimates of relatively small climate changes. For example, an extended heat wave in an un-instrumented region accompanied by a compensating cold period in a well-instrumented region may be interpreted

as a "global" cold period when it was not. Such biases can result from either spatial or temporal data gaps (Agudelo and Curry, 2004). Second, instrumental errors, particularly biases that change over time, can create erroneous trends. The seriousness of each problem depends not only on the data available but also on how they are analyzed. Finally, even if global climate is known accurately at all times and places, there remains the issue of what measures to use for quantifying climate change; different choices can sometimes create different impressions, *e.g.,* linear trends versus low frequency filtered analyses that retain some information beyond a straight line.

Temperature variations of upper air layers are characterized by large, coherent anomaly features on annual time scales, especially in the east-west direction (Wallis, 1998, Thorne *et al.,* 2005b). As a result, a given accuracy for the global mean value over, say, a year, can be attained with fewer, if reasonably spaced, upper air measurement locations than at the surface (Hurrell *et al.,* 2000). Thus, knowledge of global, long-term changes in upper-air temperature is likely limited more by instrumental errors than spatial coverage. However, for some regional changes (*e.g.,* over sparsely observed ocean areas) sampling problems may compete with or exceed instrumental ones.

### a) Climate Records
Various groups have developed long time series of climate records, often referred to as Climate Data Records (CDRs) (NRC, 2000a; 2000b; 2004) from the raw measurements generated by each observing system. Essentially, climate records are time series that include estimates of error characteristics so as to enable the study of climate variation and change on decadal and longer time scales with a known accuracy and precision.

Long-term temperature changes occur within the context of shorter-term variations, which are listed in Table 2.2. These shorter changes include: periodic cycles such as day-night and seasonal changes; fairly regular changes due to synoptic weather systems, the Quasi-Biennial Oscillation (QBO), and the El Niño-Southern Oscillation (ENSO); and longer-term variations due to volcanic eruptions or internal climate

Various groups have developed long time series of climate records, often referred to as Climate Data Records, that include estimates of error characteristics.

**Table 2.1  Data sources for climate monitoring related to the vertical temperature structure of the atmosphere.**

| DATA SOURCE | Measured Variables | Start date of main availability | Temporal Sampling | Geographic Completeness |
|---|---|---|---|---|
| Radiosondes (Balloons) | Upper Air Temperature | 1958 | 2x Day | |
| | Upper Air Humidity | 1958 | 2x Day | |
| | Upper Air Wind | 1958 | 2x Day | |
| Microwave Radiometers Space-based | Upper air Temperature | 1978 | P | |
| | Sea Surface Temperature | 1997 | P | |
| | Total Column Vapor (ocean) | 1987 | P | |
| Surface-based sounders and profilers | Upper air Temperature | ~1985 | Hrly | |
| Infrared Radiometers Space-based | Upper Air Temperature | 1973 | P, G | |
| | Land Surface Temperature | 1976 | P, G | |
| | Sea Surface Temperature | 1981 | P, G | |
| | Upper Air Humidity | 1973 | P, G | |
| Visible and Infrared Radiometers | Radiative Fluxes | 1979 | P, G | |
| GPS Satellites | Temperature | 2001 | quasi-P | |
| Surface Stations Land | Land Surface Air Temperature | ~1850 | Hrly | |
| | Land Surface Air Humidity | ~1880 | Hrly since 1973 | |
| Surface Instruments Ocean | Sea Surface Temperature | ~1850 | Syn | |
| | Marine Air Temperature | 1856 | Syn | |
| Reanalyses | All | 1950 | Syn | |

Global or near-global distribution of observations
Large regions not regularly sampled
Low spatial distribution of observations

P: Polar orbiter, twice per day per orbiter per ground location except in swath gaps 40°S – 40°N.
G: Geostationary, many observations per day per ground location
2x Day:  Twice daily at site
Hrly:  Up to several times per day, many report hourly
Syn: Synoptic or generally up to 8 times per day.  (Buoys continuous)
Quasi-P: requires transmitter and receiver (at least one of which is satellite-mounted) to be appropriately positioned to sample atmosphere. Opportunities are not spatially/temporally systematic to date but are expected to be in the future.

dynamics. These changes have different vertical temperature signatures, and the magnitude of each signal may be different at the surface, in the troposphere, and in the stratosphere. Some of these signals can complicate the identification of temperature trends in climate records.

Our survey of known atmospheric temperature variations, how well they are measured, and their impact on trend estimates suggests that most observing systems are generally able to quantify well the magnitudes of change associated with shorter time scales. For longer time scale changes, where the magnitudes of change are smaller and the stability requirements more rigorous, the observing systems face significant challenges (Seidel *et al.,* 2004).

### b) Measuring Temperature Change

Over the last three to five decades, global surface temperature records show increases of about +0.15°C per decade. Explaining atmospheric and surface trends therefore demands relative accuracies of a few hundredths of a degree C per decade in global time series of both surface and upper-air observations. As this and subsequent chapters will show, the effects of instrumental biases on the global time series are significantly larger than a few hundredths of a degree for the upper-air data, though the global surface temperature compilations do appear to reach this level of accuracy in recent decades (Folland *et al.,* 2001b). These biases, especially those of the upper air, must therefore be understood and quantified rather precisely (see section 3 below). For this fundamental reason, reliable assessment of lapse rate changes remains a considerable challenge.

Natural modes of climate variability on regional scales are manifested in decadal fluctuations in (a) the tropical Pacific, e.g., ENSO, and (b) the northern latitudes, *e.g.,* the North Atlantic, Pacific-North American, and the Arctic atmospheric oscillations (Table 2.2). Even fluctuations on longer time scales have been proposed, e.g., the Atlantic Multidecadal Oscillation/60-80 year variation (Schlesinger and Ramankutty, 1994; Enfield *et al.,* 2001; Knight *et al.,* 2005). Each of these phenomena is associated with regions of both warming and cooling. Distinguishing slow, human-induced changes from such phenomena requires identifying

the patterns and separating the influences of such modes from the warming signal (*e.g.,* as attempted for SST by Folland *et al.,* 1999.) In addition, these oscillations could themselves be influenced by human-induced atmospheric changes (Hasselmann, 1999).

## 3. LIMITATIONS

A key question addressed in this report is whether climate records built by investigators using various components of the observing system can meet the needs for assessing climate variations and trends with the accuracy and representativeness which allows any human attribution to be reliably identified. Climate record builders have usually underestimated the overall uncertainty in their products by relying on traditional sources of uncertainty that can be quantified using standard statistical methods. For example, published linear trend values exist of the same temperature product from the same observing system whose error estimates do not overlap, indicating serious issues with error determination. Thus, in 2003, three realizations of $T_2$ (or MSU channel 2) 1979-2002 global trends were published as +0.03 ±0.05, +0.12 ±0.02, and +0.24 ±0.02 °C per decade (Christy *et al.,* 2003; Mears *et al.,* 2003; and Vinnikov and Grody, 2003, respectively.) Over 40% of the difference between the first two trends is due to the treatment of a single satellite in the 1984-1986 period, with a combination of lesser differences during later satellite periods. The third data set has more complex differences, though it is being superseded by a version whose trend is now lower (Grody *et al.,* 2004, Vinnikov *et al.,* 2006).

This situation illustrates that it is very challenging to determine the true error characteristics of data sets (see Chapter 4), although considerably less attention has been paid to this than to the construction of the data sets themselves. In this report, we refer to systematic errors in the climate data records as "construction errors." Such errors can be thought of as having two fundamentally different sources, parametric and structural (see Appendix A). Parametric uncertainty, which results from finite sample sizes, is much less important than structural uncertainty.

Over 40% of the difference between two of the MSU data sets is due to the treatment of a single satellite in the 1984-1986 period.

**Table 2.2 Global atmospheric temperature variations: their time scales, sources, and properties.**

| Variation | Description | Dominant Period | Approx. Magnitude | Detectibility | Effect on Trend Estimates |
|---|---|---|---|---|---|
| Diurnal[2]<br><br>[2] Christy et al., 2003; Mears et al., 2003; Vinnikov and Grody. 2003; Dai and Trenberth, 2004; Jin, 2004; Seidel et al., 2005. | Warmer days than nights, due to Earth's rotation on its axis affecting solar heating. | Daily (outside of polar regions) | Highly variable. Surface skin T changes up to 35°C. Boundary layer changes <10°C. Free tropospheric changes <1°C. Stratospheric changes ~0.1-1°C. | Well detected in surface data. Poorly detected globally in the troposphere and stratosphere due to infrequent sampling (once or twice daily) and potential influence of measurement errors with their own diurnal signal. A few ground-based systems detect signal well. | Satellite data require adjustment of drift in the local equatorial crossing time of spacecraft orbits. Inadequate quantification of the true diurnal cycle hinders this adjustment. Different diurnal adjustments by different groups may partly account for differences in trend estimates. |
| Synoptic[3]<br><br>[3] Palmen and Newton, 1969 | Temperature changes associated with weather events, such as wave and frontal passages, due to internal atmospheric dynamics. | 3-7 days | Up to ~15°C or more at middle latitudes, ~3°C in Tropics. | Well detected by observing systems designed to observe meteorological variability. | Not significant, but contributes to noise in climate data records. |
| Intraseasonal[4]<br><br>[4] Duvel et al., 2004. | Most notably, an eastward-and vertically- propagating pattern of disturbed weather in the tropical Indo-Pacific ocean region, related to organized convection. Also, atmospheric "blocking" and wet/dry land surface can cause intraseasonal variations at mid-latitudes. | 40-60 days (Tropics), < 180 days (mid-latitudes) | 1-2°C at surface, less aloft (tropics), larger in mid-latitudes. | Temperature signals moderately well detected, with tropical atmosphere limited by sparse radiosonde network and IR-based surface temperature limited by cloud. Reanalysis data are useful. | Not significant due to short duration, but may be important if character of the oscillation changes over time. |
| Annual[5]<br><br>[5] Wallace and Hobbs, 1977 | Warmer summers than winters, and shift in position of major precipitation zones, due to tilt of the Earth's axis of rotation affecting solar heating. | Yearly | ~2-30°C; greater over land than sea, greater at high than low latitudes, greater near the surface and tropopause than at other heights. | Well observed. | Trends are often computed from "anomaly" data, after the mean annual cycle has been subtracted. Changes in the nature of the annual cycle could affect annual-average trends. |
| Quasi-Biennial Oscillation (QBO)[6]<br><br>[6] Christy and Drouilhet, 1994; Randel et al., 1999; Baldwin et al., 2001 | Nearly periodic wind and temperature changes in the equatorial stratosphere, due to internal atmospheric dynamics. | Every 23-28 months (average of 27 months because occasionally periods of up to 36 months occur.) | Up to 10°C locally, ~0.5°C averaged over the tropical stratosphere. | Fairly well observed by equatorial radiosonde stations and satellites. | Like ENSO, can influence trends in short data records, but it is relatively easy to remove this signal. |

| Variation | Description | Dominant Period | Approx. Magnitude | Detectibility | Effect on Trend Estimates |
|---|---|---|---|---|---|
| Interannual[7]<br><br>[7] Parker and Brownscombe, 1983; Pan and Oort, 1983; Christy and McNider, 1994; Parker et al., 1996; Angell, 2000; Robock, 2000; Michaels and Knappenberger, 2000; Santer et al., 2001; Free and Angell, 2002; Trenberth et al., 2002; Seidel et al., 2004; Seidel and Lanzante, 2004 | Multiannual variability due to interaction of the atmosphere with dynamic ocean and possibly land surfaces; most notably, ENSO. Can also be caused by volcanic eruptions. | ENSO events occur every 3-7 years and last 6-18 months; major volcanic eruptions, irregular but approximately every 5-20 years with effects lasting at least 2 years. | Up to 3°C in equatorial Pacific (ENSO), smaller elsewhere. Volcanic warming of stratosphere can exceed 5°C. In tropics, cooling of surface <2°C. | Fairly well observed, although the vertical structure of ENSO is not as well documented, due to sparseness of the tropical radiosonde network. | ENSO affects surface global mean temperatures by ±0.4°C, and more in the tropical troposphere. Large ENSO events near the start or end of a data record can strongly affect computed trends, as was the case for the 1997-98 event. Changes in ENSO frequency or strength affect (and may be coupled with) long-term trends. |
| Decadal to interdecadal oscillations and shifts[8]<br><br>[8] Labitzke, K., 1987; Trenberth and Hurrell, 1994; Lean et al., 1995; Zhang et al., 1997; Thompson et al., 2000; Douglass and Clader, 2002; Seidel and Lanzante, 2004; Hurrell et al., 2003; Folland et al., 1999; Power et al., 1999; Folland et al., 2002; Scaife et al., 2005. | Like interannual, but longer time scales. Prominent example is the PDO/Interdecadal Pacific Oscillation. Despite long time scale, changes can occur as abrupt shifts, for example, a warming shift around 1976. Others include regional changes in the North Atlantic, Pacific-North American, Arctic, and the Antarctic oscillations. Some changes also caused by 11-year solar cycle. | Poorly known; multi-decadal PDO cycle suggested by 20th-century observations; others a decade or two; solar 11-year cycle detectable also. | Not well studied. The 1976-77 shift associated with a sharp warming of at least 0.2°C globally, though difficult to distinguish from anthropogenic warming. 11-year cycle leads to stratospheric temperature changes of ~2°C, and interacts with the Quasi-Biennial Oscillation (QBO). | Relatively large regional changes are well observed, but global expression is subject to data consistency issues over time and possible real changes. | Can account for a significant fraction of linear trends calculated over periods of a few decades or less regionally. Such trends may differ significantly from one such period to the next. |
| Sub-centennial 60-80 year fluctuation or "Atlantic Multidecadal Oscillation"[9]<br><br>[9] Schlesinger and Ramankutty, 1994; Mann et al., 1998; Folland et al., 1999; Andronova and Schlesinger, 2000; Goldenberg et al., 2001; Enfield et al., 2001; Knight et al., 2005 | Fluctuates in instrumental and paleo data at least back to c.1600. Seems to particularly affect Atlantic sector. Possible interhemispheric component. | 60-80 years | ~ ±0.5°C in parts of the Atlantic. Apparently detectable in global mean ~ ±0.1°C | Detectable globally above the noise, clear in North Atlantic SST. | Effects small globally, but probably detectable in last few decades. Readily detectable over this period in North Atlantic Ocean where it clearly affects surface temperature trends and probably climate generally. |
| Centennial and longer variations[10]<br><br>[10] Folland et al., 2001a. | Warming during 20th century due to human influences, solar, and internal variability. Earlier changes included the "little ice age" and "medieval warm period." | None confirmed, though 1500 year Bond cycle possible. | 20th century warming of ~0.6°C globally appears to be as large or larger than other changes during the late Holocene. | Surface warming during 20th century fairly well observed; proxies covering earlier times indicated 20th century warmer than the past 5 centuries | Natural temperature variations occur on the longest time scales accessible in any instrumental record. |

The human decisions that underlie the production of climate records may be thought of as forming a structure for separating real and artificial behavior in the raw data. Assumptions made by the experts may not be correct, or important factors may have been ignored; these possibilities lead to structural uncertainty (Thorne *et al.,* 2005a) in any trend or other metric obtained from a given climate record. Experts generally tend to underestimate structural uncertainty (Morgan, 1990). The $T_2$ example above shows that this type of error can considerably exceed those recognized by the climate record builders. Sorting out which decisions are better than others, given the fact many individual decisions are interdependent and often untestable, is challenging.

Structural uncertainty is difficult to quantify because this requires considering alternatives to the fundamental assumptions, rather than just to the specific sampling or bias pattern in the available data (the main source of parametric uncertainty). For example, is an apparent diurnal variation due to (a) real atmospheric temperature change, (b) diurnal solar heating of an instrument component, (c) a combination of both, or (d) something else entirely? If the answer is not known a priori, different working assumptions may lead to a different result when corrections are determined and applied.

There may be several ways to identify structural errors. First, it is well known in statistics that one should examine the variability that is left over when known effects are removed in a data analysis, to see whether the residuals appear as small and "random" as implied by the assumptions. Even when the residuals are examined, it is often difficult to identify the cause of any non-randomness. Second, one can compare the results with external or independent data (such as comparing SST and NMAT observations). However, one then encounters the problem of assessing the accuracy of the independent data; because, in the case of global atmospheric temperature data there are no absolute standards for any needed adjustment. Christy et al. (2000) demonstrate the use of internal and external methods for evaluating the error of their upper air time series. They assumed that where agreement of independent measurements exists, there is likely to be increased confidence in the trends. Third, one can try to assess the construction uncertainty by examining the spread of results obtained by multiple experts working independently (*e.g.,* the $T_2$ example, Thorne *et al.,* 2005a). Unfortunately, though valuable, this does not establish the uncertainties of individual efforts, nor is it necessarily an accurate measure of overall uncertainty. If all investigators make common mistakes, the estimate of construction uncertainty may be too optimistic; but if some investigators are unaware of scientifically sound progress made by others, the estimate can be too pessimistic.

A general concern regarding all of the data sets used in this analysis - land air temperature, sea surface temperature, radiosonde temperature, and satellite-derived temperature - is the level of information describing the operational characteristics and evolution of the associated observing system. As indicated above, the common factor that creates the biggest differences between analyses of the same source data is the homogeneity adjustments made to account for biases in the raw data. All homogeneity adjustments would improve with better metadata – that is, information about the data (see Chapter 6). For satellite-derived temperature, additional metadata such as more data points used in the pre-launch calibration would have been helpful to know, especially if done for differing solar angles to represent the changes experienced on orbit. For the *in situ* data sets, additional metadata of various sorts likely exist in one form or another somewhere in the world and could be acquired or created. These include the type of instrument, the observing environment, the observing practices and the exact dates for changes in any of the above.

We illustrate the evolution of a data set in Table 2.3 by listing the adjustments that have been discovered and applied to the UAH $T_{2LT}$ data set since the first version was published. The UAH satellite data set is the oldest of the satellite temperature data sets, and thus has the advantage of a traceable effort toward an improved data set. Improvements in data sets generally occur when they are used regularly to monitor climate change and are therefore more thoroughly scrutinized.

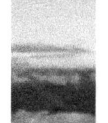

All data adjustments would improve with better metadata – that is, information about the data.

**Table 2.3 Corrections in the UAH $T_{2LT}$ data set over time. Progress occurs as data sets undergo continual and independent evaluation.**

| UAH Version of $T_{2LT}$ | Main Adjustment | Net effect on $T_{2LT}$ Trend °C/decade | Date Applied and Citation |
|---|---|---|---|
| A | Simple Bias Correction | | 1992, Spencer and Christy, 1992 |
| B | Linear diurnal drift correction for NOAA-7 | -0.03 | 1994, Christy *et al.*, 1995 |
| C | Removal of residual annual cycle related to hot target variations | +0.03 | 1997, Christy *et al.*, 1998 |
| D | Orbital Decay | +0.10 | 1998, Christy *et al.*, 2000 |
| D | Removal of dependence on time variations of hot target temperature | -0.07 | 1998, Christy *et al.*, 2000 |
| 5.0 | Non-linear diurnal correction | +0.008 | 2003, Christy *et al.*, 2003 |
| 5.1 | Tightened criteria for data acceptance | -0.004 | 2004, update file at UAH |
| 5.2 | Correction of diurnal drift adjustment | +0.035 | 2005, Spencer *et al.*, 2006 |

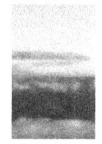

Below we identify various known issues that led to errors in the data sets examined in this report, and which have generally been addressed by the various data set builders. Note that reanalyses inherit the errors of their constituent observing systems, though they have the advantage of seeking a degree of consensus among the various observing systems through the constraint of model physics. The complex reanalysis procedure transforms these errors of output data into errors of construction methodology that are hard to quantify.

### Errors Primarily Affecting *In Situ* Observing Systems

**Spatial and temporal sampling:** The main source of this error is the poor sampling of oceanic regions, particularly in the Southern Hemisphere, and some tropical and Southern Hemisphere continental regions (see Text Box 2.1). Temporal variations in radiosonde sampling can lead to biases, (*e.g.,* switching from 00 to 12 UTC) but these are generally documented and thus potentially treatable.

**Local environmental changes:** Land-use changes, new instrument exposures, *etc.,* create new localized meteorological conditions to which the sensor responds. These issues are most important for land near-surface air temperatures but can also affect the lower elevation radiosonde data. Some changes, *e.g.,* irrigation, can act to increase nighttime minima while decreasing daytime maxima, leaving an ambiguous signal for the daily mean temperature. Such changes are sources of error only if the change in the immediate surroundings of the station is unrepresentative of changes over a larger region.

**Changes in methods of observation:** A change in the way in which an instrument is used, as in calibrating a radiosonde before launch, *i.e.,* whether it is compared against a typical outdoor sensor or against a traceable standard.

**Changes in data processing algorithms:** A change in the way raw data are converted to atmospheric information can introduce similar problems. For radiosonde data, the raw observations are often not archived and so the effects of these changes are not easily removed.

### Errors Primarily Affecting Satellite Systems

**Diurnal sampling:** It is common for polar orbiters to drift slowly away from their "sun-synchronous" initial equatorial crossing times (*e.g.,* 1:30 p.m. to 5 p.m.), introducing spurious trends related to the natural diurnal cycle of daily temperature. The later polar orbiters (since 1998) have more stable station keeping. Diurnal drift adjustments for $T_{2LT}$ and $T_2$ impact the trend by a few hundredths °C/decade. Changes in local observation time also significantly afflict *in situ* temperature observations, with a lesser impact on the global scale.

**Orbit decay:** Variations in solar activity cause expansion and contraction of the thin atmosphere at the altitudes where satellites orbit, which create variable frictional drag on spacecraft. This causes periods of altitude decay, changing the instrument's viewing geometry relative to the Earth's surface and therefore the radiation emissions observed. This issue relates most strongly to $T_{2LT}$, which uses data from multiple view angles, and is of order 0.1°C/decade.

**Calibration shifts/changes:** For satellite instruments, the effects of launch conditions or changes in the within-orbit environment (*e.g.,* varying solar shadowing effects on the spacecraft components as it drifts through the diurnal cycle) may require adjustments to the calibration equations. Adjustment magnitudes vary among the products analyzed in this report but are on the order of 0.1°C/decade for $T_{2LT}$ and $T_2$.

**Surface emissivity effects:** The intensity of surface emissions in observed satellite radiances can vary over time due to changes in surface properties, *e.g.* wet vs. dry ground, rough vs. calm seas, interannual sea ice variations *etc.*, and longer-term land cover

changes, *e.g.,* deforestation leading to higher daytime skin temperatures and larger diurnal temperature cycles.

**Atmospheric effects:** Atmospheric composition can vary over time (*e.g.,* aerosols), affecting satellite radiances, especially the infrared.

### Errors Affecting all Observing Systems

**Instrument Changes:** Systematic variations of calibration between instruments will lead to time-varying biases in absolute temperature. These involve (a) changes in instruments and their related components (*e.g.,* changes in housing can be a problem for *in situ* surface temperatures), (b) changes in instrument design or data processing (*e.g.,* radiosondes) and (c) copies of the same instrument that are intended to be identical but are not (*e.g.,* satellites).

### Errors or Differences Related to Analysis or Interpretation

**Construction Methodology:** As indicated, this is often the source of the largest differences among trends from data sets and is the least quantifiable. When constructing a homogeneous, global climate record from an observing system, different investigators often make a considerable range of assumptions as to how to treat unsampled or undersampled variability and both random and systematic instrument errors. The trends and their uncertainties that are subsequently estimated are sensitive to treatment assumptions (Free *et al.,* 2002). For example, the linear trends of the latest versions of $T_2$ from the three satellite analyses vary from +0.04 to +0.20°C/decade (Chapter 3), reflecting the differences in the combination of individual adjustments determined and applied by each team (structural uncertainty). Similarly, the $T_2$ global trends of the radiosonde-based and reanalyses data sets range from -0.04 to +0.07°C/decade indicating noticeable differences in decisions and methodologies by which each was constructed. Thus the goal of achieving a consensus with an error range of a few hundredths °C/decade is not evidenced in these results.

**Trend Methodology:** Differences between analyses can arise from the methods used

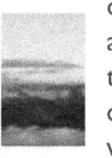

Different investigators make a range of assumptions about how to treat unsampled or undersampled variability and both random and systematic instrument errors.

## BOX 2.1: COMPARING RADIOSONDE AND SATELLITE TEMPERATURES

Attempts to compare temperatures from satellite and radiosonde measurements are hindered by a mismatch between the respective raw observations. While radiosondes measure temperatures at specific vertical levels, satellites measure radiances that can be interpreted as the temperature averaged over a deep layer. To simulate a satellite observation, the different levels of temperature in the radiosonde sounding are proportionally weighted to match the profiles shown in Figure 2.2. This can be done in one of two ways.

1. Employ a simple set of geographically and seasonally invariant coefficients or weights, called a static weighting function (Spencer and Christy, 1992). These coefficients are multiplied by the corresponding set of temperatures at the radiosonde levels and the sum is the simulated satellite temperature. Over land, the surface contributes more to the layer-average than it does over the ocean, and this difference is taken into account by slightly different sets of coefficients applied to land vs. ocean calculations. This same method may be applied to the temperature level data of global reanalyses. We have applied the "static weighting function" approach in this report.

2. Take into account the variations in the air mass temperature, surface temperature and pressure, and atmospheric moisture (Spencer *et al.*, 1990). Here, the complete radiosonde temperature and humidity profiles are ingested into a radiation model to generate the simulated satellite temperature (*e.g.*, Christy and Norris, 2004). This takes much more computing power to calculate and requires humidity information, which for radiosonde observations are generally of poorer quality than temperature information or is missing entirely. For climate applications, in which the time series of large-scale anomalies is the essential information, the output from the two methods differs only slightly (Santer *et al.*, 1999).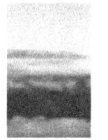

There are practical difficulties in generating long time series of simulated satellite temperatures under either approach. To produce a completely homogeneous data record, the pressure levels used in the calculation must be consistent throughout time, *i.e.*, always starting at the surface and reaching the same designated altitude. If, for example, soundings achieved higher elevations as time went on, there would likely be a spurious trend due to the effects of having measured observations during the latter period of record, while by necessity, relying on estimates for the missing values in the earlier period. We also note that HadAT2 utilizes 9 pressure levels for simulating satellite profiles while RATPAC uses 15, so differences can arise from these differing inputs.

An additional complication is that many radiosonde data sets and reanalyses may provide data at mandatory levels beginning with 1000 and/or 850 hPa, *i.e.*, with no identifiable surface. Thus, the location of the material surface, and its temperature, can only be estimated so that an additional source of error to the anomaly time series may occur. There are a number of other processing choices available when producing a time series of simulated satellite data for site-by-site comparisons between actual satellite data and radiosondes (or reanalyses) and these also have the potential to introduce non-negligible biases.

Averaging of spatially incomplete radiosonde observations for comparison of global and tropical anomalies also introduces some error (Agudelo and Curry, 2004). In this report we have first zonally averaged the data, then generated satellite-equivalent measures from these data and finally calculated global and tropical averages. The spatial coverage differs markedly between the two radiosonde data sets. However, as anomalies are highly correlated in longitude the relative poor longitudinal sampling density of RATPAC (and HadAT2 outside of the NH mid-latitudes) is not necessarily an impediment (Hurrell *et al.*, 2000). Comparing global averages estimated using only those zonally averaged grids observed at RATPAC station sites by MSU versus the globally complete fields from MSU, a sampling error of less than ±0.05°C/decade was inferred for $T_{2LT}$. Satellite and reanalyses are essentially globally complete and thus do not suffer from spatial subsampling.

to determine trends. Trends shown in this report are calculated by least squares linear regression.

**Representativeness:** Any given measure reported by climate analysts could under- or overstate underlying climatic behavior. This is not so much a source of error as a problem of interpretation. This is often called statistical error. For example, a trend computed for one time period (say, 1979-2004) is not necessarily representative of either longer or earlier periods (*e.g.*, 1958-1979), so caution is necessary in generalizing such a result. By the same token, large variations during portions of the record might obscure a small but important underlying trend (see Appendix A for a discussion of statistical uncertainties).

## 4. IMPLICATIONS

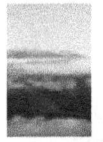

Measurements from all systems require adjustments and this report relies solely on adjusted data sets.

The observing systems deployed since the late 1950s, and the subsequent climate records derived from their data, have the capability to provide information suitable for the detection of many temperature variations in the climate system. These include temperature changes that occur with regular frequency, *e.g.,* daily and annual cycles of temperature, as well as non-periodic events such as volcanic eruptions or serious heat and cold waves. The data from these systems also have the potential to provide accurate trends in climate over the last few decades (and over the last century for surface observations), once the raw data are successfully adjusted for changes over time in observing systems, practices, and micro-climate exposure to produce usable climate records. Measurements from all systems require such adjustments and this report relies solely on adjusted data sets. The details of making such adjustments when building climate records from the uncorrected observations are examined in the following chapters.

CHAPTER 3

**W**hat do observations indicate about the changes of temperatures in the atmosphere and at the surface since the advent of measuring temperatures vertically?

**Convening Lead Author:** John R. Lanzante, NOAA
**Lead Authors:** T.C. Peterson, NOAA; F.J. Wentz, Remote Sensing Systems; K.Y. Vinnikov, Univ. of MD
**Contributing Authors:** D.J. Seidel, NOAA; C.A. Mears, Remote Sensing Systems; J.R. Christy, Univ. of AL in Huntsville; C.E. Forest, MIT; R.S. Vose, NOAA; P.W. Thorne, U. K. Met. Office; and N.C. Grody, NOAA

## KEY FINDINGS

### Observed Changes - Surface

Globally, as well as in the tropics, the temperature of the air near the Earth's surface has increased since 1958, with a greater rate of increase since 1979. All three surface temperature data sets are consistent in these conclusions.

- Globally, temperature increased at a rate of about 0.12°C per decade since 1958, and about 0.16°C per decade since 1979.
- In the tropics, temperature increased at a rate of about 0.11°C per decade since 1958, and about 0.13°C per decade since 1979.
- Most, if not all of the surface temperature increase since 1958 occured starting around the mid-1970s, a time coincident with a previously identified abrupt climate shift. However, there does not appear to be an abrupt rise in temperature at this time, rather the major part of the rise seems to occur in a more gradual fashion.

### Observed Changes - Troposphere

Globally, as well as in the tropics, both balloon-based data sets dating back to 1958 agree that the tropospheric temperature has increased slightly more than that of the surface. Since 1979, due to the considerable disagreement among tropospheric data sets, it is not clear whether the temperature of the troposphere has increased more or less than that of the surface, both globally and in the tropics.

- Globally, temperature increased at a rate of about 0.14°C per decade since 1958 according to the two balloon-based data sets. Since 1979, estimates of the increase from the two balloon and three satellite data sets range from about 0.10 to 0.20°C per decade.
- In the tropics, temperature increased at a rate of about 0.13°C per decade since 1958 according to the two balloon-based data sets. However, since 1979, estimates of the increase from the two balloon and three satellite data sets range from about 0.02 to 0.19°C per decade.
- For the balloon-based estimates since 1958, the major part of the temperature increase appears in the form of an abrupt rise in the mid-1970s, apparently in association with a climate shift that occurred at this time.

### Observed Changes - Lower Stratosphere

Globally, the temperature of the lower stratosphere has decreased both since 1958 and since 1979. The two balloon-based data sets yield reasonably consistent estimates of the rates of cooling for both time periods. However, since 1979 the two balloon data sets estimate a considerably greater rate of cooling than the two satellite data sets, which themselves disagree.

- Globally, the rate of cooling since 1958 is about 0.37°C per decade based on the two balloon data sets. Since 1979, estimates of this decrease are about 0.65°C per decade for the two balloon data sets, and from about 0.33 to 0.45° C per decade for the two satellite data sets.
- The bulk of the stratospheric temperature decrease occurred from about the late 1970s to the middle 1990s. It is unclear whether the decrease was gradual or occurred in abrupt steps in the first few years after each major volcanic eruption.

## CHAPTER 3: Recommendations

- Although considerable progress has been made in explaining the causes of discrepancies between upper-air datasets, both satellite and balloon-based, continuing steps should be taken to thoroughly assess and improve methods used to remove time-varying biases that are responsible for these discrepancies.
- New observations should be made available in order to provide more redundancy in climate monitoring. Activities should include both the introduction of new observational platforms as well as the necessary processing of data from currently under-utilized platforms. For example, Infraed Radiation (IR) and GPS satellite observations have not been used to any great extent, the former owing to complications when clouds are present and the latter owing to a short period of record. Additionally, the introduction of a network of climate-quality reference stations that include reference radiosondes, would place future climate monitoring on a firmer basis.

## I. BACKGROUND

In this chapter we describe changes in temperature at the surface and in the atmosphere based on four basic types of products derived from observations: surface, radiosonde, satellite and reanalysis. However, we limit our discussion of reanalysis products given their more problematic nature for use in trend analysis (see Chapter 2); only a few such trend values are presented for illustrative purposes.

Each of these four generic types of measurements consists of multiple data sets prepared by different teams of data specialists. The data sets are distinguished from one another by differences in the details of their construction. Each type of measurement system as well as each particular data set has its own unique strengths and weaknesses. Because it is difficult to declare a particular data set as being "the best," it is prudent to examine results derived from more than one "credible" data set of each type. Also, comparing results from more than one data set provides a better idea of the uncertainties or at least the range of results. In the interest of clarity and conciseness, we have

*Each type of measurement system as well as each particular dataset has its own unique strengths and weaknesses.*

chosen to display and perform calculations for a representative subset of all available data sets. We consider these to be the "state of the art" data sets of their type, based on our collective expert judgment.

In selecting data sets for use in this report, we limit ourselves to those products that are being actively updated and for which temporal homogeneity is an explicit goal in the construction, as these are important considerations for their use in climate change assessment. By way of a literature review, we discuss additional data sets not used in this report. Since some data sets are derivatives of earlier ones, we mention this where appropriate. One should not misconstrue the exclusion of a data set from this report as an invalidation of that product. Indeed, some of the excluded data sets have proved to be quite valuable in the past and will continue to be so into the future.

Most of the analyses that we have performed involve data that were averaged over a large region, such as the entire globe or the tropics. The spatial averaging process is complicated by the fact that the locations (gridpoints or stations) at which data values are available can vary fundamentally by data type (see Chapter 2 for details) and, even for a given type, between data production teams. In an effort towards more consistency, the spatial averages we use represent the weighted average of zonal averages[1] (*i.e.*, averages around an entire latitude line or zone), where the weights are the cosine of latitude[2]. This insures that the different latitude zones are given equal treatment across all data sets.

This chapter begins with a discussion of the four different data types, introducing some temperature data sets for each type, and then discussing their time histories averaged over the globe. Later we present more detail, concentrating on the analysis of temperature trends for two eras: (1) the period since the widespread availability of radiosonde observations in 1958, and (2) since the introduction of satellite data in 1979. We compare overall temperature trends from different measurement systems and then go into more detail on trend variations in the horizontal and vertical.

## 2. SURFACE TEMPERATURES

### 2.1 Land-based Temperature Data
Over land, temperature data come from fixed weather observing stations with thermometers housed in special instrument shelters. Records of temperature from many thousands of such stations exist. Chapter 2 outlines the difficulties in developing reliable surface temperature data sets. One concern is the variety of changes that may affect temperature measurements at an individual station. For example, the thermometer or instrument shelter might change, the time of day when the thermometers are read might change, or the station might move. These problems are addressed through a variety of procedures (see Peterson *et al.*, 1998 for a review) that are generally quite successful at re-

moving the effects of such changes at individual stations (*e.g.*, Vose *et al.*, 2003 and Peterson, 2006) whether the changes are documented in the metadata or detected via statistical analysis using data from neighboring stations as well (Aguilar *et al.*, 2003). Subtle or widespread impacts that might be expected from urbanization or the growth of trees around observing sites might still contaminate a data set. These problems are addressed either actively in the data processing stage (*e.g.*, Hansen *et al.*, 2001) or through data set evaluation to ensure as much as possible that the data are not biased[3] (*e.g.*, Jones *et al.*, 1990; Peterson, 2003; Parker, 2004; Peterson and Owen, 2005).

### 2.2 Marine Temperature Data
Data over the ocean come from moored buoys, drifting buoys, volunteer observing ships, and satellites. Historically, ships have provided most of the data, but in recent years an increasing number of buoys have been used, placed primarily in data-sparse areas away from shipping lanes. In addition, satellite data are often used after 1981. Many of the ships and buoys take both air temperature observations and sea surface temperature (SST) observations. Night marine air temperature (NMAT) observations have been used to avoid the problem that the Sun's heating of the ship's deck can make the thermometer reading greater than the actual air temperature. Where there are dense observations of NMAT and SST, over the long term they track each other very well. However, since marine observations in an area may only be taken a few times per month, SST has the advantage over air temperature in that water temperature changes much more slowly than that of air. Also, there are twice as many SST observations as NMAT from the same platforms as SSTs are taken during both the day and night and SST data are supplemented in data sparse

Data over the ocean come from moored buoys, drifting buoys, volunteer observing ships, and satellites.

---

1   The zonal averages, which were supplied to us by each data set production team, differ among data sets. We allowed each team to use their judgment as how to best produce these from the available gridpoint or station values in each latitude zone.
2   The cosine factor weights lower latitudes more than higher ones, to account for the fact that lines of longitude converge towards the poles. As a result, a zonal band in lower latitudes encompasses more area than a comparably sized band (in terms of latitude/longitude dimensions) in higher latitudes.

---

3   Changes in regional land use such as deforestation, aforestation, agricultural practices, and other regional changes in land use are not addressed in the development of these data sets. While modeling studies have suggested over decades to centuries these affects can be important on regional space scales (Oleson *et al.*, 2004), we consider these effects to be those of an external forcing to the climate system and are treated as such by many groups in the simulation of climate using the models described in Chapter 5. To the extent that these effects could be large enough to have a measurable influence on global temperature, these changes will be detected by the land-based surface network.

areas by drifting buoys which do not take air temperature measurements. Accordingly, only having a few SST observations in a grid box for a month can still provide an accurate measure of the average temperature of the month.

## 2.3 Global Surface Temperature Data

Currently, there are three main groups creating global analyses of surface temperature (see Table 3.1), differing in the choice of available data that are utilized as well as the manner in which these data are synthesized. Since the network of surface stations changes over time, it is necessary to assess how well the available observations monitor global or regional temperature. There are three ways in which to make such assessments (Jones, 1995). The first is using "frozen grids" where analysis using only those grid boxes with data present in the sparsest years is used to compare to the full data set results from other years (*e.g.*, Parker *et al.*, 1994). The results generally indicate very small errors on multi-annual timescales (Jones, 1995). The second technique is sub-sampling a spatially complete field, such as model output, only where *in situ* observations are available. Again the errors are small (*e.g.*, the standard errors are less than 0.06°C for the observing period 1880 to 1990; Peterson *et al.*, 1998b). The third technique is comparing optimum averaging, which fills in the spatial field using covariance matrices, eigenfunctions or structure functions, with other analyses. Again, very small differences are found (Smith *et al.*, 2005). The fidelity of the surface temperature record is further supported by work such as Peterson *et al.* (1999) which found that a rural subset of

global land stations had almost the same global trend as the full network and Parker (2004) that found no signs of urban warming over the period covered by this report.

### 2.3.1 NOAA NCDC

The National Oceanic and Atmospheric Administration (NOAA) National Climatic Data Center (NCDC) integrated land and ocean data set (see Table 3.1) is derived from *in situ* data. The SSTs come from the International Comprehensive Ocean-Atmosphere Data Set (ICOADS) SST observations release 2 (Slutz *et al.*, 1985; Woodruff *et al.*, 1998; Diaz *et al.*, 2002). Those that pass quality control tests are averaged into monthly 2° grid boxes (Smith and Reynolds, 2003). The land surface air temperature data come from the Global Historical Climatology Network (GHCN) (Peterson and Vose, 1997) and are averaged into 5° grid boxes. A reconstruction approach is used to create complete global coverage by combining together the faster and slower time-varying components of temperature (van den Dool *et al.*, 2000; Smith and Reynolds, 2005).

### 2.3.2 NASA GISS

The NASA Goddard Institute for Space Studies (GISS) produces a global air temperature analysis (see Table 3.1) known as GISTEMP using land surface temperature data primarily from GHCN and the U.S. Historical Climatology Network (USHCN; Easterling, *et al.*, 1996). The NASA team modifies the GHCN/USHCN data by combining at each location the time records of the various sources and adjusting the non-rural stations in such a way that their long-term trends are consistent with those from neighboring rural stations (Hansen *et al.*, 2001). These meteorological station measurements over land are combined with *in situ* sea surface temperatures and Infrared Radiation (IR) satellite measurements for 1982 to the present (Reynolds and Smith, 1994; Smith *et al.*, 1996) to produce a global temperature index (Hansen *et al.*, 1996).

### 2.3.3 UK HadCRUT2v

The UK global land and ocean data set (Had-CRUT2v, see Table 3.1) is produced as a joint effort by the Climatic Research Unit of the University of East Anglia and the Hadley Centre of the UK Meteorological (Met) Office.

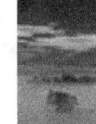

The fact that a rural subset of global land stations had almost the same trend as the full set of stations, indicates that urbanization is not a significant contributor to the global temperature trend.

The land surface air temperature data are from Jones and Moberg (2003) of the Climatic Research Unit. The global SST fields are produced by the Hadley Centre using a blend of COADS and Met Office data bank *in situ* observations (Rayner, *et al.*, 2003). The integrated data set is known as HadCRUT2v (Jones and Moberg, 2003)[4]. The temperature anoma-

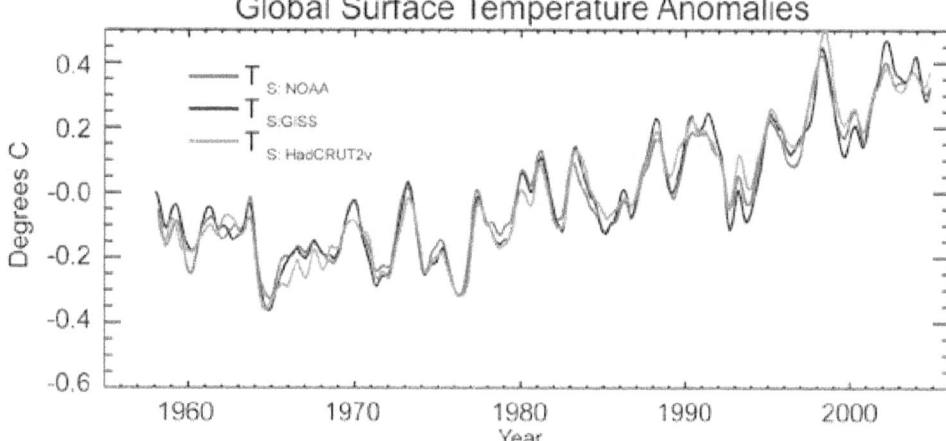

Figure 3.1

**Figure 3.1** - Time series of globally averaged surface temperature ($T_S$) for NOAA (violet), NASA (black), and HadCRUT2v (green) datasets. All time series are 7-month running averages (used as a smoother) of original monthly data, which were expressed as a departure (°C) from the 1979-97 average.

lies were calculated on a 5°x5° grid box basis. Within each grid box, the temporal variability of the observations has been adjusted to account for the effect of changing the number of stations or SST observations in individual grid-box temperature time series (Jones *et al.*, 1997, 2001). There is no reconstruction of data gaps because of the problems of introducing biased interpolated values.

2.3.4 SYNOPSIS OF SURFACE DATA SETS

Since the three chosen data sets utilize many of the same raw observations, there is a degree of interdependence. Nevertheless, there are some differences among them as to which observing sites are utilized. An important advantage of surface data is the fact that at any given time there are thousands of thermometers in use that contribute to a global or other large-scale average. Besides the tendency to cancel random errors, the large number of stations also greatly facilitates temporal homogenization since a given station may have several "near-neighbors" for "buddy-checks." While there are fundamental differences in the methodology used to create the surface data sets, the differing techniques with the same data produce almost the same results (Vose *et al.*, 2005a). The small differences in deductions about climate change derived

from the surface data sets are likely to be due mostly to differences in construction methodology and global averaging procedures.

**2.4 Global Surface Temperature Variations and Differences Between the Data Sets**

Examination of the three global surface temperature anomaly time series ($T_S$) from 1958 to the present shown in Figure 3.1 reveals that the three time series have a very high level of agreement. They all show some temperature decrease from 1958 to around 1976, followed by a strong increase. That most of the temperature change occurs after the mid 1970s has been previously documented (Karl *et al.*, 2000; Folland *et al.*, 2001b; Seidel and Lanzante, 2004). The variability of the three time series is quite similar, as are their trends. The signature of the El Niño-Southern Oscillation (ENSO), whose origin is in the tropics, is responsible for many of the prominent short-term (several year) up and down swings of temperature (Trenberth *et al.*, 2002). The strong El Niño of 1997-98 stands out as an especially large warm event within an overall upward trend.

**3. RADIOSONDE TEMPERATURES**

**3.1 Balloon-borne Temperature Data**

Since the beginning of the radiosonde era, several thousand sites have been used to launch balloons. However, many of these were in operation for only short periods of

The three global surface temperature data sets have a very high level of agreement.

---

4    Although global and hemispheric temperature time series created using a technique known as optimal averaging (Folland *et al.*, 2001a; Parker *et al.*, 2004), which provides estimates of uncertainty in the time series, including the effects of data gaps and uncertainties related to bias corrections or uncorrected biases, are available, we have used the data in their more basic form, for consistency with the other data sets.

**Table 3.1: Temperature datasets utilized in this report.  The versions of these data used in this report (*i.e.*, the versions available November 15, 2005) are archived at NOAA's National Climatic Data Center, and are available via http://www1.ncdc.noaa.gov/pub/data/ccsp.  The web sites listed below provide links to the latest versions of these data sets, which may incorporate changes made after November 2005.**

| Our Name<br>Web Page | Name Given by Producers | Producers |
|---|---|---|
| **Surface** | | |
| NOAA<br>http://www.ncdc.noaa.gov/oa/climate/monitoring/gcag/gcag.html | ER-GHCN-ICOADS | NOAA's National Climatic Data Center (NCDC) |
| NASA<br>http://www.giss.nasa.gov/data/update/gistemp/graphs/ | Land+Ocean Temperature | NASA's Goddard Institute for Space Studies (GISS) |
| HadCRUT2v<br>http://www.cru.uea.ac.uk/cru/data/temperature | HadCRUT2v | Climatic Research Unit of the University of East Anglia and the Hadley Centre of the UK Met Office |
| **Radiosonde** | | |
| RATPAC<br>http://www.ncdc.noaa.gov/oa/cab/ratpac/ | RATPAC | NOAA's: Air Resources Laboratory (ARL), Geophysical Fluid Dynamics Laboratory (GFDL), and National Climatic Data Center (NCDC) |
| HadAT2<br>http://www.hadobs.org/ | HadAT2 | Hadley Centre of the UK Met Office |
| **Satellite** | | |
| *Temperature of the Lower Troposphere* | | |
| $T_{2LT}$-UAH<br>http://vortex.nsstc.uah.edu/data/msu/t2lt | TLT | University of Alabama in Huntsville (UAH) |
| $T_{2LT}$-RSS<br>http://www.remss.com/msu/msu_data_description.html | TLT | Remote Sensing System, Inc. (RSS) |
| *Temperature of the Middle Troposphere* | | |
| $T_2$-UAH<br>http://vortex.nsstc.uah.edu/data/msu/t2 | TMT | University of Alabama in Huntsville (UAH) |
| $T_2$-RSS<br>http://www.remss.com/msu/msu_data_description.html | TMT | Remote Sensing System, Inc. (RSS) |
| $T_2$-UMd<br>http://www.atmos.umd.edu/~kostya/CCSP/ | Channel 2 | University of Maryland and NOAA/NESDIS (UMd) |
| *Temperature of the Middle Troposphere minus Stratospheric Influences* | | |
| $T^*_G$ (global)     $T^*_T$ (tropics)<br>http://www.ncdc.noaa.gov/oa/climate/research/fu-mt-uah-monthly-anom.txt (UAH)<br><br>http://www.ncdc.noaa.gov/oa/climate/research/fu-mt-rss-monthly-anom.txt (RSS) | $T_{(850-300)}$ | University of Washington, Seattle (UW) and NOAA's Air Resources Laboratory (ARL) |
| *Temperature of the Lower Stratosphere* | | |
| $T_4$-UAH<br>http://vortex.nsstc.uah.edu/data/msu/t4 | TLS | University of Alabama in Huntsville (UAH) |
| $T_4$-RSS<br>http://www.remss.com/msu/msu_data_description.html | TLS | Remote Sensing System, Inc. (RSS) |
| **Reanalysis** | | |
| NCEP50<br>http://wesley.ncep.noaa.gov/reanalysis.html | NCEP50 | National Centers for Environmental Prediction (NCEP), NOAA, and the National Center for Atmospheric Research (NCAR) |
| ERA40<br>http://www.ecmwf.int/research/era | ERA40 | European Centre for Medium-Range Weather Forecasts (ECMWF) |

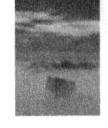

time. One approach has been to use a fixed station network consisting of a smaller number of stations having long periods of record. A complimentary approach is to grid the data, using many more stations, allowing stations to join or drop out of the network over the course of time. Since each approach has advantages and disadvantages, we utilize both. A further complication is that changes over time in instruments and recording practices have imparted artificial changes onto the temperature records. Some groups have developed methods that try to remove these artificial effects as much as possible. We employ two radiosonde data sets (see Table 3.1), one station-based and one gridded. Both data sets have been constructed using homogeneity adjustments in an attempt to minimize the effects of artificial changes.

### 3.2 Radiosonde Temperature Data Sets

#### 3.2.1 NOAA RATPAC

For several decades the 63 station data set of Angell (Angell and Korshover, 1975) was the most widely used station-based radiosonde temperature data set for climate monitoring. Recently, due to concerns regarding the effects of inhomogeneities, that network shrank to 54 stations (Angell, 2003). To better address these concerns, LKS (Lanzante, Klein, Seidel) (Lanzante *et al.*, 2003a,b) built on the work of Angell by applying homogeneity adjustments to the time series from many of his stations, as well as several dozen additional stations, to create better regional representation via a network of 87 stations. However, because of the labor-intensive nature of the homogenization process on these 87 stations, extension of the LKS data set beyond 1997 is impractical. Instead, the adjusted LKS data set is being used as the basis for a new product (see Table 3.1), Radiosonde Air Temperature Products for Assessing Climate (RATPAC), that will be updated regularly (Free *et al.*, 2003; Free *et al.*, 2005). A NOAA group (a collaboration between the ARL, GFDL, and NCDC) is responsible for the creation of RATPAC.

The RATPAC product consists of two parts: RATPAC-A and RATPAC-B[5], both of which

use the adjusted LKS data, supplemented by an extension up to present using data from the Integrated Global Radiosonde Archive (IGRA). The IGRA data used in RATPAC are based on individual soundings that have been quality controlled and then averaged into monthly station data (Durre, *et al.*, 2005). In this report we use RATPAC-B. Generally speaking, based on data averaged over large regions such as the globe or tropics, trends from RATPAC-A and RATPAC-B are closer to one another than they are to the unadjusted (IGRA) data (Free *et al.*, 2005).

#### 3.2.2 UK HadAT2

For several decades the Oort (1983) product was the most widely used gridded radiosonde data set. With the retirement of Abraham Oort, and cessation of his product, the data set produced at the Hadley Centre, UK Met Office, HadRT (Parker *et al.*, 1997) became the most widely used gridded product. Because of concern about the effects of artificial changes, this product incorporated homogeneity adjustments, although they were somewhat limited[6]. As a successor to HadRT, the Hadley Centre has created a new

> We employ two radiosonde datasets that have been adjusted in an attempt to minimize the effects of artificial changes.

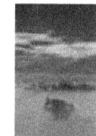

---

neities (Peterson *et al.*, 1998). However, the RATPAC-A methodology can only be used to derive homogenized temperature averaged over many stations, and thus cannot be used to homogenize temperature time series at individual stations. RATPAC-B consists of the LKS adjusted station time series that have been extended beyond 1997 by appending (unadjusted) IGRA data.

[6] Adjustments were made to upper levels only (300 hPa and above), and since they were based on satellite data, only since 1979.

---

[5] RATPAC-A uses the adjusted LKS data up through 1997 and provides an extension beyond that using a different technique to reduce the impact of inhomoge-

## Global 100-50 hPa Time Series

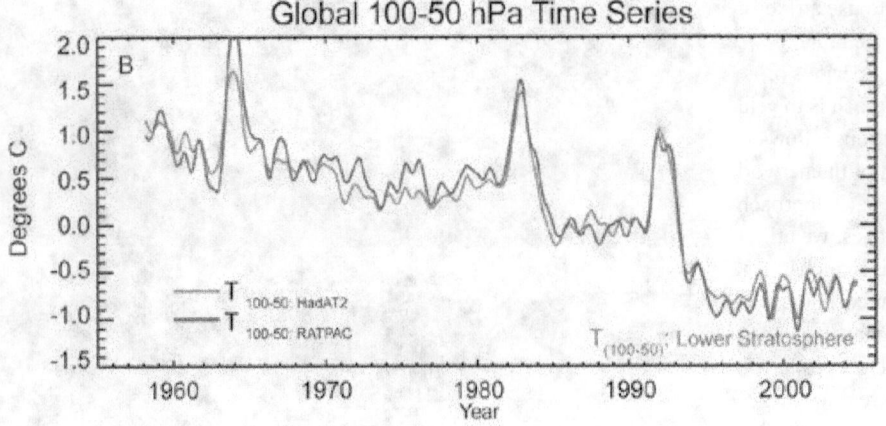

## Global 850-300 hPa Time Series

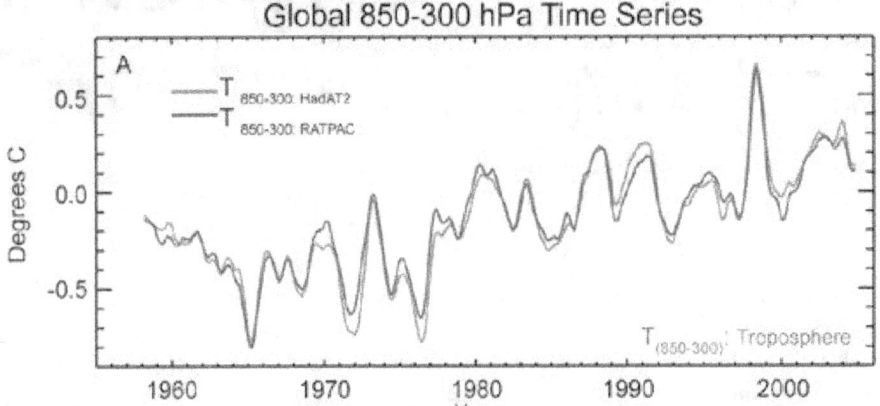

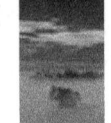

**Figure 3.2a - Bottom:** Time series of globally averaged tropospheric temperature ($T_{(850-300)}$) for RATPAC (violet) and HadAT2 (green) radiosonde datasets. All time series are 7-month running averages (used as a smoother) of original monthly data, which were expressed as a departure (°C) from the 1979-97 average.

**Figure 3.2b - Top:** Time series of globally averaged stratospheric temperature ($T_{(100-50)}$) for RATPAC (violet) and HadAT2 (green) radiosonde datasets. All time series are 7-month running averages (used as a smoother) of original monthly data, which were expressed as a departure (°C) from the 1979-97 average.

in their selection of stations in that the NOAA data set uses a relatively small number of highly scrutinized stations, while the UK data set uses a considerably larger number of stations. Compared to the surface, far fewer thermometers are in use at any given time (hundreds or less) so there is less opportunity for random errors to cancel, but more importantly, there are far fewer suitable "neighbors" to aid in temporal homogenization. While both products incorporate a common building-block data set (Lanzante *et al.*, 2003a), their methods of construction differ considerably. Any differences in deductions about climate change derived from them could be attributed to both the differing raw inputs as well as differing construction methodologies. Concerns about poor temporal homogeneity are much greater than for surface data. Indeed, it is unlikely that a recently identified cooling bias in radiosonde data (Sherwood, *et al.*, 2005; Randal and Wu, 2006) has been completely removed by the adjustment process.

product (HadAT2, see Table 3.1) that uses all available digital radiosonde data for a larger network of almost 700 stations having relatively long records[7]. Identification and adjustment of inhomogeneities was accomplished by way of comparison of neighboring stations.

### 3.2.3 SYNOPSIS OF RADIOSONDE DATA SETS
The two chosen data sets differ fundamentally

[7] High quality small station subsets, such as Lanzante *et al.* (2003a) and the Global Climate Observing System Upper Air Network, were used as a skeletal network from which to define a set of adequately similar station series used in homogenization. The data set is designed to impart consistency in both space and time and, by using radiosonde neighbors rather than satellites or reanalyses, minimizes the chances of introducing spurious changes related to the introduction of satellite data and their subsequent platform changes (Thorne *et al.*, 2005).

## 3.3 Global Radiosonde Temperature Variations and Differences Between the Data Sets
### 3.3.1 TROPOSPHERE
Figure 3.2a displays $T_{(850-300)}$ time series for the RATPAC and HadAT2 radiosonde data sets. Several noteworthy features are common to both. First, just as for the surface, ENSO signatures are clearly evident. Second, there is an apparent step-like rise of temperature around 1976-77 associated with the well-documented climate regime shift (Trenberth, 1990; Deser

*et al.,* 2004). Third, there is a long-term rise in temperatures, although a considerable amount of it may be due to the step-like change (Seidel and Lanzante, 2004). To a first approximation, both data sets display these features similarly and there is very little systematic difference between the two. Although a major component of the RATPAC product is used in the construction of the HadAT2 data set, it should be kept in mind that the former utilizes a much smaller network of stations, although the length of the station records tends to be relatively long. If the good agreement is not fortuitous, this suggests that the reduced RATPAC station network provides representative spatial sampling[8].

### 3.3.2 LOWER STRATOSPHERE

Figure 3.2b displays global temperature anomaly time series of T(100-50) from the RATPAC and HadAT2 radiosonde data sets. Several noteworthy features are common to both data sets. First is the prominent signature of three climatically important volcanic eruptions: Mt. Agung (March 1963), El Chichón (April 1982), and Mt. Pinatubo (June 1991). Temperatures rise rapidly as volcanic aerosols are injected into the stratosphere and remain elevated for about 2-3 years before diminishing. There is some ambiguity as to whether the temperatures return to their earlier values or whether they experience step-like falls in the post-volcanic period for the latter two volcanoes, particularly Mt. Pinatubo (Pawson *et al.*, 1998; Lanzante *et al.*, 2003a; Seidel and Lanzante, 2004). Second, there are small amplitude variations associated with the tropical quasi-biennial oscillation (QBO) with a period of ~ 2-3 years (Seidel *et al.*, 2004). Third, there is a downward trend, although there is some doubt as to whether the temperature decrease is best described by a linear trend over the period of record. For one thing, the temperature series prior to about 1980 exhibits little or no decrease in temperature. After that, the aforementioned step-like drops represent a viable alternative to a linear decrease (Seidel and Lanzante, 2004).

In spite of similarities among data sets, closer examination reveals some important differences. There is a rather large difference between

RATPAC and HadAT2 time series for the peak volcanic warming associated with Mt. Agung in 1963. This may be a reflection of differences in spatial sampling because the horizontal pattern of the response is not uniform (Free and Angell, 2002). More noteworthy for estimates of climate change are some subtle systematic differences between the two data sets that vary over time. A closer examination reveals that the RATPAC product tends to have higher temperatures than the HadAT2 product from approximately 1963-85, with the RATPAC product having lower values before and after this time period[9]. As we will see later, this yields a slightly greater decreasing trend for the RATPAC product. Poorer agreement between the RATPAC and HadAT2 products in the stratosphere compared to the troposphere is not unexpected because of the fact that artificial jumps in temperature induced by changes in radiosonde instruments and measurement systems tend to increase in magnitude from the near-surface upwards (Lanzante *et al.*, 2003b). More details on this issue are given in Chapter 4, Section 2.1.

---

[9]   It is worth noting that prominent artificial step-like drops, many of which were associated with the adoption of a particular type of radiosonde (Vaisala), were found in stratospheric temperatures at Australian and western tropical Pacific stations in the mid to late 1980s by Parker *et al.* (1997), Stendel *et al.* (2000), and Lanzante *et al.* (2003a). Differences in consequent homogeneity adjustments around this time could potentially explain a major part of the difference between the NOAA and UK products, although this has not been demonstrated.

---

[8]   This result is consistent with the relatively large spatial scales represented by a single radiosonde station at this level on an annual time scale demonstrated by Wallis (1998) and Thorne *et al.* (2005).

## 4. SATELLITE-DERIVED TEMPERATURES

### 4.1 Microwave Satellite Data

Three groups, employing different methodologies, have developed satellite Microwave Sounding Unit (MSU) climate data sets (see Table 3.1) derived from NOAA polar-orbiting satellites. We do not present results from a fourth group (Prabhakara *et al.,* 2000), which developed yet another methodology, since they are not continuing to work on MSU climate analyses and are not updating their time series. One of the main issues that is addressed differently by the groups is the inter-calibration between the series of satellites, and is discussed in Chapters 2 and 4.

### 4.2 Microwave Satellite Data Sets

#### 4.2.1 UNIV. OF ALABAMA IN HUNTSVILLE (UAH)

The first group to produce MSU climate products, by adjusting for the differences between satellites and the effects of changing orbits (diurnal drift), was UAH. Their approach (Christy *et al.,* 2000; Christy *et al.,* 2003) uses both an offset adjustment to allow for the systematic average differences between satellites and a non-linear hot target temperature[10] calibration to create a homogeneous series. The UAH data set has products corresponding to three temperature measures: $T_{2LT}$, $T_2$, and $T_4$ (see Chapter 2 for definitions of these measures). In this report, we use the most up-to-date versions available to us at the time, which is version 5.1 of the UAH data set for $T_2$, and $T_4$, and version 5.2 for $T_{2LT}$[11].

#### 4.2.2 REMOTE SENSING SYSTEMS (RSS)

After carefully studying the methodology of the UAH team, another group, RSS, created their own data sets for $T_2$ and $T_4$ using the same input data but with modifications to the adjustment procedure (Mears *et al.,* 2003), two of which are particularly noteworthy: (1) the method of inter-calibration from one satellite to the next

and (2) the computation of the needed correction for the daily cycle of temperature. While the second modification has little effect on the overall global trend differences between the two teams, the first is quite important in this regard. Recently, the RSS team has created its own version of $T_{2LT}$ (Mears and Wentz, 2005) and in doing so discovered a methodological error in the corresponding temperature measure of UAH. The UAH $T_{2LT}$ product used in this report is based on their corrected method. In this report, we use version 2.1 of the RSS data.

#### 4.2.3 UNIVERSITY OF MARYLAND (UMD)

A very different approach (Vinnikov *et al.,* 2004) was developed by a team involving collaborators from the University of Maryland and the NOAA National Environmental Satellite, Data, and Information Service (NESDIS) and was used to estimate globally averaged temperature trends (Vinnikov and Grody, 2003). After further study, they developed yet another new method (Grody *et al.,* 2004; Vinnikov *et al.,* 2006). As done by the other two groups, the UMd team's methodology also recalibrates the instruments based on overlapping data between the satellites. However, the manner in which they perform this recalibration differs. Also, in both versions they do not adjust for diurnal drift directly, but average the data from ascending and descending orbits. In their second approach, they substantially altered the manner in which target temperatures are used in their recalibration. The effect of their revision was to reduce the global temperature trends derived from their data from 0.22-0.26 to 0.20°C/decade. In this most recent version of their data set, which we use in this report, they apply the nonlinear adjustment of Grody *et al.* (2004) and estimate the diurnal cycle as described in Vinnikov *et al.* (2006). The UMd group produces only a measure of $T_2$, hence there is no stratospheric product ($T_4$) or one corresponding to the lower troposphere ($T_{2LT}$).

### 4.3 Synopsis of Satellite Data Sets

The relationship among satellite data sets is fundamentally different from that for surface or radiosonde products. For satellites, different data sets use virtually the same raw inputs so that any differences in derived measures are due to construction methodology. The excellent coverage provided by the orbiting sensors,

For satellites, different data sets use virtually the same raw inputs so that any differences in derived measures are due to construction methodology.

---

[10]  In fact, two targets are used, both with temperatures that are presumed to be well known. These are cold space, pointing away from the Earth, Moon, or Sun, and an onboard hot target.

[11]  The version number for $T_{2LT}$ differs from that for $T_2$, and $T_4$ because an error, which was found to affect the former (and was subsequently corrected), does not affect the latter two measures. This error was discovered by Mears and Wentz (2005).

more than half the Earth's surface daily, is a major advantage over *in situ* observations. The disadvantage is that while *in situ* observations rely on data from many hundreds or thousands of individual thermometers every day, providing a beneficial redundancy, the satellite data typically come from only one or two instruments at a given time. Therefore, any problem impacting the data from a single satellite can adversely impact the entire climate record. The lack of redundancy, compounded by occasional premature satellite failure that limits the time of overlapping measurements from successive satellites, elevates the issue of temporal homogeneity to the overwhelming explanation for any differences in deductions about climate change derived from the three data sets.

## 4.4 Global Satellite Temperature Variations and Differences Between the Data Sets

4.4.1 TEMPERATURE OF THE TROPOSPHERE Two groups (UAH and RSS) produce lower tropospheric temperature data sets, $T_{2LT}$ (see Chapter 2 for definition of this and related temperature measures) directly from satellite measurements. Their time series are shown in Figure 3.3a along with an equivalent measure constructed from the HadAT2 radiosonde data set (see Box 2.1 for an explanation as to how these equivalent measures were generated). The three temperature series have quite similar behavior, with ENSO-related variations accounting for much of the up and down meanderings, for example the historically prominent 1997-1998 El Niño. But over the full period of record, the amount of increase indicated by the data sets varies considerably. A closer look reveals that as time goes on, the RSS product indicates a noticeably greater increase of temperature than the other two. For comparison purposes, in Figure 3.3b we show an alternate measure of lower tropospheric temperatures, $T^*_G$, derived from products produced by the same three groups. From comparison of Figures 3.3a and 3.3b we see that both measures of lower tropospheric temperature agree remarkably well, even with regard to the more subtle differences relating to the longer-term changes. We will return to the issue of agreement between $T_{2LT}$ and $T^*_G$ later when we discuss trends (section 6).

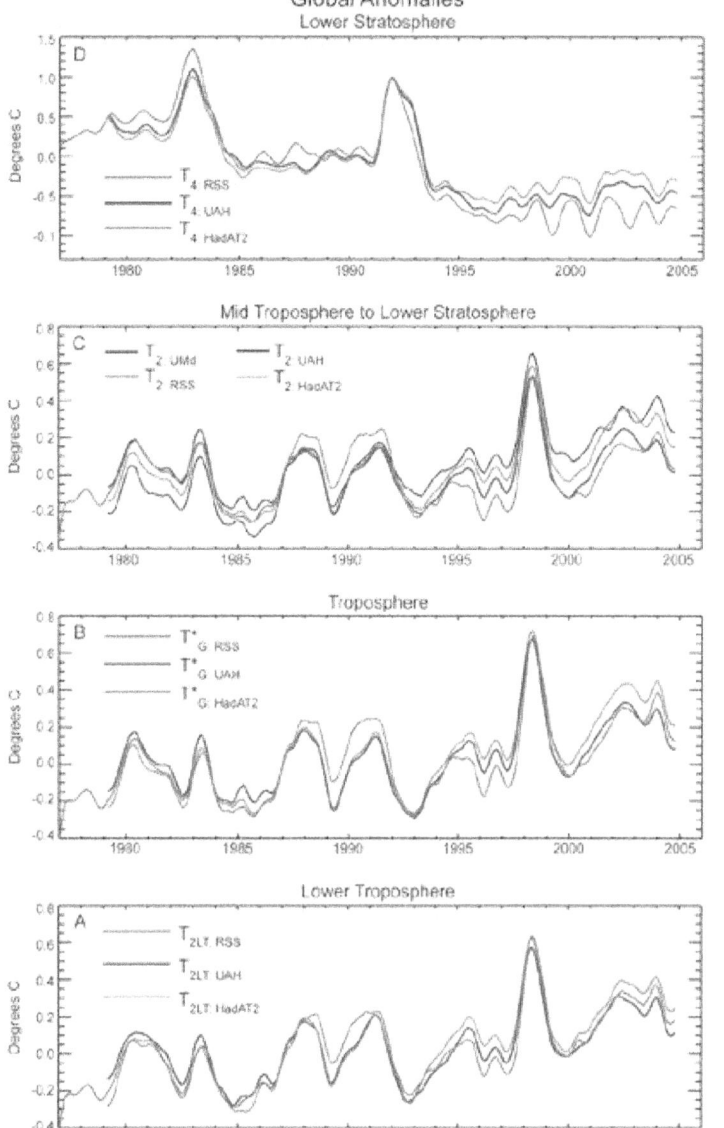

**Figure 3.3a- Bottom:** Time series of globally averaged lower tropospheric temperature ($T_{2LT}$) as follows: UAH (blue) and RSS (red) satellite datasets, and HadAT2 (green) radiosonde data. All time series are 7-month running averages (used as a smoother) of original monthly data, which were expressed as a departure (°C) from the 1979-97 average.

**Figure 3.3b- Third:** Time series of globally averaged middle tropospheric temperature ($T^*_G$) as follows: UAH (blue) and RSS (red) satellite datasets, and HadAT2 (green) radiosonde data. All time series are 7-month running averages (used as a smoother) of original monthly data, which were expressed as a departure (°C) from the 1979-97 average.

**Figure 3.3c - Second:** Time series of globally averaged upper middle tropospheric temperature ($T_2$) as follows: UAH) (blue), RSS (red), and UMd (black) satellite datasets, and HadAT2 (green) radiosonde data. All time series are 7-month running averages (used as a smoother) of original monthly data, which were expressed as a departure (°C) from the 1979-97 average.

**Figure 3.3d - Top:** Time series of globally averaged lower stratospheric temperature ($T_4$) as follows: UAH (blue) and RSS (red) satellite datasets, and HadAT2 (green) radiosonde data. All time series are 7-month running averages (used as a smoother) of original monthly data, which were expressed as a departure (°C) from the 1979-97 average.

Time series corresponding to the temperature of the upper middle troposphere ($T_2$) are shown in Figure 3.3c. The products represented in this figure are the same as for the lower troposphere, except that an additional product, that from the UMd group is available. Again, all of the time series have similar behavior with regard to the year-to-year variations. However, closer examination shows that two of the products (UMd and RSS satellite data) indicate considerable temperature increase over the period of record, whereas the other two (UAH satellite and HadAT2 radiosonde) indicate slight warming only. A more detailed discussion of the differences between the various products can be found in Chapter 4.

We note that all of the curves for the various tropospheric temperature series (Figures 3.3a-c) exhibit remarkably similar shape over the period of record. For the common time period, the satellite measures are similar to the tropospheric layer-averages computed from radiosonde data. The important differences between the various series are with regard to the more subtle long-term evolution over time, which manifests itself as differences in linear trend, discussed later in more detail.

### 4.4.2 TEMPERATURE OF THE LOWER STRATOSPHERE

Figure 3.3d shows the temperature of the lower stratosphere ($T_4$); note that there is no product from the UMd team for this layer. The dominant features for this layer are the major volcanic eruptions: El Chichón in 1982 and Mt. Pinatubo in 1991. As discussed above, the volcanic aerosols tend to warm the stratosphere for about 2-3 years before diminishing. In contrast, ENSO events have little influence on the stratospheric temperature. Both products show that the stratospheric temperature has decreased considerably since 1979, as compared to the lesser amount of increase that is seen in the troposphere. The $T_4$-RSS product shows somewhat less overall decrease than the $T_4$-UAH product, in large part as a result of the fact that the former increases relative to the latter from about 1992-94. As was the case for the troposphere, the radiosonde series show a greater decrease than the satellite data. Again, the satellite and radiosonde series for the lower-stratosphere exhibit the same general behavior over time.

## 5. REANALYSIS TEMPERATURE "DATA"

A number of agencies from around the world have produced reanalyses based on different schemes for different time periods. We focus on two of the most widely referenced, which cover a longer time period than the others (see Table 3.1). The NCEP50 reanalysis represents a collaborative effort between NOAA's National Centers for Environmental Prediction (NCEP) and the National Center for Atmospheric Research (NCAR). For the NCEP50 reanalysis, gridded air temperatures at the surface and aloft are available from 1958 to present. Using a completely different system, the European Center for Medium-Range Weather Forecasts (ECMWF) has produced similar gridded data from September 1957 to August 2002 called ERA40. Reanalyses are "hybrid products," utilizing raw input data of many types, as well as complex mathematical models to combine these data. For more detailed information on the reanalyses, see Chapter 2. As the reanalysis output does not represent a different observing platform, a separate assessment of reanalysis data will not be made.

## 6. COMPARISONS BETWEEN DIFFERENT LAYERS AND OBSERVING PLATFORMS

### 6.1 During the Radiosonde Era, 1958 to the Present

#### 6.1.1 GLOBAL

As shown in earlier sections, globally averaged temperature time series indicate increasing temperature at the surface and in the troposphere with decreases in the stratosphere over the course of the last several decades. It is desirable to derive some estimates of the magnitude of the rate of these changes. The widely-used, least-squares, linear trend technique is adopted for this purpose with the explicit caveat that long-term changes in temperature are not necessarily linear, as there may be departures in the form of periods of enhanced or diminished change, either linear or nonlinear, as well as abrupt, step-like changes[12]. While it has been

---

[12] For example, the tropospheric linear trends in the periods 1958-1979 and 1979-2003 were shown to be much less than the trend for the full period (1958-2003), based on one particular radiosonde data set (Thorne *et al.*, 2005), due to the abrupt rise in tem-

*Global average time series indicate increasing temperature at the surface and in the troposphere with decreases in the stratosphere over the course of the last several decades.*

shown that such constructs are plausible, it is nevertheless difficult to prove that they provide a better fit to the data, over the time periods addressed in this report, than the simple linear model (Seidel and Lanzante, 2004). Additional discussion on this topic can be found in Appendix A.

Trends computed for the radiosonde era are given in Table 3.2 for the surface as well as various tropospheric and stratospheric layer averages[13]. The surface products are quite consistent with one another, as are the radiosonde products in the troposphere. In the stratosphere, the radiosonde products differ somewhat, although there is an inconsistent relationship involving the two stratospheric measures ($T_{(100-50)}$ and $T_4$) regarding which product indicates a greater decrease

in temperature[14]. The reanalysis products, which are "hybrid-measures," agree better with the "purer" surface and radiosonde measures at and near the surface. Agreement degrades with increasing altitude such that the reanalyses indicate more tropospheric temperature increase and considerably less stratospheric decrease than do the radiosonde products. The disparity between the reanalyses and other products is not surprising given the suspect temporal homogeneity of the reanalyses (see Chapter 2, Section 1c).

Perhaps the most important result shown in Table 3.2 is that both the radiosonde and reanalysis trends indicate that the tropospheric temperature has increased as fast as or faster than the surface over the period 1958 to present. For a given data set, the 3 measures ($T_{2LT}$, $T_{(850-300)}$, and $T*_G$) always indicate more increase in

Both the radiosonde and reanalysis trends indicate that the tropospheric temperature has increased as fast as or faster than the surface over the period from 1958 to present.

___

perature in the mid 1970s.

[13] Note that it is instructive to examine the behavior of radiosonde and reanalysis temperatures averaged in such a way as to correspond to the satellite layers ($T_{2LT}$, $T*_G$, $T_2$, and $T_4$) even though there are no comparable satellite measures prior to 1979.

___

[14] The reason for this inconsistency is that the HadAT2 product records data at fewer vertical levels than the RATPAC product, so the comparison is not one-to-one.

**Table 3.2 - Global temperature trends in °C per decade from 1958 through 2004 (except for ERA40 which terminates September 2001) calculated for the surface or atmospheric layers by data source. The trend is shown for each, with the approximate 95% confidence interval (2 sigma) below in parentheses. The levels/layers, from left to right, go from the lowest to the highest in the atmosphere. Bold values are estimated to be statistically significantly different from zero (at the 5% level). A Student's t-test, using the lag-1 autocorrelation to account for the non-independence of residual values about the trend line, was used to assess significance (see Appendix A for discussion of confidence intervals and significance testing).**

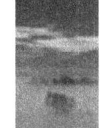

| | $T_S$ | $T_{2LT}$ | $T_{(850-300)}$ | $T*_G$ | $T_2$ | $T_{(100-50)}$ | $T_4$ |
|---|---|---|---|---|---|---|---|
| **Surface:** | | | | | | | |
| NOAA | **0.11** (0.02) | | | | | | |
| NASA | **0.11** (0.02) | | | | | | |
| HadCRUT2v | **0.13** (0.02) | | | | | | |
| **Radiosonde:** | | | | | | | |
| RATPAC | **0.11** (0.02) | **0.13** (0.03) | **0.13** (0.03) | **0.13** (0.03) | **0.07** (0.03) | **-0.41** (0.09) | **-0.36** (0.08) |
| HadAT2 | **0.12** (0.03) | **0.16** (0.04) | **0.14** (0.04) | **0.15** (0.04) | **0.08** (0.04) | **-0.39** (0.08) | **-0.38** (0.08) |
| **Reanalyses:** | | | | | | | |
| NCEP50 | **0.12** (0.03) | **0.15** (0.05) | **0.17** (0.05) | **0.17** (0.06) | **0.13** (0.06) | -0.18 (0.23) | -0.18 (0.22) |
| ERA40 | **0.11** (0.03) | **0.15** (0.04) | **0.15** (0.04) | **0.14** (0.04) | **0.10** (0.04) | **-0.21** (0.13) | -0.17 (0.13) |

the troposphere than at the surface, although this is usually not true when the $T_2$ measure is considered. The reason for the inconsistency involving $T_2$ is because of contributions to the layer that it measures from stratospheric cooling, an effect first recognized by Spencer and Christy (1992) (see discussion of this issue in Chapters 2 and 4). The development of $T^*_G$ as a global measure, and its counterpart, $T^*_T$ for the tropics (Fu *et al.,* 2004; Fu and Johanson, 2005; Johanson and Fu, 2006) was an attempt to remove the confounding effects of the stratosphere using a statistical approach (see Chapter 2).

### 6.1.2 Land vs. Ocean

The annual average temperature of most of the land and ocean surface increased during the radiosonde era, with the exception of parts of the North Atlantic Ocean, the North Pacific Ocean, and a few smaller areas. With a few exceptions, such as the west coast of North America, trends in land air temperature in coastal regions are generally consistent with trends in SST over neighboring ocean areas (Houghton *et al.,* 2001). Because bias adjustments are performed separately for land and ocean areas, before merging to create a global product, it is unlikely that the land-ocean consistency is an artifact of the construction methods used in the various surface analyses. However, land air temperatures did increase somewhat more rapidly than SSTs in some regions during the past two decades. Possibly related to this is the fact that since the mid-1970s, whether due to anthropogenic or natural causes, El Niño has frequently been in its "warm" phase, which tends to bring higher than normal temperatures to much of North America, among other regions, which have had strong temperature increases over the past few decades (Hurrell, 1996). Also, when global temperatures are rising or falling, the global mean land temperature tends to both rise and fall faster than the ocean, which has a tremendous heat storage capacity (Waple and Lawrimore, 2003). The physical reasons for these differences between land and ocean are given in Chapter 1.

### 6.1.3 Marine Air vs. Sea Surface Temperature

In ocean areas, it is natural to consider whether the temperature of the air and that of the ocean

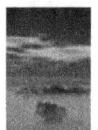

The surface temperature increase has accelerated in recent decades while the tropospheric increase has decelerated.

surface (SST) increases or decreases at the same rate. Several studies have examined this question. Overall, on seasonal and longer scales, the SST and marine air temperature generally move at about the same rate globally and in many ocean basin scale regions (Bottomley *et al.,* 1990; Parker *et al.,* 1995; Folland *et al.,* 2001b; Rayner *et al.,* 2003). However, differences between SST and marine air temperature in the tropics were noted by Christy *et al.* (1998) and then examined in more detail by Christy *et al.* (2001). The latter study found that tropical SST increased more than NMAT from 1979-1999 derived from the Tropical Atmosphere Ocean (TAO) array of tropical buoys and transient marine ship observations. Over the satellite era, some unexplained differences in these trends were also noted by Folland *et al.* (2003) in parts of the tropical south Pacific using the Rayner *et al.* (2003) NMAT data set which incorporates new corrections for the effect on NMAT of increasing deck (and hence measurement) heights.

### 6.1.4 Minimum vs. Maximum Temperatures Over Land

Daily minimum temperature increased about twice as fast as daily maximum temperature over global land areas during the radiosonde era (Karl *et al.,* 1993; Easterling *et al.,* 1997; Folland et al, 2001b). Vose *et al.* (2005b) confirmed this using a more spatially complete data set, but also found that during the satellite era maximum and minimum temperatures have been rising at nearly the same rate. In addition, their rate of warming increased near the start of the satellite era, consistent with the evolution of surface temperatures as depicted in Fig. 3.1. The causes of this asymmetric warming during the radiosonde era are still debated, but many of the areas with greater increases of minimum temperatures correspond to those where cloudiness appears to have increased over the period as a whole (Dai *et al.,* 1999; Henderson-Sellers, 1992; Sun and Groisman, 2000; Groisman *et al.,* 2004). This makes physical sense since clouds tend to cool the surface during the day by reflecting incoming solar radiation, and warm the surface at night by absorbing and reradiating infrared radiation back to the surface.

## 6.2  During the Satellite Era, 1979 to the Present

### 6.2.1  GLOBAL

A comparable set of global trends for the satellite era is given in Table 3.3. Comparison between Tables 3.2 and 3.3 reveals that some of the relationships between levels and layers, as well as among data sets, are different during the two eras. Comparing satellite era trends with the radiosonde era trends for data sets that have both periods in common, it is clear that the surface temperature increase (see Figure 3.1) has accelerated in recent decades while the tropospheric increase (see Figure 3.2a) has decelerated. Since most of the stratospheric decrease has

occurred since 1979 (see Figure 3.2b) the rate of temperature decrease there is close to twice as large as during the full radiosonde era. Thus, care must be taken when interpreting results from only the most recent decades. Agreement among different surface and radiosonde data sets is reasonable and about as good as during the longer radiosonde era. The reanalysis data sets show poorer agreement with surface data and especially with stratospheric radiosonde data for the ERA40 product.

Comparisons of trends between different satellite products and between satellite and radiosonde products yields a range of results

**Table 3.3 - Global temperature trends in °C per decade from 1979 through 2004 (except for ERA40 which terminates September 2001) calculated for the surface or atmospheric layers by data source. The trend is shown for each, with the approximate 95% confidence interval (2 sigma) below in parentheses. The levels/ layers, from left to right, go from the lowest to the highest in the atmosphere. Bold values are estimated to be statistically significantly different from zero (at the 5% level). A Student's t-test, using the lag-1 auto-correlation to account for the non-independence of residual values about the trend line, was used to assess significance (see Appendix A for discussion of confidence intervals and significance testing).**

| | $T_S$ | $T_{2LT}$ | $T_{(850-300)}$ | $T^*_G$ | $T_2$ | $T_{(100-50)}$ | $T_4$ |
|---|---|---|---|---|---|---|---|
| **Surface:** | | | | | | | |
| NOAA | **0.16** (0.04) | | | | | | |
| NASA | **0.16** (0.04) | | | | | | |
| HadCRUT2v | **0.17** (0.04) | | | | | | |
| **Radiosonde:** | | | | | | | |
| RATPAC | **0.17** (0.05) | **0.13** (0.06) | **0.10** (0.07) | **0.11** (0.08) | 0.02 (0.07) | **-0.70** (0.24) | **-0.65** (0.21) |
| HadAT2 | **0.18** (0.05) | **0.14** (0.07) | **0.12** (0.08) | **0.12** (0.08) | 0.03 (0.08) | **-0.63** (0.24) | **-0.64** (0.24) |
| **Satellite:** | | | | | | | |
| UAH | | **0.12** (0.08) | | **0.12** (0.09) | 0.04 (0.08) | | **-0.45** (0.42) |
| RSS | | **0.19** (0.08) | | **0.19** (0.09) | **0.13** (0.08) | | -0.33 (0.38) |
| UMd | | | | | **0.20** (0.07) | | |
| **Reanalyses:** | | | | | | | |
| NCEP50 | **0.12** (0.07) | **0.12** (0.10) | **0.11** (0.10) | 0.06 (0.11) | -0.04 (0.10) | **-0.76** (0.45) | **-0.74** (0.44) |
| ERA40 | **0.11** (0.06) | **0.11** (0.10) | 0.10 (0.10) | **0.13** (0.11) | 0.07 (0.10) | -0.31 (0.53) | -0.34 (0.49) |

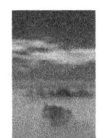

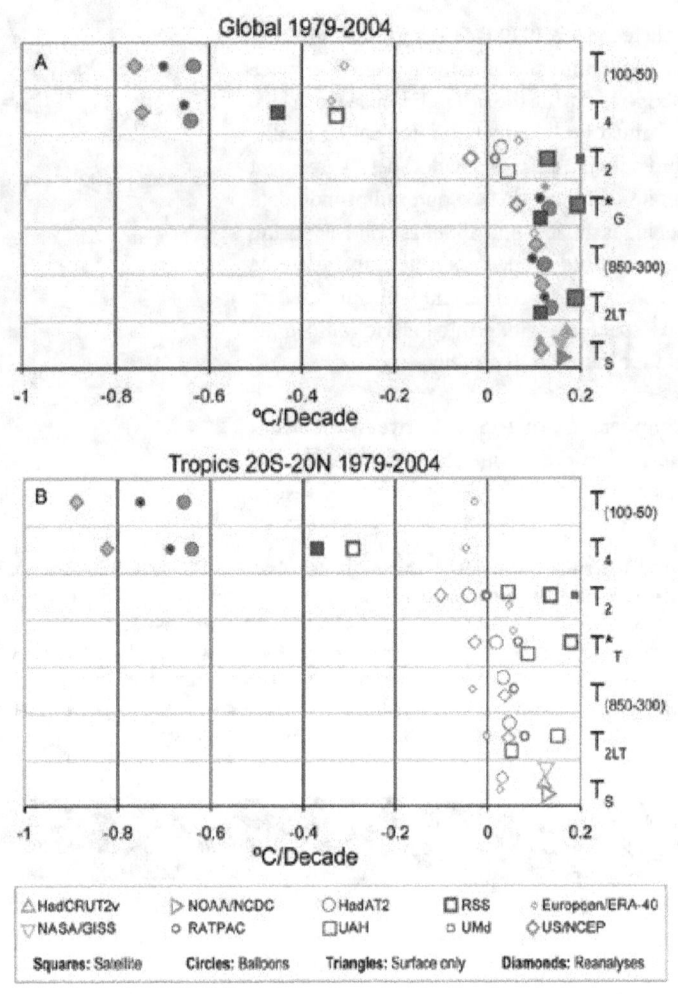

**Figure 3.4a (top)** - Global temperature trends (°C/decade) for 1979-2004 from Table 3.3 plotted as symbols. See figure legend for definition of symbols. Filled symbols denote trends estimated to be statistically significantly different from zero (at the 5% level). A Student's t-test, using the lag-1 autocorrelation to account for the non-independence of residual values about the trend line, was used to assess significance (see Appendix A for discussion of confidence intervals and significance testing).

**Figure 3.4b (bottom)** - Tropical (20°N-20°S) temperature trends (°C/decade) for 1979-2004 from Table 3.4 plotted as symbols. See figure legend for definition of symbols. Filled symbols denote trends estimated to be statistically significantly different from zero (at the 5% level). A Student's t-test, using the lag-1 autocorrelation to account for the non-independence of residual values about the trend line, was used to assess significance (see Appendix A for discussion of confidence intervals and significance testing).

as indicated by examination of the numerical trend values found in Table 3.3, which are also graphed in Figure 3.4a. While the tropospheric satellite products from the UAH team have trends that are not too dissimilar from the corresponding radiosonde trends, the two other satellite data sets show a considerably greater increase in tropospheric temperature. In the stratosphere, there is a large disagreement between satellite and radiosonde products, with the latter indicating much greater decreases in temperature. Here too, the reanalyses are quite inconsistent, with the ERA40 product closer to the satellites and the NCEP50 product closer to the radiosondes.

Perhaps the most important issue is the relationship between trends at the surface and in the troposphere. As shown in Table 3.3 and Figure 3.4a, both radiosonde data sets as well as the UAH satellite products indicate that, in contrast to the longer radiosonde era, during the satellite era the temperature of the surface has increased more than that of the troposphere. However, tropospheric trends from the RSS satellite data set, based on both measures of temperature having little or no stratospheric influence ($T_{2LT}$ and $T^*_G$) yield an opposing conclusion: the tropospheric temperature has increased as much or more than the surface. For the third satellite data set, comparisons with surface temperature are complicated by the fact that the UMd team produces only $T_2$, which is influenced by stratospheric cooling (see Chapter 2). Nevertheless, we can infer that it too suggests more of a tropospheric temperature increase than that at the surface[15].

Since climate change theory suggests more warming of the troposphere than the surface only in the tropics (see Chapter 1), much of the interest in observed trends has been in this region. Therefore, to compliment the global trends (Figure 3.4a and Table 3.3), we present a similar plot of tropical trends in Figure 3.4b (with corresponding trend values in Table 3.4).

---

[15] The difference in trends, $T^*_G$ minus $T_2$, for the UAH and RSS data sets is about 0.06 to 0.08 C/decade. Adding this amount to the UMd $T_2$ trend (0.20 C/decade) yields an estimate of the UMd trend in $T^*_G$ of about 0.26 to 0.28 C/decade. In this calculation we are assuming that the effects of the stratospheric cooling trend on the UMd product are the same as from the UAH and RSS data sets.

**Table 3.4 – Tropical (20°N-20°S) temperature trends in °C per decade from 1979 through 2004 (except for ERA40 which terminates September 2001) calculated for the surface or atmospheric layers by data source. The trend is shown for each, with the approximate 95% confidence interval (2 sigma) below in parentheses. The levels/layers, from left to right, go from the lowest to the highest in the atmosphere. Bold values are estimated to be statistically significantly different from zero (at the 5% level). A Student's t-test, using the lag-1 autocorrelation to account for the non-independence of residual values about the trend line, was used to assess significance (see Appendix A for discussion of confidence intervals and significance testing).**

| | $T_S$ | $T_{2LT}$ | $T_{(850-300)}$ | $T^*_G$ | $T_2$ | $T_{(100-50)}$ | $T_4$ |
|---|---|---|---|---|---|---|---|
| **Surface:** | | | | | | | |
| NOAA | 0.13 (0.15) | | | | | | |
| NASA | 0.13 (0.15) | | | | | | |
| HadCRUT2v | 0.12 (0.17) | | | | | | |
| **Radiosonde:** | | | | | | | |
| RATPAC | **0.13** (0.07) | 0.08 (0.12) | 0.06 (0.14) | 0.07 (0.15) | 0.00 (0.14) | **-0.75** (0.36) | **-0.69** (0.29) |
| HadAT2 | **0.15** (0.12) | 0.05 (0.15) | 0.03 (0.16) | 0.02 (0.18) | -0.04 (0.17) | **-0.66** (0.30) | **-0.64** (0.31) |
| **Satellite:** | | | | | | | |
| UAH | | 0.05 (0.18) | | 0.09 (0.19) | 0.05 (0.17) | | **-0.37** (0.28) |
| RSS | | 0.15 (0.19) | | 0.18 (0.20) | 0.14 (0.18) | | -0.29 (0.30) |
| UMd | | | | | **0.19** (0.16) | | |
| **Reanalyses:** | | | | | | | |
| NCEP50 | 0.03 (0.16) | 0.05 (0.17) | 0.04 (0.17) | -0.03 (0.18) | -0.10 (0.17) | **-0.89** (0.41) | **-0.83** (0.34) |
| ERA40 | 0.03 (0.21) | 0.00 (0.23) | -0.03 (0.25) | 0.06 (0.26) | 0.05 (0.23) | -0.03 (0.45) | -0.05 (0.42) |

Compared to the global trends, the tropical trends show even more spread among data sets, particularly in the lower stratosphere[16]. The result of the greater spread is that the range of plausible values for the difference in trends between the surface and troposphere is larger than that for the globe as a whole. Similar to the global case, in the tropics the UAH satellite plus the two radiosonde data sets (RATPAC and HadAT2) suggest more warming at the surface than in the troposphere, while the opposite

conclusion is reached based on the other two satellite products (RSS and UMd). Resolution of this issue would seem to be of paramount importance in the interpretation of observed climate change central to this Report.

### 6.2.2 LATITUDE BANDS

Globally averaged temperatures paint only part of the picture. Different layers of the atmosphere behave differently depending on the latitude. Furthermore, even the processing of the data can make for latitudinal difference in long-term trends. Figure 3.5 shows the trends in temperature for different data sets and levels averaged over latitude bands. Each of these trends was created by making a latitudinally

---

[16] The larger spread may be partially an artifact of the fact that when averaging over a smaller region, there is less cancellation of random variations. In addition, the fact that the networks of *in situ* observations are much sparser in the tropics than in the extratropics of the Northern Hemisphere may also contribute.

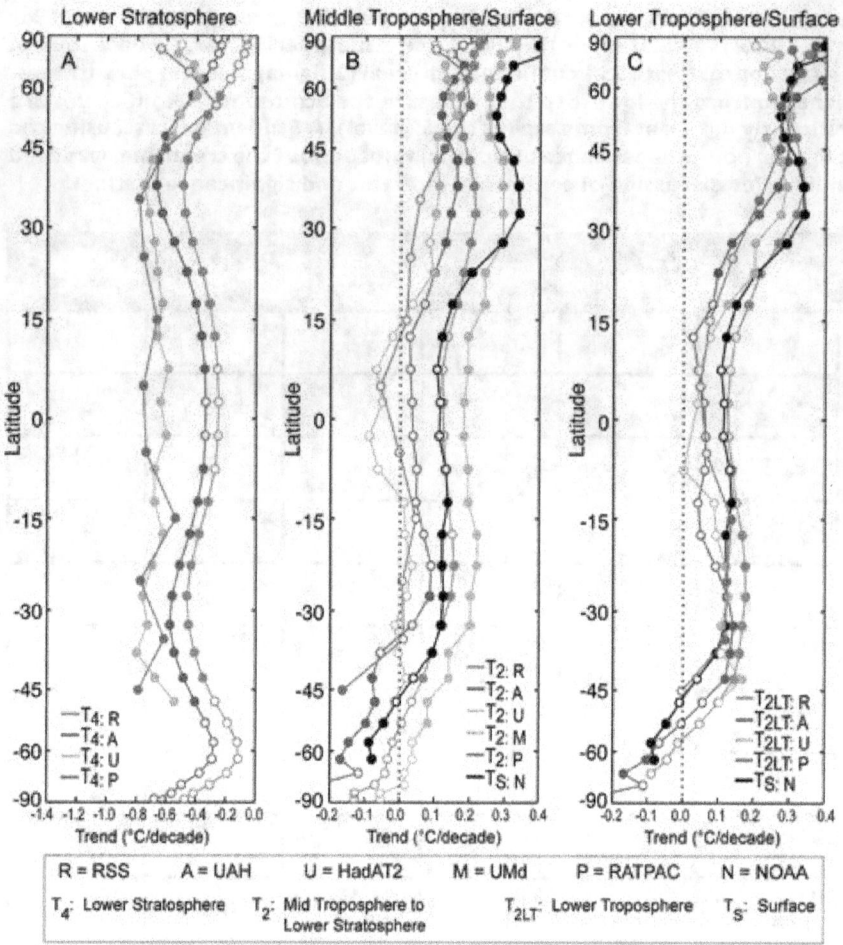

**Figure 3.5 --** Temperature trends for 1979-2004 (°C/decade) by latitude.
**Left:** stratospheric temperature ($T_4$) based on RSS (red) and UAH (blue) satellite datasets, and RATPAC (violet) and HadAT2 (green) radiosonde datasets.
**Middle:** mid-tropospheric temperature ($T_2$) based on UMd (orange), RSS (red) and UAH (blue) satellite datasets, and RATPAC (violet) and HadAT2 (green) radiosonde datasets; and surface temperature ($T_S$) from NOAA data (black).
**Right:** surface temperature ($T_S$) from NOAA data (black) and lower tropospheric temperature ($T_{2LT}$) from RSS (red) and UAH satellite data (blue), and from RATPAC (violet) and HadAT2 (green) radiosonde data.

Filled circles denote trends estimated to be statistically significantly different from zero (at the 5% level). A Student's t-test, using the lag-1 autocorrelation to account for the non-independence of residual values about the trend line, was used to assess significance (see Appendix A for discussion of confidence intervals and significance testing).

decrease occurs in the deep tropics. By contrast, the RATPAC and HadAT2 radiosonde data sets are quite different from the satellite products, with much flatter profiles. It is worth noting that there is a fundamental disagreement between the radiosonde and satellite products. Except for the mid-latitudes of the Northern Hemisphere[17], at most other latitudes the radiosonde products show more of a temperature decrease than the satellite products, with the largest discrepancy in the tropics[18].

For the middle troposphere (middle panel of Figure 3.5) there is general agreement among the radiosonde and satellite data sets in depicting the same basic structure. The largest temperature increase occurs in the extratropics of the Northern Hemisphere, with a smaller increase or slight decrease in the tropics, and even lesser increase or more decrease in the extratropics of the Southern Hemisphere. At most latitudes, $T_{2-UMd}$ indicates the most increase (least decrease), followed next by $T_{2-RRS}$, then $T_{2-UAH}$, and finally the radiosonde products with the least increase (most decrease).

For the lower troposphere and surface (right panel of Figure 3.5) the profiles are roughly similar in shape to those for the middle

averaged time series of monthly anomalies and then fitting that time series with a standard least-squares linear regression slope.

In the stratosphere (left panel of Figure 3.5), trend profiles for the two satellite data sets are fairly similar, with a greater temperature decrease everywhere according to $T_{4-UAH}$ than $T_{4-RRS}$. Some of the largest temperature decrease occurs in the south polar region, where ozone depletion is largest. A broad region of weaker

[17] The apparently better radiosonde-satellite agreement in the midlatitudes of the Northern Hemisphere may be the result of spurious stratospheric warming at stations located in countries of the former Soviet Union, offsetting the more typical spurious cooling bias of radiosonde temperatures (Lanzante *et al.*, 2003a,b).

[18] We note that in the tropics, where the radiosonde and satellite products differ the most, abrupt artificial drops in temperature appear to be particularly problematic for radiosonde data (Parker *et al.*, 1997; Lanzante *et al.*, 2003a,b). Other studies (Sherwood *et al.*, 2005; Randel and Wu, 2006) also suggest spurious cooling for radiosonde temperatures, especially in the tropics. For further discussion see Chapter 4.

troposphere with one major exception: the higher-latitude temperature increase of the Northern Hemisphere is more pronounced compared to the other regions. Comparing the surface temperature trend profile (black) with that from the various tropospheric products in the middle and right panels of Figure 3.5 suggests that the sign and magnitude of this difference is highly dependent upon which tropospheric measure is used.

### 6.2.3 MAPS

Trend maps represent the finest spatial granularity with which different levels/layers and observing platforms can be compared. However, since maps may not be the optimal way in which to examine trends[19], we present only a limited number of such maps for illustrative purposes. Figure 3.6 presents maps of trends for the surface (bottom), lower troposphere (second from bottom), upper middle troposphere (second from top), and stratosphere (top). The surface map is based on the NOAA data set[20] while those for the troposphere and stratosphere are based on the RSS satellite data set[21]. In examining these maps it should be kept in mind that based on theory we expect the difference in trend between the surface and troposphere to vary by location. For example, as shown in Chapter 1, climate model projections typically indicate that human induced changes should lead to more warming of the troposphere than the surface in the tropics, but the opposite in the Arctic and Antarctic. In addition, land and ocean respond differently, as discussed in Chapter 1 as well.

---

19  Averaging over space (*e.g.,* over latitudes, the tropics or the globe, as presented earlier) tends to reduce noise that results from the statistical uncertainties inherent to any observational measurement system. Furthermore, models that are used to study climate change have limited ability to resolve the smallest spatial scales and therefore there is little expectation of detection at the smallest scales (Stott and Tett, 1998). The formal methodology that is used to compare models with observations ("fingerprinting," see Chapter 5) concentrates on the larger-scale signals in both models and observations in order to optimize the comparisons.

20  Trend maps from other surface data sets (not shown) tend to be fairly similar to that of the NOAA map, differing mostly in their degree of spatial smoothness, which is a function of data set construction methodology.

21  A comparison between UAH and RSS trend maps for tropospheric layers is given in Chapter 4.

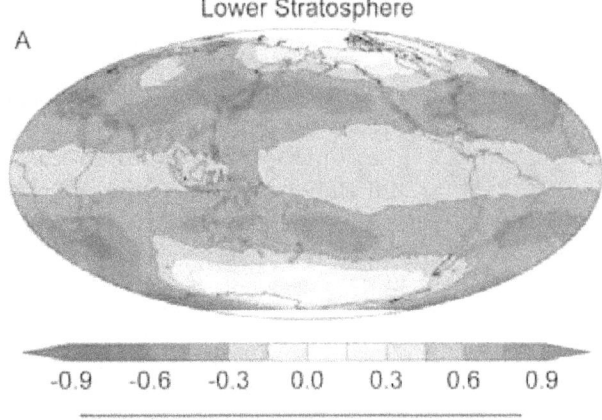

Lower Stratosphere

A

-0.9   -0.6   -0.3   0.0   0.3   0.6   0.9

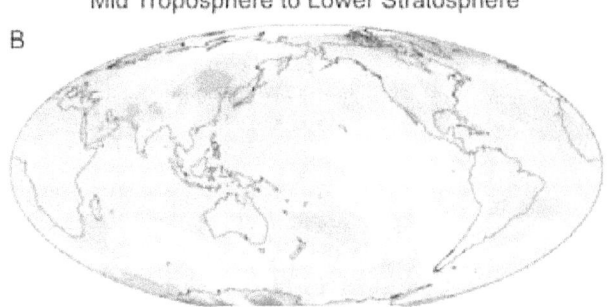

Mid Troposphere to Lower Stratosphere

B

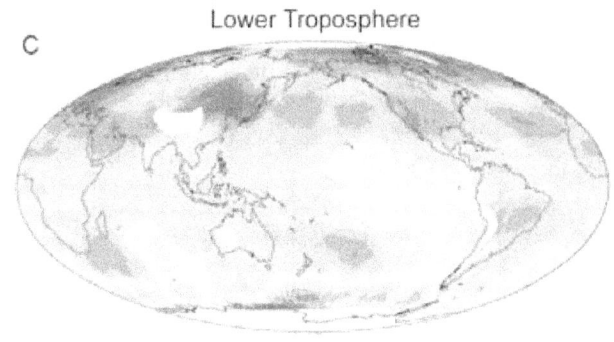

Lower Troposphere

C

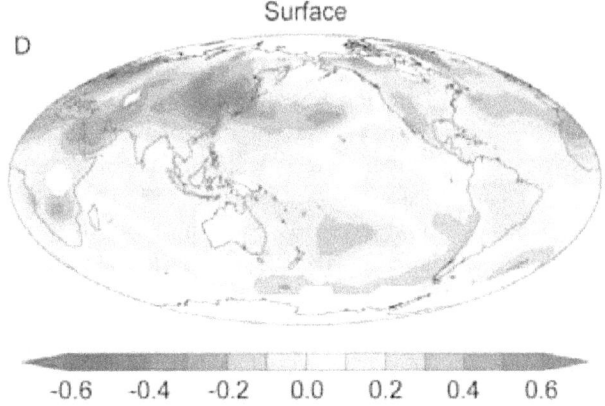

Surface

D

-0.6   -0.4   -0.2   0.0   0.2   0.4   0.6

**Figure 3.6 -** Temperature trends for 1979-2004 (°C /decade).
**Bottom (d)**: NOAA surface temperature ($T_{S-NOAA}$).
**Third (c):** RSS lower tropospheric temperature ($T_{2LT-RRS}$).
**Second  (b):** RSS upper middle tropospheric temperature ($T_{2-RRS}$).
**Top (a):** RSS lower stratospheric temperature ($T_{4-RRS}$).

The trend maps indicate both similarities and differences between the surface and tropospheric trend patterns. There is a rough correspondence in patterns between the two. The largest temperature increase occurs in the extra-tropics of the Northern Hemisphere, particularly over landmasses. A decrease or smaller increase is found in the high latitudes of the Southern Hemisphere as well as in the eastern tropical Pacific. Note the general correspondence between the above noted features in Figures 3.6c,d and the zonal trend profiles (middle and right panels of Figure 3.5). Note that the temperature of the mid troposphere to lower stratosphere is somewhat of a hybrid measure, being affected most strongly by the troposphere, but with a non-negligible influence by the stratosphere.

In contrast to the surface and troposphere, a temperature decrease is found almost everywhere in the stratosphere (Figure 3.6a). The largest decrease is found in the midlatitudes of the Northern Hemisphere and the South Polar Region, with a smaller decrease in the tropics. Again note the correspondence between the main features of the trend map (Figure 3.6a) and the corresponding zonal trend profiles (left panel of Figure 3.5).

## 7. CHANGES IN VERTICAL STRUCTURE

### 7.1 Vertical Profiles of Trends

Up to this point, our vertical comparisons have contrasted trends of surface temperature with trends based on different layer-averaged temperatures. Layers are useful because the averaging process tends to reduce noise. The use of layer-averages is also driven by the limitations of satellite measurement systems that are unable to provide much vertical detail. However, as illustrated in Chapter 1, changes in various forcing agents can lead to more complex changes in the vertical. Radiosonde data, because of their greater vertical resolution, are much better suited for this than currently available satellite data.

Figure 3.7 shows vertical profiles of trends from the RATPAC and HadAT2 radiosonde data sets for temperature averaged over the globe (top) or tropics (bottom) for the radiosonde (left) and satellite (right) eras. The trend values of Figure

3.7 are also given in Table 3.5. Each graph has profiles for the two radiosonde data sets. The tropics are of special interest because many climate models suggest that under global warming scenarios trends should increase from the lower troposphere upwards, maximizing in the upper troposphere (see Chapters 1 and 5).

For the globe, the figure indicates that during the longer period the tropospheric temperature increased slightly more than that of the surface. By contrast, for the globe during the satellite era, the surface temperature increased more than that of the troposphere. Both data sets agree reasonably well in these conclusions. For the tropics, the differences between the two eras are more pronounced. For the longer period there is good agreement between the two data sets in that the temperature increase is smaller at the surface and maximized in the upper troposphere. The largest disagreement between data sets and least amount of tropospheric temperature increase is seen in the tropics during the satellite era. For the RATPAC product, the greatest temperature increase occurs at the surface with a slight increase (or decrease) in the lower and middle troposphere followed by somewhat larger increase in the upper troposphere. The HadAT2 product also shows largest increase at the surface, with a small increase in the troposphere, however, it lacks a distinct return to increase in the upper troposphere. In summary, the two data sets have fairly similar profiles in the troposphere with the exception of the tropics during the satellite era[22]. For the stratosphere, the decrease in temperature is noticeably greater for both the globe and the tropics during the satellite than radiosonde era as expected (see Figure 3.2b). Some of the largest discrepancies between data sets are found in the stratosphere.

### 7.2 Lapse Rates

Temperature usually decreases in the troposphere going upward from the surface. Lapse rate is defined as the rate of decrease in temperature with increasing altitude and is a measure of the stability of the atmosphere[23]. Most of the

---

[22]  However, the differences between data sets may not be meaningful since they are small compared to the statistical uncertainty estimates (see Table 3.5 and discussion in Appendix A).

[23]  A larger lapse rate implies more unstable conditions and a greater tendency towards vertical mixing of

From 1958 to the present, the tropospheric temperature increased slightly more than that of the surface, but from 1979 to the present, the surface temperature increased more than that of the troposphere.

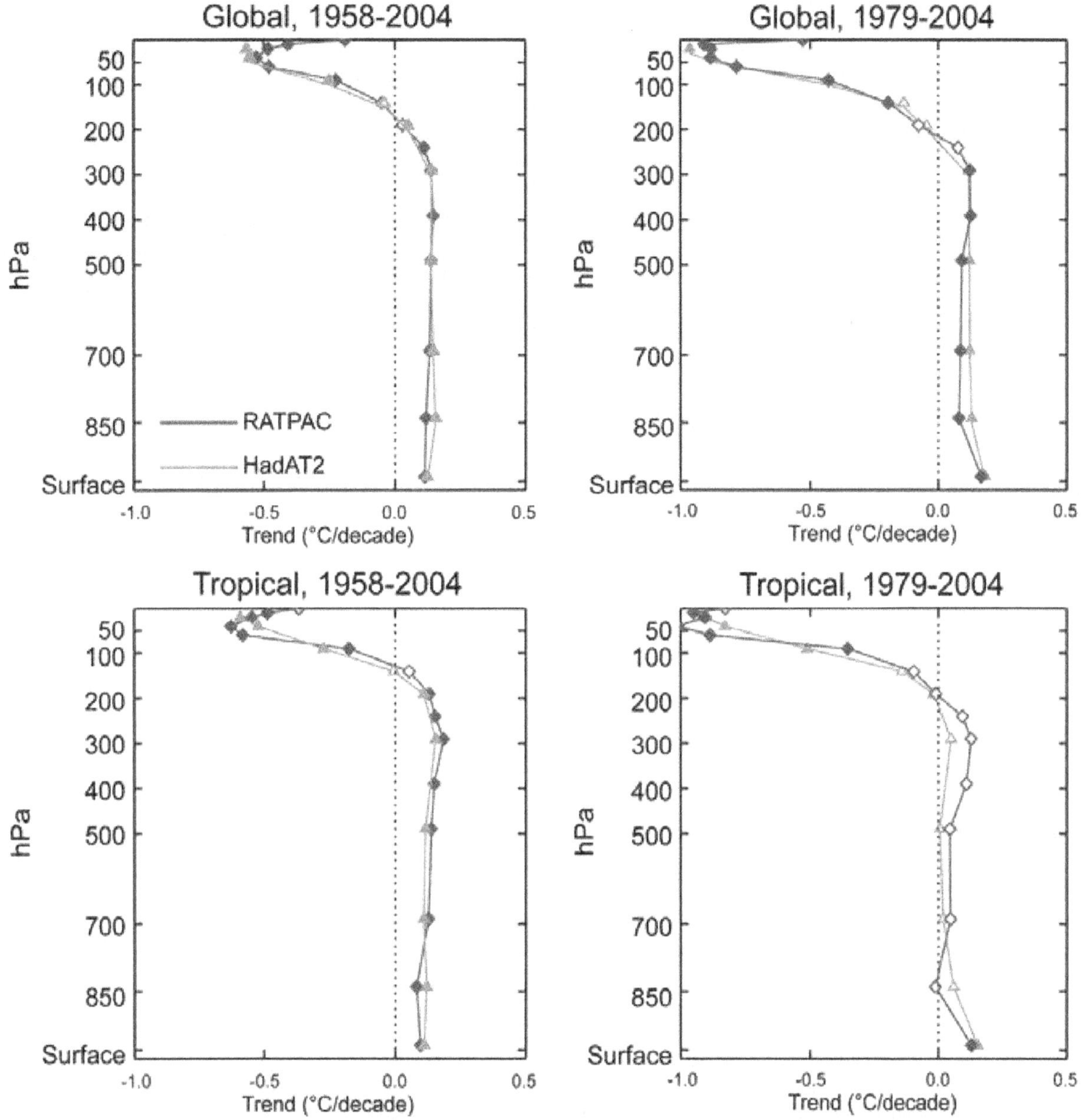

**Figure 3.7** -- Vertical profiles of temperature trend (°C/decade) as a function of altitude (*i.e.*, pressure in hPa) computed from the RATPAC (violet) and HadAT2 (green) radiosonde datasets. Trends (which are given in Table 3.5) have been computed for 1958-2004 (left) and 1979-2004 (right) based on temperature that has been averaged over the globe (top) or the tropics, 20°N-20°S (bottom). Surface data for the HadAT2 product is taken from HadCRUT2v since the HadAT2 dataset does not include values at the surface; the surface values have been averaged so as to match their observing locations with those for the radiosonde data. By contrast, the surface temperatures from the RATPAC product are those from the RATPAC dataset, which are surface station values reported with the radiosonde data. Note that these differ from the NOAA surface dataset values (ER-GHCN-ICOADS) as indicated in Table 3.1. Filled symbols denote trends estimated to be statistically significantly different from zero (at the 5% level). A Student's t-test, using the lag-1 autocorrelation to account for the non-independence of residual values about the trend line, was used to assess significance (see Appendix A for discussion of confidence intervals and significance testing).

observational work to date has not examined lapse rates themselves, but instead has used an approximation in the form of a vertical temperature difference[24]. This difference has taken on the form of the surface temperature minus some tropospheric temperature, either layer-averaged (in the case of satellite data) or

---

air.

[24] The reasons for this are two-fold: (1) satellite measurement systems are only able to resolve temperatures in deep layers rather than at specific levels, and (2) radiosonde measurements are consistently recorded at a fixed number of constant pressure rather than height levels.

**Table 3.5 – Temperature trends in °C per decade from the RATPAC and HadAT2 radiosonde datasets corresponding to the plots in Figure 3.7 (see figure caption for further details). Global and tropical trends are given for 1958 through 2004 and 1979 through 2004 (except for ERA40 which terminates September 2001). The HadAT2 dataset does not have temperatures for some of the levels, hence the empty table cells. The trend is shown for each vertical level (hPa), with the approximate 95% confidence interval (2 sigma) below in parentheses. Bold values are estimated to be statistically significantly different from zero (at the 5% level). A Student's t-test, using the lag-1 autocorrelation to account for the non-independence of residual values about the trend line, was used to assess significance (see Appendix A for discussion of confidence intervals and significance testing).**

| Level (hPa) | 1958-2004 | | | | 1979-2004 | | | |
|---|---|---|---|---|---|---|---|---|
| | RATPAC Global | HadAT2 Global | RATPAC Tropical | HadAT2 Tropical | RATPAC Global | HadAT2 Global | RATPAC Tropical | HadAT2 Tropical |
| 20 | **-0.41** (0.08) | | **-0.49** (0.14) | | **-0.91** (0.14) | | **-0.95** (0.32) | |
| 30 | **-0.48** (0.09) | **-0.57** (0.10) | **-0.55** (0.18) | **-0.59** (0.20) | **-0.88** (0.23) | **-0.96** (0.25) | **-0.91** (0.52) | **-0.90** (0.59) |
| 50 | **-0.53** (0.12) | **-0.55** (0.12) | **-0.63** (0.22) | **-0.52** (0.23) | **-0.89** (0.33) | **-0.88** (0.35) | **-1.01** (0.57) | **-0.83** (0.59) |
| 70 | **-0.48** (0.11) | | **-0.58** (0.22) | | **-0.79** (0.26) | | **-0.89** (0.45) | |
| 100 | **-0.23** (0.06) | **-0.25** (0.06) | **-0.18** (0.06) | **-0.27** (0.07) | **-0.43** (0.16) | **-0.43** (0.15) | **-0.36** (0.17) | **-0.51** (0.16) |
| 150 | -0.05 (0.06) | -0.04 (0.06) | 0.05 (0.07) | -0.01 (0.06) | **-0.19** (016) | -0.13 (0.14) | -0.10 (0.19) | -0.14 (0.16) |
| 200 | 0.03 (0.05) | **0.05** (0.05) | **0.13** (0.08) | **0.11** (0.09) | -0.08 (0.11) | -0.05 (0.11) | -0.01 (0.20) | -0.02 (0.22) |
| 250 | **0.11** (0.04) | | **0.15** (0.08) | | 0.08 (0.10) | | 0.09 (0.20) | |
| 300 | **0.14** (0.04) | **0.14** (0.04) | **0.18** (0.07) | **0.15** (0.08) | **0.12** (0.09) | **0.12** (0.09) | 0.13 (0.18) | 0.05 (0.21) |
| 400 | **0.15** (0.04) | | **0.15** (0.06) | | **0.13** (0.08) | | 0.11 (0.15) | |
| 500 | **0.14** (0.03) | **0.14** (0.04) | **0.14** (0.06) | **0.11** (0.06) | **0.09** (0.07) | **0.12** (0.07) | 0.05 (0.12) | 0.01 (0.14) |
| 700 | **0.13** (0.03) | **0.15** (0.04) | **0.13** (0.05) | **0.11** (0.06) | **0.09** (0.05) | **0.12** (0.07) | 0.05 (0.12) | 0.02 (0.13) |
| 850 | **0.12** (0.02) | **0.15** (0.03) | **0.08** (0.03) | **0.12** (0.05) | **0.08** (0.05) | **0.13** (0.06) | -0.01 (0.06) | 0.06 (0.11) |
| Surface | **0.11** (0.02) | **0.12** (0.03) | **0.10** (0.03) | **0.11** (0.04) | **0.17** (0.05) | **0.18** (0.05) | **0.13** (0.07) | **0.15** (0.12) |

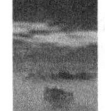

at some specific pressure level (in the case of radiosonde data)[25].

Much of the interest in lapse rate variations has focused on the tropics. Several studies (Brown

---

[25] When constant pressure level data from radiosondes are used, the resulting lapse rate quantity may be influenced by changes in the thickness (*i.e.*, average temperature) of the layer. However, some calculations by Gaffen *et al.* (2000) suggest that thickness changes do not have very much influence. Therefore, we consider vertical temperature differences to be a suitable approximation of lapse rate

*et al.*, 2000; Gaffen *et al.*, 2000; Hegerl and Wallace, 2002; Lanzante *et al.*, 2003b) present time series related to tropical lapse rate based on either satellite or radiosonde measures of tropospheric temperature. As examples, we present some such time series in Figure 3.8, based on measures of lower tropospheric temperature from three different data sets. Some essential low-frequency characteristics are common to all. A considerable proportion of the variability of the tropical lapse rate is associ-

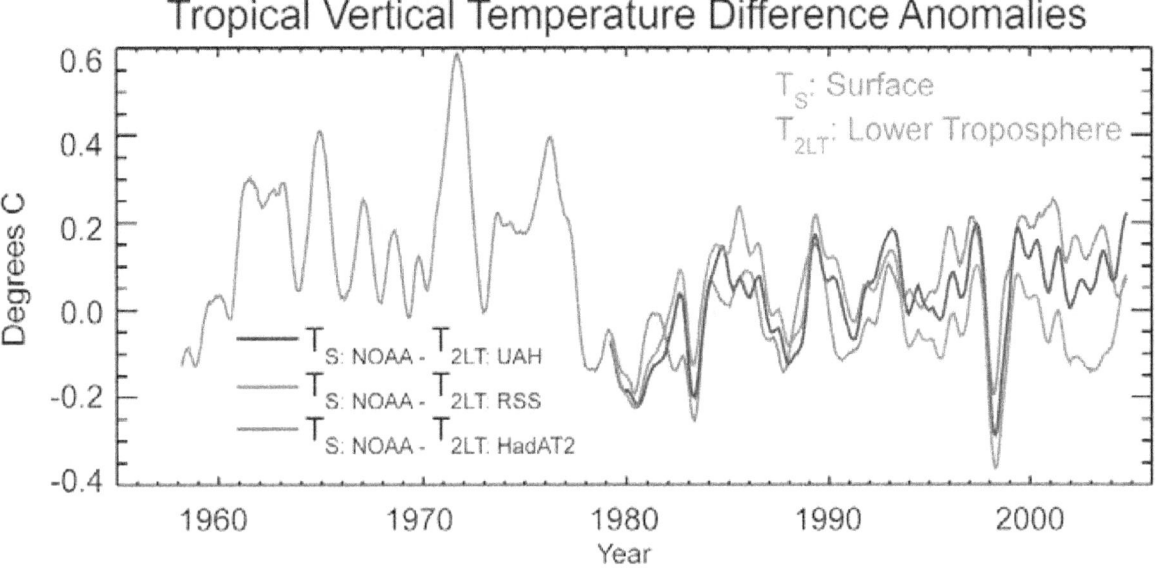

**Figure 3.8 -** Time series of vertical temperature difference (surface minus lower troposphere) for the tropics (20°N-20°S). NOAA surface temperatures ($T_{S-NOAA}$) are used in each case to compute differences with lower tropospheric temperature ($T_{2LT}$) from three different groups: HadAT2 radiosonde (green), RSS satellite (red), and UAH satellite (blue). All time series are 7-month running averages (used as a smoother) of original monthly data, which were expressed as a departure (°C) from the 1979-97 average.

ated with ENSO[26], a manifestation of which is the up and down swings of about 3-7 years in the series shown in Figure 3.8. Another feature evident in the four studies cited above, and seen in Figure 3.8 as well, is an apparent strong association with the climate regime shift that occurred ~1976-77 (Trenberth and Hurrell, 1994). There is a rather sharp drop in tropical lapse rate at this time[27], coincident with an abrupt change in a measure of convective stability (Gettelman *et al.*, 2002). Overall, the variation in tropical lapse rate can be characterized as highly complex, with rapid swings over a few years, superimposed upon persistent periods of a decade or more, as well as longer-term drifts or trends evident during some time periods.

The feature of the tropical lapse rate series that has drawn the most interest is the linear trend component during the satellite era. From a long historical perspective (see also Figure 3.8), this trend is a rather subtle feature, being overshadowed by both the ENSO-related variations as well as the regime shift of the late 1970s. Several studies (Brown *et al.*, 2000; Gaffen *et al.*, 2000; Hegerl and Wallace, 2002; Lanzante *et al.*, 2003b) have estimated trends in lower tropospheric lapse rate while another (Christy *et al.*, 2001) has estimated trends in the difference between SST and surface air temperature.

The different trend estimates vary considerably among the above-cited studies, being dependent upon the details of the calculations[28]. From the cited studies, satellite-era trends in lapse rate based on temperatures averaged over the tropics range from nearly zero (no change) to about 0.20°C/decade (surface warms more than the troposphere). The time series of Figure 3.8 also exhibit a wide range of satellite-era trends[29]. During the longer radiosonde era, the

The feature of the tropical lapse rate series that has drawn the most interest is the linear trend component during the satellite era.

---

[26] Lapse rate changes occur about five to six months after a particular change in ENSO (Hegerl and Wallace, 2002; Lanzante *et al.*, 2003b). During a tropical warming event (El Niño) the tropical troposphere warms relative to the surface; the opposite is true during a tropical cooling event (La Niña).

[27] Lanzante *et al.* (2003b) also noted an apparent decrease in the amplitude of ENSO-related tropical lapse rate variations after the ~1976-77 regime shift.

[28] These details include: time period, latitude zone, data sets utilized, station network vs. grid, time of day of observations, use of homogeneity adjustment, and whether or not measurements in the troposphere and surface were taken from the same locations. Particularly noteworthy is the fact that Lanzante *et al.* (2003b) found that during the satellite era, use of adjusted data could, depending the other details of the analysis, either halve or eliminate the positive tropical lapse rate trend found using the unadjusted data.

[29] Trends from 1979 to 2004 ( C /decade) for the three time series in Figure 3.8 are: 0.11 (HadAT2 radiosonde), 0.08 (UAH satellite), and -0.02 (RSS satellite).

various studies found trends of opposite sign (*i.e.,* air temperature at the surface increases more slowly than that of air aloft) and show less sensitivity, with a range of values of near-zero to about -0.05°C/decade[30].

Spatial variations in lapse rate trends have also been examined. During the satellite era, some have found predominantly increasing trends in the tropics (Gaffen *et al.,* 2000; Brown *et al.,* 2000) while others have found a greater mixture, with more areas of negative trends (Hegerl and Wallace, 2002; Lanzante *et al.,* 2003b). Outside of the tropics, both Hegerl and Wallace (2002) and Lanzante *et al.* (2003b) found complex spatial patterns of trend. Lanzante *et al.* (2003b) also found considerable local sensitivity to homogeneity adjustment in the tropics and even more so over the extra-tropics of the Southern Hemisphere, which is quite sparsely sampled.

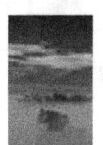

Satellite-era trends in lapse rate based on temperatures averaged over the tropics range from nearly zero (no change) to about 0.20°C/decade (surface warms more than the troposphere).

While the first two of these trends are statistically significant at the 5% level, the third is not (see Appendix A for discussion of significance testing).

[30] The trend from 1958 to 2004 for the HadAT2 radiosonde series shown in Figure 3.8 is -0.02 C/decade. This trend is not statistically significant at the 5% level (see Appendix A for discussion of significance testing).

CHAPTER 4

**W**hat is our understanding of the contribution made by observational or methodological uncertainties to the previously reported vertical differences in temperature trends?

***Convening Lead Author:*** Carl A. Mears, Remote Sensing Systems

***Lead Authors:*** C.E. Forest, MIT; R.W. Spencer, Univ. of AL in Huntsville; R. S. Vose, NOAA; R.W. Reynolds, NOAA

***Contributing Authors:*** P.W. Thorne, U. K. Met. Office; J.R. Christy, Univ. of AL in Huntsville

## KEY FINDINGS

### Surface

It is likely that errors in the homogenized surface air temperature data do not contribute substantially to the large-scale differences between trends for different levels because these errors are very likely to be smaller than those for the upper air data.

- Systematic local biases in surface trends may exist due to changes in station exposure or instrumentation over land, and due to the small number of measurements over a number of regions of the Earth, including parts of the oceans, sea ice areas, and some land areas. Such biases have been documented at the local and regional scale, but no such effect (including any significant urban bias) has been identified in the zonal and global averages presented in this Report. On large spatial scales, sampling studies suggest that these local biases in trends are likely to mostly cancel through the use of many observations with differing instrumentation.

- Since all known bias adjustments have not yet been applied to sea surface temperature data, it is likely that errors remain in these data, though it is generally agreed that these errors are likely to be small compared to errors in radiosonde and satellite measurements of the upper air, especially for the satellite era.

### Troposphere

While all data sets indicate that the troposphere has warmed over both the radiosonde era and the satellite era, uncertainties in the tropospheric data make it difficult to determine whether the troposphere has warmed more than or less than the surface. Some tropospheric data sets indicate that the troposphere has warmed more than the surface, while others indicate the opposite.

- It is very likely that errors remain in the adjusted radiosonde data sets in the troposphere since the methods used to produce them are only able to detect and remove the more obvious errors, and involve many subjective decisions. It is likely that a net spurious cooling corrupts the area-averaged adjusted radiosonde data in the tropical troposphere in at least one and probably both of the data sets, causing the data to indicate less warming than has actually occurred.

- For tropospheric satellite data ($T_2$ and $T_{2LT}$), the primary cause of trend discrepancies between different versions of the data sets is differences in how the data from the different satellites are merged together.

- A secondary contribution to the differences between these data sets is the difference between the diurnal adjustments that are used to account for drifting measurement times. These differences in the diurnal adjustment are more important for regional trends than for global trends, though regional trend differences are also partly influenced by differences in merging methods.

- Each tropospheric satellite data set has strengths and weaknesses that are coming into better focus. Improvements have occurred in several data sets even during the drafting of this Report, each moving it closer to the others, suggesting that further convergence in the not-too-distant future is a strong possibility.
- Comparisons between radiosonde data and satellite data for $T_2$ are very likely to be corrupted by the excessive cooling in the radiosonde data from the stratosphere which are used to help construct the radiosonde-derived $T_2$ data. Trend discrepancies between radiosonde and satellite data sets are reduced by considering a multi-channel retrieval that estimates and removes the stratospheric influence ($T^*_G$).

### Stratosphere
Despite their large discrepancies, all data sets indicate that the stratosphere has cooled considerably over both the radiosonde era and the satellite era.
- The largest discrepancies between data sets are in the stratosphere, particularly between the radiosonde and satellite-based data sets. It is very likely that the satellite-sonde discrepancy arises primarily from uncorrected errors in the radiosonde data.
- There are also substantial discrepancies between the satellite data sets in the stratosphere, indicating that there remain unresolved issues with these data sets as well.

## CHAPTER 4: Recommendations

All of the surface and atmospheric temperature datasets used in this report have undergone extensive testing and analysis in an effort to make them useful tools for investigating Earth's climate during the recent past. In order to further increase our confidence in their use as climate diagnostics, they require ongoing assessment to further quantify uncertainty and to identify and remove any possible systematic biases that remain after the appropriate homogenization methods have been applied.

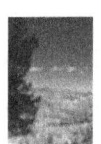

- The diurnal cycles in both atmospheric and surface temperature need to be accurately determined and validated to reduce uncertainties in the satellite data due to the diurnal adjustment. Possible approaches include examining more model or reanalysis data to check the diurnal adjustments currently in use, concerted *in situ* measurement campaigns at a number of representative locations, or operating satellite-borne sounders in a non Sun-synchronous orbit. Information about the surface skin temperature diurnal cycle may be obtained by studying data from existing satellites, or the upcoming Global Precipitation Mission.
- The relative merits of different merging methods for satellite data for all relevant layers need to be diagnosed in detail. Possible approaches include comparison with other temperature data sources (radiosondes or IR satellites) over limited time periods where the discrepancies between the satellite results are the greatest, comparison with other ancillary data sources such as winds and integrated water vapor, and comparison of trends on regional spatial scales, particularly in regions where trends are large or well characterized by radiosonde data.
- The methods used to remove radiosonde inhomogeneities and their effects on trends need to be rigorously studied. The detailed intercomparisons of the methods used by different groups to construct satellite-based climate records have been beneficial to our understanding of these products, and similar parallel efforts to create climate records from radiosonde data would be likely to provide similar benefits.
- Possible errors in trends in spatially averaged surface temperature need to be assessed further. On land these errors may arise from local errors due to changes in instrumentation or local environment that do not completely cancel when spatial averaging is performed. Over the ocean, these errors may arise from the small number of samples available in many regions, and long-term changes in measurement methods. For historical data, these assessments may benefit from the recovery of additional metadata to better characterize possible non-climatic signals and from efforts to assess the self-consistency of historical data.
- Tools and methods need to be developed to help reduce structural uncertainty by providing methods to objectively differentiate between different datasets and construction methods. To the extent possible, such tools should be based on generally accepted physical principles, such as consistency of the temperature changes at adjacent levels in the atmosphere, include physically-based comparisons with external ancillary data, and take account of the consistency of intermediate data generated while producing the datasets.

## 1. BACKGROUND

In the previous Chapter, we have discussed a number of estimates of vertically resolved global temperature trends. Different sources of data (*e.g.,* surface measurements, vertical profiles from radiosondes, and data from satellite borne sounding radiometers), as well as different analysis methods applied to the same data, can yield long term (multi-decadal) temperature trends that differ by as much as several tenths of a °C per decade. This is of comparable magnitude to the actual climate change signal being searched for. In this chapter we discuss these discrepancies in light of the observing system capabilities and limitations described in Chapter 2. We note the degree to which estimates of uncertainty can account for the differences in reported values for the temperature trends in given layers, and differences in the trends of adjacent layers. Most of the time our focus will be on the period from 1979-2004 during which atmospheric temperatures were observed using multiple observing systems.

We begin our discussion in the stratosphere, and move to successively lower layers until we reach the Earth's surface. We proceed in this order because the largest discrepancies in trends between data sources occur in upper atmospheric layers, especially the stratosphere. As mentioned in Box 2.1 (in Chapter 2), when satellite-equivalent measures are made from vertically resolved radiosonde data to facilitate comparisons between the two systems, large stratospheric errors can significantly influence measures centered much lower in the atmosphere.

## 2. UNCERTAINTY IN STRATOSPHERIC TEMPERATURE TRENDS

Long-term observations of the stratosphere have been made by two observing systems: radiosondes and satellite-borne sounders. On both the global and the zonally averaged scale, there is considerably less variation between data sets derived from the same type of observing system for this layer than between those from different observing systems. This can be seen in the leftmost panel of Fig. 3.5, which shows the zonally averaged trends over the satellite

era (1979-2004) for two radiosonde-based data sets, and two satellite-based data sets. The radiosonde data ($T_{4\text{-HadAT2}}$ and $T_{4\text{-RATPAC}}$) show more cooling than data sets based on satellite data ($T_{4\text{-UAH}}$ and $T_{4\text{-RSS}}$), and also do not show the reduced cooling in the tropics relative to the mid-latitudes that is seen in the satellite data.

### 2.1 Radiosonde Uncertainty in the Stratosphere

Radiosonde data are plagued by numerous spurious discontinuities in measured temperature that must be detected and removed in order to construct a homogenized long-term record of atmospheric temperature, a task that is particularly difficult in the absence of reliable metadata describing changes in instrumentation or observing practice. A number of physical sources of such discontinuities have larger effects in the stratosphere. The lower atmospheric pressure in the stratosphere leads to reduced thermal contact between the air and the temperature sensor in the radiosonde package. This in turn leads to increased errors due to daytime solar heating and lags between the real atmospheric temperature and the sensor response as the instrument rises through atmospheric layers with rapidly varying temperatures. Such systematic errors are not important for trend studies provided that they do not change over the time period being studied. In practice, as noted in Chapter 2, radiosonde design, observing practices, and procedures used to attempt to correct for radiation and lag errors have all changed over time.

Past attempts to make adjustments to radiosonde data using detailed physical models of the in-

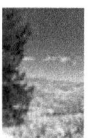

struments (Luers and Eskridge, 1998) improved data homogeneity in the stratosphere, but not in the troposphere (Durre *et al.*, 2002). Since it is important to use the same methods for all radiosonde levels for consistency, scientists have tended to instead use empirical methods to deduce the presence and magnitude of any suspected discontinuity. Both of the homogenized radiosonde data sets used in this report make these estimates using retrospective statistical analyses of the radiosonde data without input from other measurements. The investigators who constructed these data sets have attempted to identify and to adjust for the effects of suspected change points, either by examination of station time series in isolation (RATPAC), or by comparison with nearby stations (HadAT2). Both approaches can most successfully identify changes that are large and step-like. While based in statistics, both these methods also include significant subjective components. As a result, different investigators with nominally the same sets of radiosonde data can calculate different trend estimates because of differences in adjustment procedures (Free *et al.*, 2002). The lack of sensitivity to small or gradual changes may bias the resulting homogenized products if such changes are numerous and predominantly of one sign or the other[1]. The relative frequency of large step-like changes and smaller changes that may be statistically indistinguishable from natural variability remains an open question.

Since the adjustments needed to remove the resulting discontinuities tend to be larger for the stratosphere than for lower levels (Parker *et al.*, 1997; Christy *et al.*, 2003; Lanzante *et al.*, 2003), the uncertainty associated with the homogenization procedures is very likely to be larger in the stratosphere than at lower levels, as has been shown for the HadAT2 radiosonde data set (Thorne *et al.*, 2005a). The best estimate of the size of this source of uncertainty is obtained by comparing the statistics (*e.g.*, the trends) from the two adjusted radiosonde data sets that are currently available. However, the HadAT2 group analysis is partly based upon the RATPAC data set, so we may be under-estimating the uncertainty. Only through increasing

the number of independently produced data sets under different working assumptions can we truly constrain the uncertainty (Thorne *et al.*, 2005b).

Differences in trends between daytime and nighttime observations in the uncorrected radiosonde data used in constructing the RATPAC and HadAT2 radiosonde data sets, suggest that the biases caused by solar heating[2] have been reduced over time, leading to a spurious cooling trend in the raw daytime data (Sherwood *et al.*, 2005). Many of the changes in observing practice will affect both day and night time observations; e.g., a change in practice may yield a spurious 0.5°C daytime cooling and 0.4°C night time cooling, so day-night differences cannot be used in isolation to correct the observations. Whether the RATPAC and HadAT2 methods have successfully removed day-night and other effects, or if sufficiently targeted are capable of doing so, is a matter for ongoing research. Randel and Wu (2005) have shown for a subset of tropical stations in the RATPAC data set, there is strong evidence for step-like residual cooling biases following homogenization, which will cause a spurious cooling in the tropical area-averaged RATPAC time series considered here. They find that the effect is not limited to daytime launches, as would be expected from discussions above, and that it is likely to affect at least the upper-troposphere as well as the stratosphere. Finally, the balloons that carry the instruments aloft have improved over time, so they are less likely to burst at high altitudes or in extreme cold. This could also lead to a warm sampling bias within the stratosphere in early radiosondes which has gradually ameliorated with time, introducing a spurious stratospheric cooling signal (Parker and Cox, 1995). Taken together these results imply that any residual systematic errors in the homogenized radiosonde products will likely lead to a spurious cooling bias.

Different investigators with nominally the same sets of radiosonde data can calculate different trend estimates because of differences in adjustment procedures.

---

[1]  It is speculated that gradual changes could result from the same changes in instrumentation or practices that cause the step like changes, provided that these changes are implemented gradually (Lanzante et al., 2003).

---

[2]  For some types of radiosondes, radiation adjustments based on information provided by the manufacturer are made as part of routine processing of radiosonde data by the observing station. The findings cited here refer to data that has already had these corrections performed. The reduction in daytime biases is likely to be due to a combination of improvements in instrument design, and improvements in the radiation adjustment procedure.

Since the radiosonde stations selected for inclusion in the adjusted data sets do not cover the entire globe[3], there can be a bias introduced in to the global mean trend depending on the locations of the chosen stations. On a global scale, this bias has been estimated to be less than 0.02 to 0.03°C/decade for $T_4$ by sub-sampling globally complete satellite or reanalysis data sets at the station locations[4], and thus it is not an important cause of the differences between the data sets on large spatial scales (Hurrell *et al.,* 2000; Free and Seidel, 2005). Though they have not been explicitly calculated, sampling errors are likely to be more important for the zonal radiosonde trends plotted in Figure 3.5, and may account for some of the zone-to-zone variability seen in the radiosonde data in that figure that is not duplicated in the smoother satellite data. The sampling effects also permeate in the vertical - above 100hPa there is a significant reduction in the number of valid measurements whereas below this level the number of measurements is relatively stable. Because the trends vary with height, this can lead to errors, particularly when calculating satellite-equivalent measures.

## 2.2 Satellite Uncertainty in the Stratosphere

The two satellite-based stratospheric data sets ($T_{4\text{-UAH}}$ and $T_{4\text{-RSS}}$) have received considerably less attention than their tropospheric counterparts (see section 4.3 below), though they differ in estimated trend by roughly the same absolute amount (~0.1°C/decade) as the corresponding tropospheric data sets produced by the same institutions. However the importance of the differences is perceived to be much less because the trend is much larger (a cooling over 1979-2004 of approximately 0.8°C). A detailed comparison of the methods used to construct the two data sets has not yet been performed. Despite the lack of such a study, it is very likely that in the stratosphere, like the troposphere (discussed in section 4.3), structural uncertainty is the most important source of uncertainty.

Two important types of structural uncertainty are likely to dominate: those associated with the method of correcting for drifts in diurnal sampling time, and those associated with the method of correcting calibration drifts associated with the temperature of the hot calibration target. Section 3 discusses how these uncertainty sources are treated in the troposphere.

Despite unresolved problems in the satellite data sets, the similarity of the satellite measurement and homogenization methods suggest that the satellite measurements of the stratosphere are no more uncertain than those of the midtroposphere, where satellites and radiosondes are in much closer agreement. This assessment, coupled with the evidence presented above that residual artificial cooling is likely to exist in the stratospheric radiosonde data, particularly in the tropics, implies that the discrepancy between radiosonde and satellite estimates of stratospheric trends (see Table 3.3) during the satellite era is very likely to be mostly due to uncorrected biases in the radiosonde measurements.

*The discrepancy between radiosonde and satellite estimates of stratospheric trends during the satellite era is very likely to be mostly due to uncorrected biases in the radiosonde measurements.*

# 3. UNCERTAINTY IN TROPOSPHERIC TRENDS

In contrast to the stratosphere, differences in reported tropospheric trends from the same type of measurement are as large or larger than differences in trends reported from different data sources. This can be seen in Figure 3.5 and Tables 3.3 and 3.4. Also note that the radiosonde data for the two tropospheric layers show the

---

[3]   In the Southern Hemisphere, not even all latitude bands are represented

[4]   This estimate is valid for the RATPAC data set and a previous version of the HadAT2 data set. The estimated bias increases to about 0.05 C for a tropical average. In the cited work the tropics were defined to be 30 S to 30 N – we would expect the sampled error to be a few hundredths of a degree per decade larger for the 20 S to 20 N definition of the tropics used in this report.

same general north-south pattern (*i.e.*, more temperature increase in the mid-latitudes than at the poles or in the tropics) as the satellite data, in contrast to the stratospheric results.

### 3.1 Radiosonde Uncertainty in the Troposphere

The main sources of error in tropospheric radiosonde trends are similar to those encountered in the stratosphere. The challenge is to assess to what extent these types of errors, which in the stratosphere likely result in artificial cooling even in homogenized data sets, extend down into the troposphere. Another important issue is that when performing calculations to directly compare radiosonde data with satellite trends for the $T_2$ layer, the contribution of errors in the stratospheric trends to the results for this layer become important, since 10% to 15% of the weight for this layer comes from the stratosphere.

#### 3.1.1 REMOVING NON-CLIMATIC INFLUENCES.

There are several pieces of evidence that suggest that any residual bias in tropospheric radiosonde data will be towards a cooling. First, the more obvious step-like inhomogeneities that have been found tend to predominantly introduce spurious cooling into the raw time series, especially in the tropics. This suggests that any undetected change points may also favor spurious cooling (Lanzante *et al.,* 2003). Second, solar-heating-induced errors, while largest in the stratosphere have been found to bias daytime measurements to higher temperatures at all levels, particularly in the tropics. Periodic radiosonde intercomparisons (most recently at Mauritius in Feb. 2005) undertaken under the auspices of the World Meteorological Organization (WMO) imply that the magnitude of these errors has been reduced over time, and that radiosondes from independent manufacturers have become increasingly similar (and presumably more accurate) over time[5] (da Silveira *et al.,* 2003; Nash *et al.,* 2005). If these effects have on average been uncorrected by the statistical procedures used to construct the homogenized radiosonde data sets discussed in this report, they would have introduced an artificial cooling

Any residual bias in tropospheric radiosonde data is likely to be towards a cooling.

signal into the radiosonde records. Of course on an individual station basis the picture is likely to be much more ambiguous and many stations records, even following homogenization efforts, are likely to retain large residual warm or cold biases. But on average, the evidence outlined above suggests that if there is a preferred sign it is likely to be towards a residual cooling. It is important to stress that to date the quantitative evidence to support such an argument, at least away from a small number of tropical stations (Randel and Wu, 2006), is at best ambiguous.

#### 3.1.2 SAMPLING UNCERTAINTY

The fact that most radiosonde data are primarily collected over Northern Hemispheric land areas naturally leads to uncertainties about whether or not averages constructed from radiosonde data can faithfully represent global trends. However, Wallis (1998), Hurrell *et al.* (2000), and Thorne et al. (2005a) show that stations can be representative of much larger scale averages above the boundary layer, particularly within the deep tropics. Spatial and temporal sampling errors for the radiosonde data sets have been assessed by sub-sampling trends in reanalyses or satellite data at the locations of radiosonde stations used in the production of global data sets, and comparing the results to the full global average of the reanalysis or satellite data (Hurrell *et al.,* 2000; Free and Seidel, 2004). Typically, errors of a few hundredths of a °C per decade have been estimated for global averages, too small to fully account for the differences between radiosonde and satellite trends, though it has been suggested that the existing sampling could lead to a warm bias in the radiosonde record (Agudelo and Curry, 2004). As is the case for the stratosphere, sampling errors may be part of the cause for the zone-to-zone variability seen in the radiosonde data. Residual differences between the global means of the two radiosonde data sets are assessed to be approximately equally caused by sampling error, choice of raw data, and choice of adjustments made[6].

---

5   These intercomparisons provide a source of data about the differences between different type of sondes that has not yet been used to homogenize sonde data.

6   This comparison was made using a previous version of the UK data set (HadRT), which uses a different set of stations than the current version. This difference is very unlikely to substantially alter these conclusions.

### 3.1.3 The Influence of Uncertainty in Stratospheric Measurements

To compare data that represent identical layers in the atmosphere, "satellite-equivalent" radiosonde data products have been constructed using a weighted average of radiosonde temperatures at a range of levels (see Box 2.1, Chapter 2). The $T_2$ radiosonde data sets have been constructed to match the weighting function for Microwave Sounding Unit (MSU) channel 2. Since 10% to 15% of the weight for this channel comes from the stratosphere (see Figure 2.1), it is important to keep in mind the suspected relatively large errors in the stratospheric measurements made by radiosondes. It is possible that stratospheric errors could cause the trends in the radiosonde-derived $T_2$ to be as much as 0.05°C/decade too cool, particularly in the tropics, where the suspected stratospheric errors are the largest (Randel and Wu, 2005) and therefore have a large impact on area-weighted averages. This error source may be partly eliminated by considering the multi-channel tropospheric retrievals discussed in section 5 below.

## 3.2. Satellite Uncertainty in the Troposphere

Satellite-derived temperature trends in the middle and upper troposphere have received considerable attention. In particular, the causes of the differences between $T_{2-UAH}$ and $T_{2-RSS}$ have been examined in detail; less work has been done concerning $T_{2-UMd}$ because this data set is newer. There are two potentially important contributions to the residual uncertainty in satellite estimates of global trends for the satellite-based data sets: (1) corrections for drifts in diurnal sampling, and (2) different methods of merging data from the series of different satellites.

### 3.2.1 Diurnal Sampling Corrections

During the lifetime of each satellite, the orbital parameters tend to drift slowly with time. This includes both a slow change of the local equator crossing time (LECT), and a decay of orbital height over time due to drag by the upper atmosphere. The LECT is the time at which the satellite passes over the equator in a northward direction. Changes in LECT indicate corresponding changes in local observation time for the entire orbit. Because the temperature changes with the time of day (*e.g.,* the cycle of daytime heating and nighttime cooling), slow changes in observation time can cause a spurious long-term trend. These diurnal sampling effects must be estimated and removed in order to produce a climate-quality data record.

The three research groups that are actively analyzing data from microwave satellite sounders first average together the ascending and descending orbits, which has the effect of removing most of the first harmonic of the diurnal cycle. For the purposes of this report, "diurnal correction" means the removal of the second and higher harmonics. Each group uses a different method to perform the diurnal correction.

The UAH group calculates mean differences by subtracting the temperature measurements on one side of the satellite track from the other (Christy *et al.,* 2000). This produces an estimate of how much, on average, the temperature changes due to the difference in local observation times from one side of the satellite swath to another, typically about 40 minutes. This method has the advantage of not relying on data from other sources to determine the diurnal cycle, but it has been shown to be sensitive to satellite attitude errors (Mears and Wentz, 2005), and is too noisy to produce a diurnal adjustment useable on small spatial scales.

The RSS group uses hourly output from a climate model in a microwave radiative transfer algorithm to estimate the diurnal cycle in brightness temperature at each grid point in the satellite data set (Mears *et al.,* 2003). This method has the advantage that a diurnal adjustment can be made at the data resolution. However, it is likely that the climate model-based adjustment contains errors, both because models are often unable to accurately represent the diurnal cycle[7] (Dai and Trenberth, 2004),

*Satellite-derived temperature trends in the middle and upper troposphere have received considerable attention.*

---

[7]  Dai and Trenberth found that the CCSM2 climate model (whose atmospheric component is similar to the CCM3 model used by the RSS group) often underestimated the surface diurnal cycle over the oceans relative to the observational data set they used, with the model indicating that the diurnal amplitude is in the range of 0.0 to 0.4 C, while their observations, derived from ship data, indicate a range of 0.4 to 1.0 C. However, the model range is more consistent with satellite observations of diurnal skin temperature (Gentemann, *et al*). It is possible that spurious diurnal signals due to solar heating of the measurement apparatus have not been completely removed from the ship data. Dai and Trenberth found that the

and because the parameterization of the ocean surface temperature used as a lower boundary for the atmospheric component of the climate model used does not include diurnal variability. The model has been shown to represent the first harmonic of the diurnal cycle for MSU channel 2 with less than 10% error, but less is known about the accuracy of the second and higher harmonics that are more important for adjusting for the diurnal sampling errors (Mears *et al.*, 2003).

Both groups use their diurnal cycle techniques to adjust the satellite data before merging the data from the different satellites. In contrast, the Maryland group averaged the ascending and descending satellite data to remove only the first harmonic in the diurnal cycle before merging, and used a fitting procedure to account for both the first and second harmonic diurnal components when performing the trend analysis after merging the data from different satellites (Vinnikov and Grody, 2003; Vinnikov *et al.*, 2006). Since they only accounted for the first harmonic diurnal component during the merging of satellite data, errors in the diurnal cycle can cause errors in the data analysis following the merging procedure. Although the removal of the diurnal cycle before merging may also introduce some error into UAH and RSS merging procedures if the assumed diurnal cycle is inaccurate, the removal of the diurnal harmonics before merging seems to be a more logical approach as the diurnal harmonics will tend to cause errors unless removed.

On a global scale, the total impact of the diurnal correction applied by the RSS and UAH groups to the microwave sounding data for the RSS data is to increase the decadal trend by about 0.03°C/decade for $T_2$ (Christy *et al.*, 2003; Mears *et al.*, 2003). The impact of the Maryland group's adjustment is almost negligible. For the RSS $T_2$ data, when a diurnal correction is applied that is 50% or 150% as large as the best estimate, these adjustments significantly worsen the magnitude of the intersatellite differences. Changes of this magnitude in the diurnal cycle lead to temperature trends that differ by 0.015°C; so we estimate that the uncertainty in trends due to uncertainty in the diurnal correc-

tion is about 0.015°C/decade for $T_2$. The UAH group estimates that the diurnal correction for $T_2$ is known to 0.01°C/decade (Christy *et al.*, 2000). These estimates of residual uncertainty are relatively small, and are considerably less than the structural uncertainties associated with the satellite merging methodology described in the next section. Despite the global agreement for the diurnal adjustment for the RSS and UAH results, significant differences in the adjustments exist as a function of location (Mears and Wentz, 2005), which may explain some of the difference on smaller spatial scales between these two data sets that can be seen in Fig. 3.5 and Fig. 4.3.

### 3.2.2 SATELLITE MERGING METHODOLOGY

It is very likely that the most important source of uncertainty in microwave sounding temperature trends is due to inter-satellite calibration offsets, and calibration drifts that are correlated with the temperature of the calibration target (Christy *et al.*, 2000; Mears *et al.*, 2003). When results from supposedly identical co-orbiting satellites are compared, intersatellite offsets are immediately apparent. These offsets, typically a few tenths of a °C, must be identified and removed or they will produce errors in long-term trends of several tenths of a °C per decade. When constant offsets are used to remove the inter-satellite differences, the UAH group found that significant differences still remain that are strongly correlated with the temperature of the calibration target[8] (Christy *et al.*, 2000). This effect has since been confirmed by the RSS group (Mears *et al.*, 2003). Both the UAH and RSS groups now remove the calibration target temperature effect using a model that includes a constant offset for each satellite, and an additional empirical "target factor" multiplied by the calibration target temperature.

Despite the similarity in methods, the RSS and UAH groups obtain significantly different values for the global temperature trends (see Table 3.3). In particular, the difference between the trends for $T_2$ has received considerable attention. A close examination of the procedures suggests that about 50% of the discrepancy in trends is accounted for by a difference be-

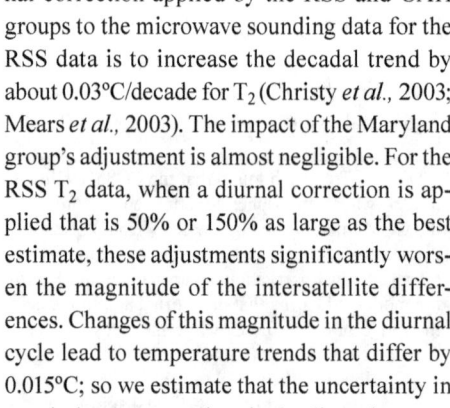

It is very likely that the most important source of uncertainty in microwave sounding temperature trends is due to inter-satellite calibration offsets and calibration drifts.

---

model accurately represents the diurnal pressure tide, suggesting that upper air temperatures are reliable.

[8]   The calibration target can change temperature by tens of C over the course of the life of the satellite due to orbit- and season-dependent solar heating.

tween the target factor for the NOAA-09 instrument deduced by the two groups. This difference mainly arises from the subsets of data used by the two groups when determining the satellite merging parameters (*i.e.,* offsets and target factors). The UAH group emphasizes pairs of satellites that have long periods of overlap, and thus uses data from six pairs of satellites, while RSS uses all available (12) overlapping pairs of satellites. Most of the remainder of the difference is due to a smaller difference in the calibration target temperature proportionality constant for NOAA-11, and to small differences in the diurnal correction. Both these differences primarily affect the measurements made by NOAA-11 and NOAA-14, due to their large drifts in local measurement time, which in addition to their direct effect on the diurnal correction, also lead to large changes in the temperature of the calibration target.

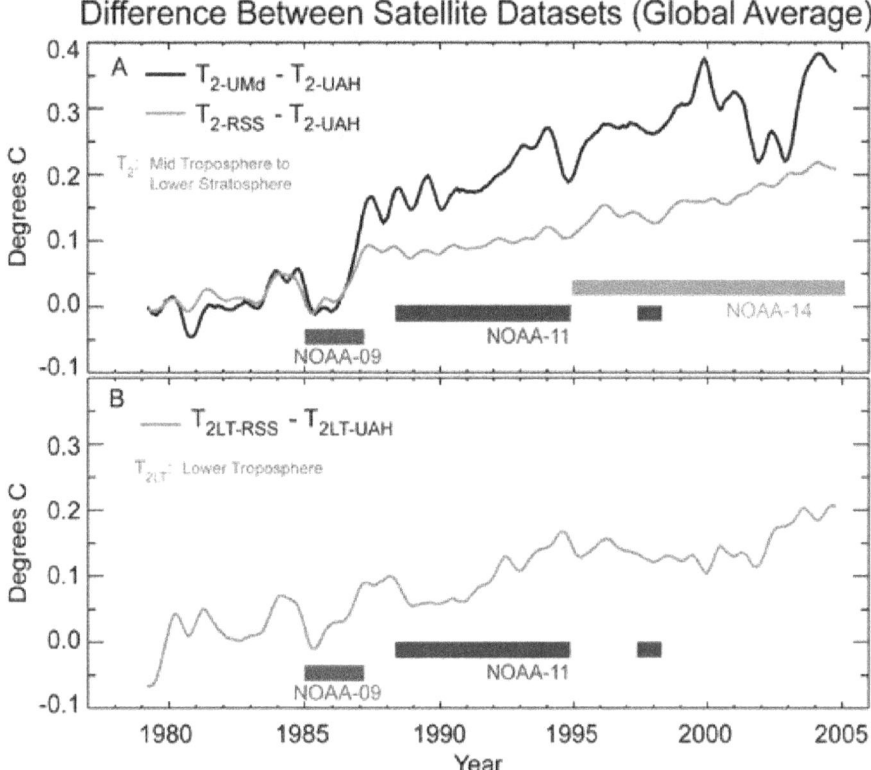

**Figure 4.1** (a) Time series of the difference between global averages of satellite-derived $T_2$ datasets. Both the RSS and UMd datasets show a step-like feature relative to the UAH dataset during the lifetime of NOAA-09. The difference between the RSS and the UAH datasets shows a slow drift during the NOAA-11 and NOAA-14 lifetimes. Both these satellites drifted more than 4 hours in observations time. (b) Time series difference between global averages of satellite derived $T_{2LT}$ datasets. A slow drift is apparent during the lifetime of NOAA-11, but the analysis during the NOAA-14 lifetime is complicated because the $T_{2LT-RSS}$ dataset does not include data from the AMSU instruments on NOAA-15 and NOAA-16, while the $T_{2LT-UAH}$ dataset does. All time series have been smoothed using a Gaussian filter with width = 7 months.

In Fig. 4.1a, we plot the difference ($T_{2-RSS}$ - $T_{2-UAH}$) between the RSS and UAH time series. There is an obvious step that occurs in 1986, near the end of the NOAA-09 observation period, and a gradual slope that occurs during the observation periods of NOAA-11 and NOAA-14. Note that the trend difference between these two data sets is statistically significant at the 1% level, even though the error ranges quoted in Table 3.3 overlap, due to the presence of nearly identical short term fluctuations in the two data sets (see Appendix A for more details).

The Maryland group data set ($T_{2-UMd}$), in its most recent version (Grody *et al.,* 2004; Vinnikov *et al.,* 2006), implemented a more detailed, physically based error model to describe the errors that correlated with a nonlinear combination of the observed brightness temperature measurements and the warm target temperature used for calibration[9]. They use a substantially different merging procedure to deduce values of the parameters that describe the intersatellite differences. First, they use

---

[9]   The Maryland group accounted for uncertainties in the radiometers non-linearity parameter as well as errors in the warm target radiation temperature (due to uncertainties in its emissivity and physical temperature) and errors in the cold space radiation temperature (due to uncertain antenna side lobe contributions for example). However, while all of these error sources are accounted for, they are assumed to be constant during the lifetime of a given instrument and thus do not take into account the possibility of contributions to the side lobe response from the Earth or warm parts of the satellites whose temperature varies with time. These error sources lead, when globally averaged and linearized, to an expression where the target temperature is the most important factor. Thus while the exact physical cause of the observed effect is not known precisely, it is possible to accurately model and remove it on a global scale from the data using either method

## Mid to Upper Troposphere Temperature Trend
### 1979-2004 (°C/Decade)

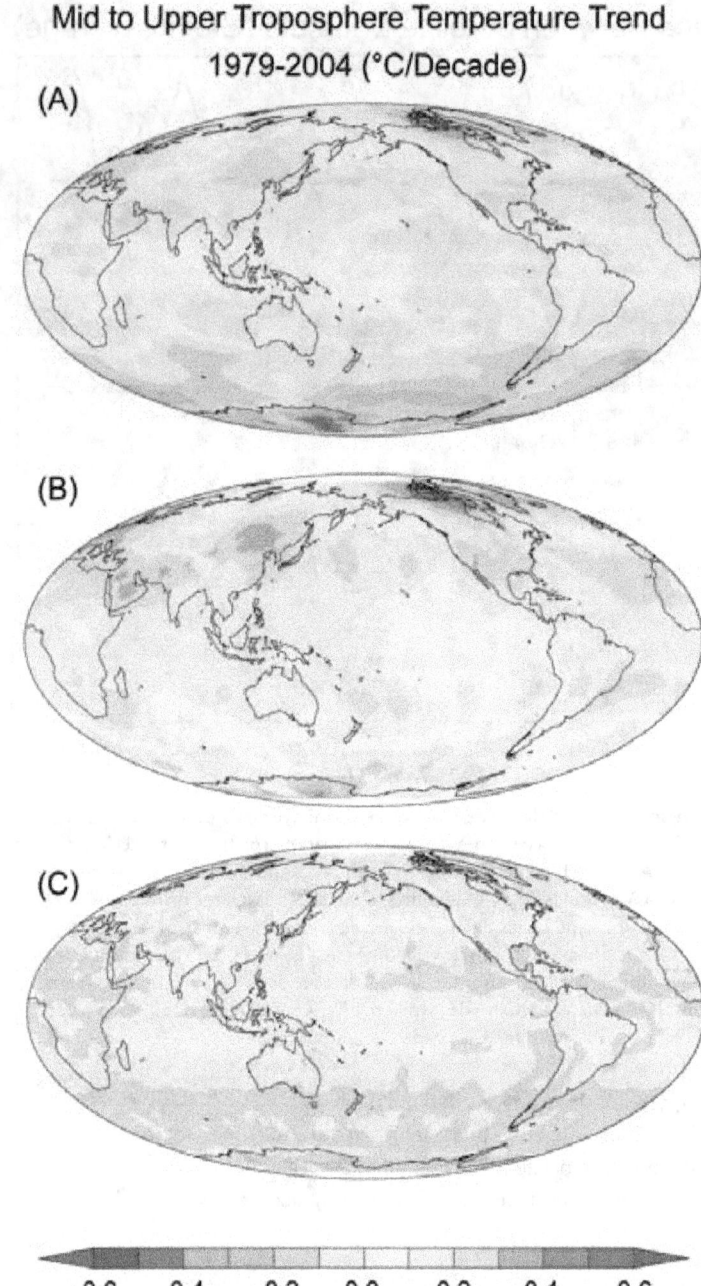

**Figure 4.2** Global maps of trends from 1979-2004 for (a) $T_{2\text{-UAH}}$ and (b) $T_{2\text{-RSS}}$. Except for an overall difference between the two results, the spatial patterns are very similar. A map of the difference $T_{2\text{-UAH}} - T_{2\text{-RSS}}$ between trends for the two products shown in (c) reveals more subtle differences in the trend.

measurements only from the nadir view, in contrast to the central 5 views used by the other groups. Second, as noted above, only the first harmonic diurnal component is accounted for during the satellite merging, possibly causing errors in the retrieved parameters. Third, they only use the spatial variation seen by the differ-

ent MSU instruments to derive the calibration adjustments and perform long-time-scale temporal averaging of the measured temperatures to reduce the noise in the overlapping satellite measurements. This averaging procedure may attenuate the time dependent signal that the UAH empirical error model was introduced to explain. The large step in the $T_{2\text{-UMd}} - T_{2\text{-UAH}}$ difference time series that occurs in 1986 (see Fig. 4.1a) suggests that uncertainty in the parameters for the NOAA-09 satellite are also important for this data set[10]. The cause of the large fluctuations in the difference during the 2000-2004 time period is not known, but may be related to the absence of Advanced MSU (AMSU) data in the $T_{2\text{-UMd}}$ data set. Due to its relatively recent appearance, considerably less is known about the reasons for the differences between the Maryland data set and the RSS and UAH data sets, thus the comments about these differences should be viewed as more speculative than the statements about the RSS-UAH differences.

These differences are an excellent example of structural uncertainty, where identical input data and three seemingly reasonable methodologies lead to trends that differ significantly more than the amount expected given their reported internal uncertainties. Since methodological differences yield data products showing differences in trends in $T_2$ of about 0.1°C per decade, it is clear that the most important source of uncertainty for satellite data are structural uncertainties and that these need to be included in any overall assessment of uncertainties in the estimates of tropospheric temperature trends and lapse rates.

### 3.2.3 DIFFERENCES IN SPATIAL PATTERN
Only $T_{2\text{-UAH}}$ and $T_{2\text{-RSS}}$ have provided gridded results. Maps of gridded trends for these products are shown in Figure 4.2, along with a map of the differences between the trends. The overall pattern in the trends is very similar between the two data sets, aside from a difference in the globally averaged trends. Differences in the latitude dependence are due to the use of zonally varying intersatellite offsets in the construction of $T_{2\text{-UAH}}$ (in contrast to the constant offsets in $T_{2\text{-RSS}}$) and to differences

---

[10] The trend in this difference time series is statistically significant at the 1% level.

in the applied diurnal adjustment as a function of latitude. Other differences may be caused by the spatial smoothing applied to the $T_{2\text{-UAH}}$ during the construction of the data set, and to differences in spatial averaging performed on the diurnal adjustment before it was applied. This last difference will be discussed in more detail in section 4.2 below because the effects are more obvious for the $T_{2LT}$ layer.

## 4. UNCERTAINTY IN LOWER TROPOSPHERIC TRENDS

### 4.1 Radiosonde Uncertainty in the Lower Troposphere

Uncertainties in lower tropospheric trends measured by radiosondes are very similar to those discussed above for the middle-upper troposphere. The most important difference is that when comparing to the $T_{2LT}$ satellite product, the contribution of the stratospheric radiosonde trends, which is suspected to be erroneous to some extent, is substantially less than for the $T_2$ data records. This decreases the likelihood that $T_{2LT}$ data products constructed from radiosonde data are biased toward excess cooling. However, it is possible that undetected negative trend bias remains in all tropospheric levels (see Section 3.1 above for more details), so radiosonde trends may still be contaminated by spurious cooling.

### 4.2 Satellite Uncertainty in the Lower Troposphere

Currently, there are two lower tropospheric satellite data records, $T_{2LT\text{-UAH}}$ and $T_{2LT\text{-RSS}}$. As mentioned in the Preface, both data sets are relatively recent, thus little is known about the specific reasons for their differences. Because of the noise amplification effects of the differencing procedure[11] used to construct the data record (Spencer and Christy, 1992), the merging parameters tend to be more sensitive

to the methods used to deduce them. A number of different methods were explored in the creation of $T_{2LT\text{-RSS}}$, leading to an estimate of the structural uncertainty of 0.08°C/decade for global trends. When combined with internal uncertainty, the estimated total global trend uncertainty for this data set is 0.09°C/decade (Mears and Wentz, 2005). Note that the difference between the global trends for $T_{2LT\text{-RSS}}$ (0.19°C/decade) and $T_{2LT\text{-UAH}}$ (0.12°C/decade) shown in Table 3.3 is less than this estimated uncertainty. The estimated global trends in the radiosonde data sets are also within the $T_{2LT\text{-RSS}}$ error range. In Figure 4.1b we plot the difference ($T_{2LT\text{-RSS}}$ - $T_{2LT\text{-UAH}}$) between the RSS and UAH time series. This time series shows more variability than the corresponding $T_2$ difference time series, making it more difficult to speculate about the underlying causes of the differences between them. The step-like feature during the 1985-1987 period is less obvious, and while there appears to be a slow drift during the NOAA-11 lifetime, a corresponding drift during the NOAA-14 lifetime is less obvious, perhaps because the RSS data do not yet include data from the more recent AMSU satellites. We speculate that the drift during NOAA-11 is in part due to differences in the diurnal correction applied. The UAH diurnal correction is based on a parameterization of the diurnal cycle that is constrained by measurements made during a time period with 3 co-orbiting satellites (Spencer *et al.*, 2006), while RSS uses a model-based diurnal correction analogous to that used for $T_2$.

In Figure 4.3, we show global maps of the gridded trends for $T_{2LT\text{-UAH}}$ and $T_{2LT\text{-RSS}}$, along with a map of the trend differences. The spatial variability in the trend differences between the two data sets is much larger than the variability for $T_2$, though both data sets show similar patterns in general, with the greatest temperature increase occurring in the Northern Hemisphere, particularly over Eastern Asia, Europe, and Northern Canada. The two data sets are in relatively good agreement north of 45°N latitude. In the tropics and subtropics, the largest differences occur over land, particularly over arid regions.

We speculate that this may be in part due to differences in how the diurnal adjustment is

The two satellite data sets are in relatively good agreement north of 45°N latitude. In the tropics and subtropics, the largest differences occur over land, particularly over arid regions.

---

[11] The $T_{2LT}$ data sets are constructed by subtracting 3 times the average temperature measured by the outermost 4 (near-limb) views from 4 times the average temperature measured by the 4 adjacent views, which are closer to nadir. This has the effect of removing most of the stratospheric signal, and moving the effective weighting function lower in the troposphere (Spencer and Christy, 1992). Assuming that the errors is each measurement are uncorrelated, this have the effect of amplifying these errors by a factor of about 5 relative to $T_2$ (Mears and Wentz, 2005). Even if some of the error is correlated between views, this argument still applies to the uncorrelated portion of the error.

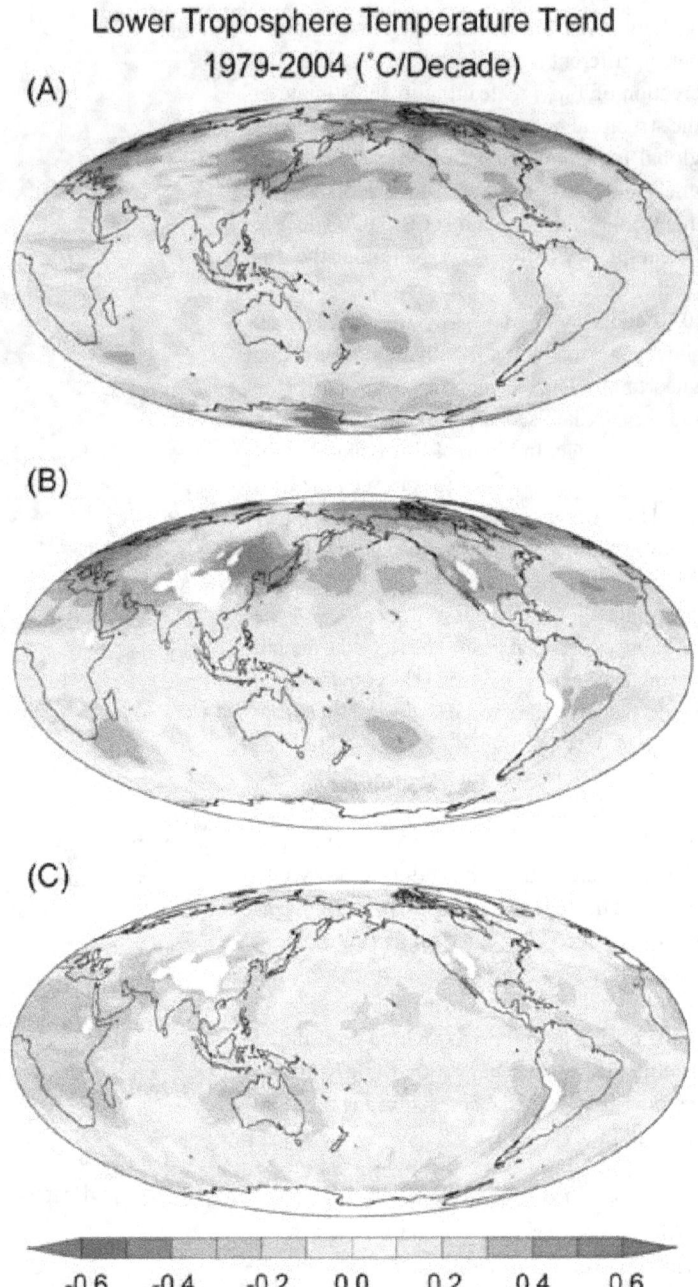

Lower Troposphere Temperature Trend
1979-2004 (˚C/Decade)

(A)

(B)

(C)

-0.6    -0.4    -0.2    0.0    0.2    0.4    0.6

**Figure 4.3** Global maps of trends from 1979-2004 for (a) $T_{2LT-UAH}$ and (b) $T_{2LT-RSS}$. Except for an overall difference between the two results, the spatial patterns are similar. A map of the difference $T_{2LT-UAH}$ - $T_{2LT-RSS}$ between trends for the two products shown in (c) shows that the largest differences are over tropical and subtropical land areas. Data from land areas with elevation higher than 2000 meters are excluded from the $T_{2LT-RSS}$ dataset and shown in white.

the diurnal cycle varies strongly with longitude. More arid regions (*e.g.,* subtropical Africa), which typically have much larger surface diurnal cycles, may be under-adjusted when the zonally averaged correction is applied, leading to long-term trends that are too low. Problems over Africa in the UAH data set were first identified by (Hurrell and Trenberth, 1998). Correspondingly, more humid regions and oceans may be over-adjusted, in some cases making up for the overall difference between the two data sets, perhaps accounting for the good agreement in regions such as Southeast Asia, Southern India, and Northern South America. Further analysis is required using a range of alternative diurnal correction estimation techniques for definitive conclusions to be reached. Other differences, such the north-south streaking seen in the RSS data, may be caused by differences in spatial smoothing, and by the inclusion of AMSU data in $T_{2LT-UAH}$, but not in $T_{2LT-RSS}$.

The decay of orbital height over each satellite's lifetime can cause substantial errors in satellite-derived $T_{2LT}$ because changes in height lead to changes in the Earth incidence angles for the near-limb observations used to construct the data record Wentz and Schabel (1998)[12]. Both the RSS and UAH groups now correct for this error by calculating the expected change in observed temperature as a function of incidence angle, and then using this estimate to remove the effect of orbital decay[13]. The straight-forward method used to make these corrections, combined with its insensitivity to assumptions about the vertical structure of the atmosphere, leads to the conclusion that errors due to orbital decay have been accurately removed from both data sets and are not an important cause of any differences between them.

## 4.3 Comparison Between Satellite and Well-characterized Radiosonde Stations

Point-by-point comparisons between radiosonde and satellite data eliminate many sources

done by the two groups. The UAH group applies an averaged diurnal adjustment for each zonal band, based on different adjustments used for land and ocean. The RSS group uses a grid-point resolution diurnal correction. The UAH method may lead to errors for latitudes where

---

[12]  Note that the adjustment for orbital decay is only important for the $T_{2LT}$ data sets. The $T_2$ data sets only use nadir and near-nadir observations. Since changes in orbital height only lead to small changes in incidence angle for these views, the $T_2$ data sets are insensitive to the effects of orbital decay.

[13]  The UAH group began to use this adjustment in version D of their product, which is described in Christy *et al.,* (2000).

of sampling error normally present in radiosonde data. Also, since uniform global coverage is less important when using radiosondes to validate satellite data locally, stations can be chosen to minimize the contribution due to undocumented changes in radiosonde instrumentation or observing practice. For instance, if one restricts comparisons of the satellite and radiosonde data to 29 Northern Hemisphere radiosonde stations that have consistently used a single type of instrumentation (the Viz sonde) since 1979, the average difference between these radiosonde trends and $T_{2LT\text{-}UAH}$ trends since 1979-2004 is only 0.03°C/decade (Christy *et al.*, 2003). Similarly, when this set of radiosondes is extended to include a set of Southern Hemisphere stations where instrument changes were well documented, agreement between $T_{2LT\text{-}UAH}$ and radiosonde trends is almost as good (Christy and Norris, 2004). This suggests that, for the $T_{2LT}$ layer, where the stratospheric problems with radiosonde data are minimized, some level of corroboration can be attained from these two diverse measurement systems.

## 5. MULTI-CHANNEL RETRIEVALS OF TROPOSPHERIC TEMPERATURE

As mentioned above, the single channel satellite measurements commonly identified as tropospheric temperature ($T_2$) are impossible to interpret as solely tropospheric temperatures because 10% to 15% (seasonally and latitudinally varying) of the signal measured by MSU channel 2 arises from the stratosphere. In principle, it is possible to reduce the stratospheric contribution to Channel 2 by subtracting out a portion of the stratospheric Channel 4 (Fu *et al.*, 2004), though the exact values of the weights used in this procedure are controversial (see Chapter 2 for more details). Despite this controversy, there is little doubt that the resulting trends are more representative of the troposphere than the $T_2$ data sets. The reduction in stratospheric signal also reduces the difference between trends in the satellite data and the radiosonde data (see Table 3.3), because the error-prone stratospheric levels in the stratosphere have reduced (but still non-zero) weight.

The existence of a stratosphere-corrected tropospheric retrieval allows tests for consistency of temperature trends among the different data sets constructed by a research group for different atmospheric layers. One test, when applied to an earlier version (v5.1) of the UAH global average trends, did not prove inconsistency on the global scale, because the difference between the $T_{2LT\text{-}UAH}$ trend and the retrieval-calculated $T_{2LT}$ trend was well within the published margin of error. However, a clearer inconsistency was found for the tropics (Fu and Johanson, 2005). In this case, the difference between the retrieval-calculated trend and $T_{2LT\text{-}UAH}$ trend was larger than its estimated error range, an indication of uncharacterized error in at least one of the UAH products, or more generally that $T_{2LT\text{-}UAH}$, $T_{2\text{-}UAH}$ and $T_{4\text{-}UAH}$ were not strictly self-consistent as a set. This inconsistency no longer exists (within error estimates) after the introduction of version 5.2 of the $T_{2LT\text{-}UAH}$ data set in mid 2005. The RSS versions of the $T_2$, $T_4$ and T* data sets were found to be consistent for both global and tropical averages (Fu and Johanson, 2005). The trends in the RSS version of the $T_{2LT}$ data set (produced after Fu and Johanson was submitted) is also consistent with the other RSS based data sets.

## 6. UNCERTAINTY IN SURFACE TRENDS

### 6.1 Sea Surface Temperature Uncertainty

Temperature analyses over the ocean are produced from sea surface temperatures (SST) instead of marine air temperatures. This is because marine air temperatures are biased from daytime ship deck heating (Folland and Parker, 1995; Rayner *et al.*, 2003) and because satellite observations are available for SST beginning in November 1981 to augment *in situ* data (Reynolds and Smith, 1994). Spatially complete analyses of SSTs can be produced by combining satellite and *in situ* data (from ships and buoys) (Reynolds *et al.*, 2002; Rayner *et al.*, 2003), from *in situ* data alone (Smith and Reynolds, 2004), or from satellite data alone (Kilpatrick *et al.*, 2001).

#### 6.1.1 SATELLITE SST UNCERTAINTIES
Climate comparison analyses based on infrared satellite data alone are not useful because of

An inconsistency was found for the tropics in one of the satellite-derived tropospheric data sets.

possible large time-dependent biases. These biases have typically occurred near the end of a satellite's life time when the instrument no longer works properly, or during periods when assumptions made about the atmospheric profile in the satellite algorithm are no longer valid, e.g., during periods immediately following volcanic eruptions, when a large amount of dust from the eruption is present in the stratosphere (Reynolds, 1993; Reynolds *et al.*, 2004). These problems may be partially mitigated in the future by use of the microwave SST sensors that became available starting with the launch of the Tropical Rainfall Measuring Mission (TRMM) in 1987 (Wentz *et al.*, 2000), but these microwave SST data have not been available long enough to derive meaningful trends, and are difficult to calibrate absolutely due to various instrument related problems (Wentz *et al.*, 2001; Gentemann *et al.*, 2004). Thus, analyses now use multiple satellite instruments blended with or anchored to *in situ* data that reduce the overall analysis errors (*e.g.*, Reynolds *et al.*, 2002, Rayner *et al.*, 2003).

### 6.1.2 IN SITU SST UNCERTAINTIES

As discussed in Chapter 2, the primary sources of uncertainty in *in situ* SST measurements are non-climatic signals caused by changes in the mix of instrumentation over time and sampling errors. Over time the measurements have typically evolved from insulated bucket measurements to engine intake, through hull, and buoy mounted sensors - these changes are not necessarily accurately recorded in the metadata. Both non-climatic signals and sampling errors are thought to be largest in sparsely sampled regions, such as the southern oceans, where a single erroneous or unrepresentative measurement could bias the average for an entire measurement cell for the month in question. Both types of errors have been calculated for the Extended Reconstruction SST (ERSST) data set and included in the quoted error range (see figure 4.4).

### 6.2 Land Surface Air Temperature Uncertainty

The three surface temperature analyses exhibit similar warming rates since 1958. As the surface data sets have many stations in common, they are not totally independent. However, the MSU series take identical input, and radiosonde

data sets have common data also, so this issue is not unique to the surface records. The fact that the range in trends is much smaller for the surface data sets than for these other data sets implies that the structural uncertainty arising from data set construction choices is much smaller at the surface, in agreement with the arguments made in Thorne *et al.* (2005b). Also, a number of studies *e.g.*, (Peterson *et al.*, 1999; Vose *et al.*, 2004) suggest that long-term, large-scale trends are not particularly sensitive to variations in choice of station networks. But because most land networks were not designed for climate monitoring, the data contain biases that data set creators address with different detailed methods of analysis. The primary sources of uncertainty from a land-surface perspective are (a) the construction methods used in the analyses and (b) local environmental changes around individual observing stations (*e.g.*, urbanization) that may not have been addressed by the homogeneity assessments.

Because the stations are not fully representative of varying-within-area land surface, coastal, and topographical effects, global data sets are produced by analyzing deviations of temperature from station averages (anomalies) as these deviations vary more slowly with a change in location than the temperatures themselves (Jones *et al.*, 1997). Random errors in inhomogeneity detection and adjustments may result in biased trend analyses on a grid box level. However, on the relatively large space scales of greatest importance to this Report, such problems are unlikely to be significant in current data sets in the period since 1958 except where data gaps are still serious, *e.g.*, in parts of central Africa, central South America, and over parts of Antarctica. Note that for the contiguous United States, the period 1958-2004 uses the greatest number of stations per grid box anywhere on the Earth's land surface, generally upwards of 20 stations per grid box. For regions with either poor coverage or data gaps, trends in surface air temperature should be regarded with considerable caution, but do not have serious effects on the largest of scales as most of the spatial variability is well sampled.

A variety of studies have documented that urbanization has a warming effect on the local microclimate; however, no study has demonstrated

For surface air temperature data sets, the structural uncertainty arising from data set construction choices is much smaller than for SST or upper air data sets.

that urban warming imparts a significant bias to multi-decadal trends over large areas. In fact, the effect appears at most to be roughly an order of magnitude smaller than long-term trends (*e.g.*, Jones *et al.*, 1990). Several recent global (*e.g.*, Easterling *et al.*, 1997; Peterson *et al.*, 1999) and national analyses (*e.g.*, Li *et al.*, 2004; Peterson *et al.*, 2003) also indicate that urban and rural station networks had comparable trends since roughly the mid-20th century. In addition, minimum temperature trends since 1950 were similar on both windy and calm nights, the latter being more susceptible to urban warming (Parker, 2004). To insure that potential urbanization effects do not impact analyses, the NASA group adjusts the data from all urban stations so that their long-term trends are consistent with those from neighbouring rural stations (Hansen *et al.*, 2001). It is generally accepted that local biases in trends mostly cancel through the use of many stations or ocean observations. Because such a cancellation has not been rigorously proved, partly due to the lack of adequate metadata, it is conceivable that systematic changes in many station exposures of a similar kind may exist over the land during the last few decades. If such changes exist, they may lead to small amounts of spurious cooling or warming, even when the data are averaged over large land regions.

## 6.3 Combined Land-ocean Analyses Uncertainty

Global combined surface temperature products are computed by combining ocean and land gridded data sets. The latest version of the UK surface data set, HadCRUT2v, (Jones and Moberg, 2003) has been optimally averaged with uncertainties for the globe and hemispheres. The NOAA surface temperature data set produced by Smith and Reynolds (2005), uses Global Historical Climatology Network (GHCN), merged with the *in situ* ERSST analysis of Smith and Reynolds (2004). The analyses are done separately over the ocean and the land following the ERSST methods. Error estimates include the bias, random and sampling errors.

As an example of uncertainties in a combined land-ocean analysis, near-global time series (60°S to 60°N) are shown in Figure 4.4 for SST, land-surface air temperature, and the combined SST and land-surface air temperature (Smith and Reynolds, 2005). (The combined product is the GHCN-ERSST product used in Chapter

3). The SST has the tightest (95%) uncertainty limits (upper panel). The land-surface air temperature (middle panel) has a larger trend over the period since 1958, but its uncertainty limits are also larger than for SST. Land surface air temperature uncertainty is larger than the uncertainty for SST because of higher variability of surface air temperature over land (see Chapter 1), persistently un-sampled regions, including central Africa and interior South America, and because the calculations include an increasing urbanization bias-error estimate. Merged temperature anomalies and their uncer-

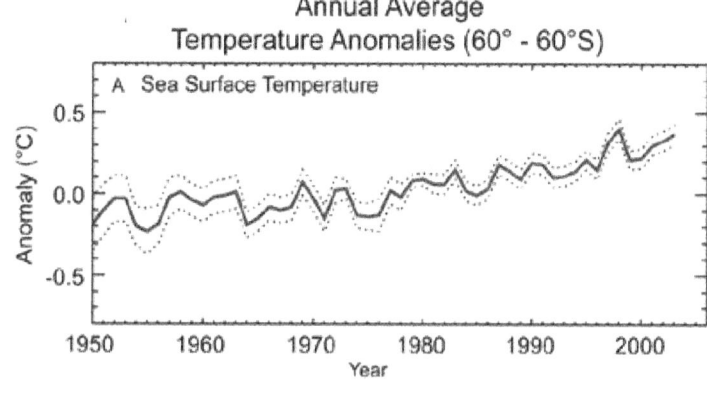

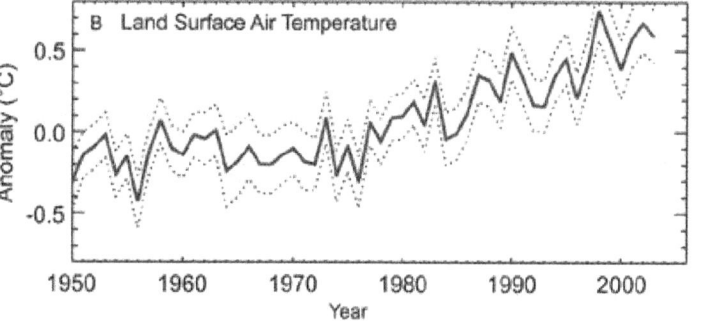

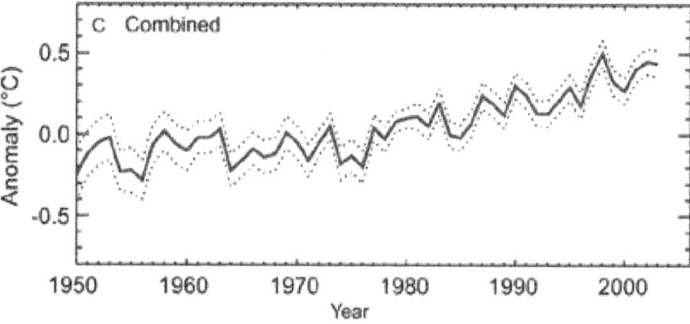

**Figure 4.4.** SST, Land Surface Air Temperature, and the Combined Temperature Data Record anomaly averaged annually and between 60°S and 60°N (purple), with its estimated 95% confidence intervals (dashed). Data are from the $T_{S-NOAA}$ dataset (Smith and Reynolds, 2005). Anomalies are relative to the 1982-2001 period for SST, and 1982-1991 for land.

tainty (lower panel) closely resemble the SST result, since oceans cover most of the surface area. Similar uncertainty was found by (Folland *et al.*, 2001) using different methods.

## 7. INTERLAYER COMPARISONS

### 7.1 Troposphere/Stratosphere
All data sources agree that on a global scale, the stratosphere has cooled substantially while the troposphere has warmed over both the 1958-2004 and the 1979-2004 time periods (note that this is not true for all 25-year time periods within the longer 1958-2004 time period). We suspect that the stratospheric cooling trends estimated from radiosondes are larger in magnitude than the actual trend. Despite the uncertainty in the exact magnitude of stratospheric cooling, we have very high confidence that the lower stratosphere has cooled by several tenths of a °C per decade over the past five decades.

### 7.2 Lower Troposphere/Mid-Upper Troposphere
The difference in trend between the lower troposphere and mid-upper troposphere is not well characterized by the existing data. On a global scale, all data sets suggest that $T_{2LT}$ is warming relative to $T_2$, but it is important to note that the $T_2$ data records have significant stratospheric contributions that reduce their warming trends. Radiosonde measurements suggest that the $T_{(850-300)}$ layer (which does not include the stratosphere) is warming at about the same rate as $T_{2LT}$, while satellite data suggest that $T^*_G$ is warming more rapidly than $T_{2LT}$. The magnitude of these inter-data set differences are typically less than their individual estimates of uncertainty, substantially reducing confidence in our ability to deduce the sign of the lower troposphere-mid-upper troposphere trend difference.

### 7.3 Surface/Lower Troposphere
On a global scale, one satellite data set ($T_{2LT-RSS}$) suggests that the troposphere has warmed more than the surface, while both radiosonde data sets and one of the satellite data sets ($T_{2LT-UAH}$) indicate the opposite. The magnitude of these differences is less than the uncertainty estimates for any one data record, thus no discrepancy is indicated. The situation is similar in the tropics. Both global and tropical averages of

the radiosonde data contain many stations with less reliable data and metadata, which may be part of the cause for the surface-tropospheric differences. In contrast, in North America and Europe the warming in the surface and lower troposphere appears to be very similar in all data sets. This may be due to a combination of the presence of more reliable radiosonde stations in these areas and the stronger correlation between the surface and the lower troposphere over land. It is also illuminating to investigate the spatial patterns in the difference in trends between these data sets. In Figure 5.5, panels E and F (in Chapter 5), we plot maps of the trend differences between the $T_{S-NOAA}$ data set and the two satellite derived $T_{2LT}$ data sets. This figure shows that the trends differences are much larger over arid tropical and subtropical land regions for the $T_{2LT-UAH}$ data set than for the $T_{2LT-RSS}$ data set. As discussed in more detail in Section 4.2, this is likely to be due to the method the UAH team uses to adjust for diurnal drifts, which is likely to under correct regions with large diurnal variability.

### 7.4 Surface/Mid Troposphere
It is also interesting to consider the trend differences between the surface and mid troposphere since more satellite data sets are available for $T_2$ than for $T_{2LT}$. Here, mostly due to the large structural uncertainty in the trends in $T_2$, the various data sets are unable to agree on the sign of the trend difference over the 1979-2004 period. On a global scale, the two radiosonde data sets and two of the satellite data sets (RSS and UAH) suggest that $T_2$ has warmed less than the surface, but the other satellite data set (UMd) suggests that the opposite is true. Similar results are found for tropical averages. It is important to remember that $T_2$ is contaminated by stratospheric cooling. $T^*_G$, which is adjusted to remove these effects, shows smaller differences between the surface and tropospheric trends, with two satellite data sets (RSS and UMd) indicating more warming than at the surface.

## 8. RESOLUTION OF UNCERTAINTY

In almost all of the tropospheric and stratospheric data records considered, our uncertainty is dominated by structural uncertainty

All data sources agree that on a global scale, the stratosphere has cooled substantially while the troposphere has warmed over both the 1958-2004 and the 1979-2004 time periods.

arising through data set construction choices (Thorne *et al.*, 2005b). Differences arising as a result of different, seemingly plausible correction models applied by different groups to create a climate-quality data record are significantly larger than the uncertainties internal to each method, in the raw data measurements, or in the sampling uncertainties. These structural uncertainties are difficult to assess in an absolute sense. The best estimates we can currently make come from examining the spread of results obtained by different groups analyzing the same type of data. This "all data sets are equal" approach has been employed in our present analysis. As outlined in Chapter 2, this estimate of uncertainty can either be too small or too large, depending on the situation. Given this caveat, it is always better to have multiple (preferably at least three) data records that purport to measure the same aspects of climate with the same data, so we can get some idea of the structural uncertainty.

In reality, all data sets are not equally plausible realizations of the true climate system evolution. The climate system has evolved in a single way, and some data sets will be closer to this truth than others. Given that the importance of structural uncertainty, particularly for trends aloft, has only recently been recognized, it is perhaps not surprising that we are unable to quantify this at present. We could make value-based judgments to imply increased confidence in certain data sets, but these would not be unambiguous, may eventually be proven wrong, and are not a tenable approach in the longer term from a scientific perspective. Therefore tools need to be developed to objectively discriminate between data sets. These may include (1) measures of the internal consistency of the construction methods, (2) assessment of the physical plausibility of the merged products, including consistency of vertically resolved trends, and (3) comparisons with vicarious data – for example, changes in temperature need to be compared with changes in water vapor, winds, clouds, and various measures of radiation to assess consistency with the expected physical relationships between these variables. Taken together such a suite of indicators can be used to provide an objectively based way of highlighting residual problems in the data sets and gaining a closer estimate of the truth. Such

an audit of current data sets should be seen as very high priority and preferably undertaken independently of the data set builders in a similar manner to the model intercomparisons performed at Lawrence Livermore National Laboratory. In addition to an agreed set of objective analysis tools, such an effort requires full and open access to all of the data sets including a full audit trail.

Some specific suggestions for resolving some of the issues brought forward in this chapter are mentioned here, but these are not exhaustive and further investigation is required.

### 8.1 Radiosondes

A significant contribution to the long-term inhomogeneity of the radiosonde record appears to be related to changes in radiative heating of the temperature sensor for various radiosonde models, and changes in the adjustments made to attempt to correct for these changes. Recent work suggests that such problems may account for much of the apparent tropical cooling shown in unadjusted data. Other recent work suggests that step-like changes in bias may still remain, even in adjusted data sets. Suitable tests on radiosonde products may therefore include: stability of day-night differences, spatial consistency, internal consistency (perhaps including wind data that to date have not been incorporated), and consistency with MSU-derived and other independent estimates.

### 8.2 Satellites

The most important contributions to satellite uncertainty are merging methodology and the diurnal adjustment. The satellite data are simple enough that considerable understanding can result from examination of intermediate results in the merging process, including intersatellite differences that remain after the merging adjustments are complete. Consistent reporting of such results can help differentiate between methods. It appears that the differences in merging methodology often result in sharp step-like features in difference time series between data sets. Other data sets, such as spatially averaged adjusted radiosonde data, might be expected to show more slowly changing errors, since their errors are due to the overlap of many different, potentially step-like errors that occur at different times. So comparisons of satellite data

We could make value-based judgments to imply increased confidence in certain data sets, but these would not be unambiguous, and are not a tenable approach in the longer term from a scientific perspective.

with radiosonde data over short time periods may help differentiate between satellite data sets. The diurnal adjustment can be improved by a more rigorous validation of model-derived diurnal cycles, or by further characterization of the diurnal cycle using the TRMM satellite or concerted radiosonde observing programs designed to characterize the diurnal cycle at a number of representative locations.

### 8.3 Surface

The uncertainty in the historical near-surface temperature data is dominated by sampling uncertainty, systematic changes in the local environment of surface observing stations, and by difficult-to-characterize biases due to changes in SST measurement methods. The relative maturity of the surface data sets suggests that to a large degree, these problems have been addressed to the extent possible for the historical data, due to the absence of the required metadata (for the bias-induced uncertainties) or the existence of any observations at all. However, it is likely that much of the relatively recent SST data can be adjusted for measurement type as some of the needed metadata is available or can be estimated.

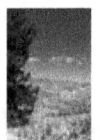

The best estimates we can currently make come from examining the spread of results obtained by different groups analyzing the same type of data.

CHAPTER 5

# How well can the observed vertical temperature changes be reconciled with our understanding of the causes of these changes?

***Convening Lead Author:*** Benjamin D. Santer, DOE LLNL
***Lead Authors:*** J.E. Penner, Univ. of MI; P.W. Thorne, U.K. Met. Office
***Contributing Authors:*** W. Collins, NSF NCAR; K. Dixon, NOAA; T.L. Delworth, NOAA; C. Doutriaux, DOE LLNL; C.K. Folland, U.K. Met. Office; C.E. Forest, MIT; J.E. Hansen, NASA; J.R. Lanzante, NOAA; G.A. Meehl, NSF NCAR; V. Ramaswamy, NOAA; D.J. Seidel, NOAA; M.F. Wehner, DOE LBNL; T.M.L. Wigley, NSF NCAR

## KEY FINDINGS

### Fingerprint Pattern Studies

Fingerprint studies use rigorous statistical methods to compare spatial and temporal patterns of climate change in computer models and observations.

**1.** Both human and natural factors have affected Earth's climate. Computer models are the only tools we have for estimating the likely climate response patterns ("fingerprints") associated with different forcing mechanisms.

To date, most formal fingerprint studies have focused on a relatively small number of climate forcings. Our best scientific understanding is that:
* Increases in well-mixed greenhouse gases (which are primarily due to fossil fuel burning) result in large-scale warming of the Earth's surface and troposphere, and cooling of the stratosphere.
* Human-induced changes in the atmospheric burdens of sulfate aerosol particles cause regional cooling of the surface and troposphere.
* Depletion of stratospheric ozone cools the lower stratosphere and upper troposphere.
* Large volcanic eruptions cool the surface and troposphere (for 3 to 5 years) and warm the stratosphere (for 1 to 2 years).
* Increases in solar irradiance warm globally throughout the atmospheric column (from the surface to the stratosphere).

**2.** Results from many different fingerprint studies provide consistent evidence of a human influence on the three-dimensional structure of atmospheric temperature over the second half of the 20th century.

Robust results are:
* Detection of greenhouse-gas and sulfate aerosol signals in observed surface temperature records.
* Detection of an ozone depletion signal in stratospheric temperatures.
* Detection of the combined effects of greenhouse gases, sulfate aerosols, and ozone in the vertical structure of atmospheric temperature changes (from the surface to the stratosphere).

**3.** Natural factors have influenced surface and atmospheric temperatures, but cannot fully explain their changes over the past 50 years.

* The multi-decadal climatic effects of volcanic eruptions and solar irradiance changes are identifiable in some fingerprint studies, but results are sensitive to analysis details.

**Trend Comparisons**

Linear trend comparisons are less powerful than "fingerprinting" for studying cause-effect relationships, but when treated with caution can highlight important differences (and similarities) between models and observations.

**4.** When run with natural and human-caused forcings, model global-mean temperature trends for individual atmospheric layers are consistent with observations.

**5.** Comparing trend differences between the surface and the troposphere exposes potential discrepancies between models and observations in the tropics.

• Differencing surface and tropospheric temperature time series (a simple measure of the temperature lapse rate) removes much of the common variability between these layers. This makes it easier to identify discrepancies between modeled and observed lapse-rate changes.
• For globally averaged temperatures, model-predicted trends in tropospheric lapse rates are consistent with observed results.
• In the tropics, most observational data sets show more warming at the surface than in the troposphere, while most model runs have larger warming aloft than at the surface.

**Amplification of Surface Warming in the Troposphere**

**6.** In the tropics, surface temperature changes are amplified in the free troposphere. Models and observations show similar amplification behavior for monthly and interannual temperature variations, but not for decadal temperature changes.

• Tropospheric amplification of surface temperature anomalies is due to the release of latent heat by moist, rising air in regions experiencing convection.
• Despite large inter-model differences in variability and forcings, the size of this amplification effect is remarkably similar in the models considered here, even across a range of timescales (from monthly to decadal).
• On monthly and annual timescales, amplification is also a ubiquitous feature of observations, and is very similar to values obtained from models and basic theory.
• For longer-timescale temperature changes over 1979 to 1999, only one of four observed upper-air data sets has larger tropical warming aloft than in the surface records. All model runs with surface warming over this period show amplified warming aloft.

• These results could arise due to errors common to all models; to significant non-climatic influences remaining within some or all of the observational data sets, leading to biased long-term trend estimates; or a combination of these factors. The new evidence in this Report (model-to-model consistency of amplification results, the large uncertainties in observed tropospheric temperature trends, and independent physical evidence supporting substantial tropospheric warming) favors the second explanation.
• A full resolution of this issue will require reducing the large observational uncertainties that currently exist. These uncertainties make it difficult to determine whether models still have common, fundamental errors in their representation of the vertical structure of atmospheric temperature change.

**Other Findings**

**7.** It is important to account for both model and observational uncertainty in comparisons between modeled and observed temperature changes.

• There are large "construction uncertainties" in the process of generating climate data records from raw observations. These uncertainties can critically influence the outcome of consistency tests between models and observations.

**8.** Inclusion of spatially variable forcings in the most recent climate models does not fundamentally alter simulated lapse-rate changes at the largest spatial scales.

- Changes in black carbon aerosols and land use/land cover (LULC) may have had significant influences on regional temperatures, but these influences have not been quantified in formal fingerprint studies.
- These forcings were included for the first time in about half the global model simulations considered here. Their incorporation did not significantly affect simulations of lapse-rate changes at very large spatial scales (global and tropical averages).

## CHAPTER 5: Recommendations

**1.** Separate the uncertainties in climate forcings from uncertainties in the climate response to forcings.

The simulations of 20th century (20CEN) climate analyzed here show climate responses that differ because of differences in:
- Model physics and resolution;
- The forcings incorporated in the 20CEN experiment;
- The chosen forcing history, and the manner in which a specific forcing was applied.
- Model initial conditions.

We consider it a priority to partition the uncertainties in climate forcings and model responses, and thus improve our ability to interpret differences between models and observations. This could be achieved by better coordination of experimental design, particularly for the 20CEN simulations that are most relevant for direct comparison with observations.

**2.** Quantify the contributions of changes in black carbon aerosols and land use/land cover to recent large-scale temperature changes.

We currently lack experiments in which the effects of black carbon aerosols and LULC are varied individually (while holding other forcings constant). Such "single forcing" runs will help to quantify the contributions of these forcings to global-scale changes in lapse rates.

**3.** Explicitly consider model and observational uncertainty.

Efforts to evaluate model performance or identify human-induced climate change should always account for uncertainties in both observations and in model simulations of historical and future climate. This is particularly important for comparisons involving long-term changes in upper-air temperatures. It is here that current observational uncertainties are largest and require better quantification.

**4.** Perform the "next generation" of detection and attribution studies.

Formal detection and attribution studies utilizing the new generation of model and observational data sets detailed herein should be undertaken as a matter of priority.

## I. INTRODUCTION

Climate models can be used to simulate the response to changes in a single forcing or a combination of forcings, and thus have real advantages for studying cause-effect relationships.

A key scientific question addressed in this report is whether the Earth's surface has warmed more rapidly than the troposphere over the past 2-3 decades (NRC, 2000). Chapter 1 noted that there are good physical reasons why we do not expect surface and tropospheric temperatures to evolve in unison at all places and on all time-scales. Chapters 2, 3, and 4 summarized our current understanding of observed changes in surface and atmospheric temperatures. These chapters identified important differences between surface and tropospheric temperatures, some of which may be due to remaining problems with the observational data, and some of which are likely to be real.

In Chapter 5, we seek to explain and reconcile the apparently disparate estimates of observed changes in surface and tropospheric temperatures. We make extensive use of computer models of the climate system. In the real world, multiple "climate forcings" vary simultaneously, and it is difficult to identify and separate the climate effects of individual factors. Furthermore, the experiment that we are performing with the Earth's climate system lacks a suitable control – we do not have a convenient "parallel Earth" on which there are no human-induced changes in greenhouse gases, aerosols, or other climate forcings. Climate models can be used to perform such controlled experiments, or to simulate the response to changes in a single forcing or combination of forcings, and thus have real advantages for studying cause-effect relationships. However, models also have systematic errors that can diminish their usefulness as a tool for interpretation of observations (Gates *et al.,* 1999; McAvaney *et al.,* 2001).

### BOX 5.1: Climate Models

Climate models provide us with estimates of how the real world's climate system behaves and is likely to respond to changing natural and human-caused forcings. Because of limitations in our physical understanding and computational capabilities, models are simplified and idealized representations of a very complex reality. The most sophisticated climate models are direct descendants of the computer models used for weather forecasting. While weather forecast models seek to predict the specific timing of weather events over a period of days to several weeks, climate models attempt to simulate future changes in the *average distribution* of weather events.

Because the climate system is chaotic, fully coupled models of the atmosphere and ocean cannot simulate exactly the same sequence of individual weather events that occurred in the real world (see Section 2). Such models can, however, capture many of the statistical characteristics of observed weather and climate variability, on timescales of days to decades. Many models have demonstrated skill in their portrayal of major modes of observed climate variability, such as the North Atlantic Oscillation (Hurrell *et al.,* 2003), the El Niño/Southern Oscillation (ENSO; AchutaRao and Sperber, 2006) or the Atlantic Multidecadal Oscillation (Knight *et al.,* 2005). This variability contributes to the background "noise" against which any signal of human effects on climate must be detected[a.] (Box 5.5).

Simulations of 21st century climate are typically based on "scenarios" of future emissions of GHGs, aerosols and aerosol precursors, which in turn derive from scenarios of population changes, economic growth, energy usage, developments in energy production technology, *etc.* Climate models are also used to "hindcast" the climate changes that we have observed over the 20th century. When run in "hindcast" mode, a climate model is not constrained by actual weather observations from satellites or radiosondes. Instead, it is driven by our best estimates of changes in some (but probably not all) of the major forcings, such as GHG concentrations, the Sun's energy output, and the amount of volcanic dust in the atmosphere. In hindcast experiments, a climate model is free to simulate the full four-dimensional (latitude, longitude, height/depth and time) distributions of temperature, moisture, *etc.* Comparing the results of such an experiment with long observational records constitutes a valuable test of model performance.

A more complete assessment of climate models and their ability to represent many different aspects of the climate system will be covered in CCSP Synthesis and Assessment Product 3.1: "Climate Models: An Assessment of Strengths and Limitations for User Applications."

[a.] There is some evidence that human-induced climate change may modulate the statistical behavior of existing modes of climate variability (Hasselmann, 1999).

We evaluate published research that has made rigorous quantitative comparisons of modeled and observed temperature changes, primarily over the satellite and radiosonde eras. Some new model experiments (performed in support of the IPCC Fourth Assessment Report) involve simultaneous changes in a wide range of natural and human-induced climate forcings. These experiments are highly relevant for direct comparison with satellite-, radiosonde-, and surface-based temperature observations. We review their key results here.

## 2. MODEL SIMULATIONS OF RECENT TEMPERATURE CHANGE

Many different types of computer model are used for studying climate change issues (Meehl, 1984; Trenberth, 1992; see Box 5.1). Models span a large range of complexity, from the one- or two-dimensional energy-balance models (EBMs) through Earth system Models of Intermediate Complexity (EMICs) to full three-dimensional atmospheric General Circulation Models (AGCMs) and coupled atmosphere-ocean GCMs (CGCMs). Each type has advantages and disadvantages for specific applications. The more complex AGCMs and CGCMs are most appropriate for understanding problems related to the atmosphere's vertical temperature structure, since they explicitly resolve that structure, and incorporate many of the physical processes (*e.g.,* convection, interactions between clouds and radiation) thought to be important in maintaining atmospheric temperature profiles. They are also capable of representing the horizontal and vertical structure of unevenly distributed climate forcings that may contribute to differential warming of the surface and troposphere. Examples include volcanic aerosols (Robock, 2000) or the sulfate and soot aerosols arising from fossil fuel or biomass burning (Penner *et al.,* 2001; Ramaswamy *et al.,* 2001a,b).

AGCM experiments typically rely on an atmospheric model driven by observed time-varying changes in sea-surface temperatures (SSTs) and sea-ice coverage. This is a standard reference experiment that many AGCMs have performed as part of the Atmospheric Model Intercomparison Project ("AMIP"; Gates *et al.,* 1999). The

AMIP-style experiments discussed here also include specified changes in a variety of natural and human-caused forcing factors (Hansen *et al.,* 1997, 2002; Folland *et al.,* 1998; Tett and Thorne, 2004).

In both observations and climate models, variations in the El Niño-Southern Oscillation (ENSO) have pronounced effects on surface and tropospheric temperatures (Yulaeva and Wallace, 1994; Wigley, 2000; Santer *et al.,* 2001; Hegerl and Wallace, 2002; Hurrell *et al.,* 2003). When run in an AMIP configuration, an atmospheric model "sees" the same changes in ocean surface temperature that the real world's atmosphere experienced. The time evolution of ENSO effects on atmospheric temperature is therefore very similar in the model and observations. This facilitates the direct comparison of modeled and observed temperature changes[1]. Furthermore, AMIP experiments reduce climate noise by focusing on the random variability arising from the atmosphere rather than on the variability of the coupled atmosphere-ocean system (which is larger in amplitude). This "noise reduction" aspect of AMIP runs has been exploited in efforts to identify human effects on year-to-year changes in atmospheric temperatures (Folland *et al.,* 1998; Sexton *et al.,* 2001) and volcanic influences on surface air temperature (Mao and Robock, 1998).

One disadvantage of the AMIP experimental set-up is that significant errors in one or more of the applied forcing factors (or omission of key forcings) are not "felt" by the prescribed SSTs. Such errors are more obvious in a CGCM experiment, where the ocean surface is free to respond to imposed forcings. The lack of an ocean response, combined with the masking effects of natural variability, make it difficult to use an AMIP-style experiment to estimate the slow response of the climate system to an imposed forcing change[2]. CGCM experiments

Climate models are also used to "hindcast" the climate changes that we have observed over the 20th century. Comparing the results of such an experiment with long observational records constitutes a valuable test of model performance.

---

[1]  This does not mean, however, that the atmospheric model will necessarily capture the correct amplitude and horizontal and vertical structure of the tropospheric temperature response to the specified SST and sea-ice changes. Even with the specification of observed ocean boundary conditions, the time evolution of modes of variability that are forced by both the ocean and the atmosphere (such as the North Atlantic Oscillation; see Rodwell *et al.,* 1999) will not be the same in the model and in the real world (except by chance).

[2]  Volcanic forcing provides an example of the signal

are more useful for this specific purpose (see Chapter 1, Figure 1.3).

The CGCM experiments of interest here involve a model that has been "spun-up" until it reaches some quasi-steady climate state[3]. The CGCM is then run with estimates of how a variety of natural and human-caused climate forcings have changed over the 20th century. We refer to these subsequently as "20CEN" experiments. Since the true state of the climate system is never fully known, the same forcing changes are applied $n$ times,[4] each time starting from a slightly different initial climate state. This procedure yields $n$ different realizations of climate change. All of these realizations contain some underlying "signal" (the climate response to the imposed forcing changes) upon which are superimposed $n$ different manifestations of "noise" (natural internal climate variability). Taking averages over these $n$ realizations yields less noisy estimates of the signal (Wigley *et al.*, 2005a).

In a CGCM, ocean temperatures are fully predicted rather than prescribed. This means that even a (hypothetical) CGCM which perfectly captured all important aspects of ENSO physics would not have the same timing of El Niño and La Niña events as the real world (except by chance). The fact that ENSO variability – and its effects on surface and atmospheric temperatures – does not "line up in time" in observations and CGCM experiments hampers direct comparisons between the two[5]. This problem

> AMIP-style experiments and CGCM runs are both useful tools for exploring the possible causes of differential warming.

can be ameliorated by statistical removal of ENSO effects (Santer *et al.*, 2001; Hegerl and Wallace, 2002; Wigley *et al.*, 2005a)[6].

The bottom line is that AMIP-style experiments and CGCM runs are both useful tools for exploring the possible causes of differential warming. We note that even if these two experimental configurations employ the same atmospheric model and the same climate forcings, they can yield noticeably different simulations of changes in atmospheric temperature profiles. These differences arise for a variety of reasons, such as AGCM-versus-CGCM differences in sea-ice coverage, SST distributions, and cloud feedbacks, and hence in climate sensitivity (Sun and Hansen, 2003)[7].

Most models undergo some adjustment of poorly-known parameters which directly affect key physical processes, such as convection and rainfall. Parameters are varied within plausible ranges, which are generally derived from direct observations. The aim of this procedure is to reduce the size of systematic model errors and improve simulations of present-day climate. Adjustment of uncertain model parameters is *not* performed over the course of a 20CEN experiment.

Several groups are now beginning to explore model parameter space, and are investigating the possible impact of parameter uncertainties on simulations of mean present-day climate and future climate change by running "perturbed physics" ensembles (Allen, 1999; Forest *et al.*, 2002; Murphy *et al.*, 2004; Stainforth *et al.*, 2005). Such work will help to quantify one component of model uncertainty. Another component of model uncertainty arises from differences in the basic structure of models[8].

---

estimation problem. The aerosols injected into the stratosphere during a massive volcanic eruption are typically removed within 2-3 years (Sato *et al.*, 1993; Hansen *et al.*, 2002; Ammann *et al.*, 2003). Because the large thermal inertia of the oceans causes a lag in response to this forcing, the cooling effect of the aerosols on the troposphere and surface persists for much longer than 2-3 years (Santer *et al.*, 2001; Free and Angell, 2002; Wigley *et al.*, 2005a). In the real world and in "AMIP-style" experiments, this slow, volcanically induced cooling of the troposphere and surface is sometimes masked by the warming effects of El Niño events (Christy and McNider, 1994; Wigley, 2000; Santer *et al.*, 2001), thus hampering volcanic signal estimation.

[3]  There are a variety of different spin-up strategies.

[4]  In most of the experiments reported on here, $n$ is between 3 and 5.

[5]  If $n$ is large enough to adequately sample the (simulated) effects of natural variability on surface and tropospheric temperatures, it is not necessarily a disadvantage that the simulated and observed variability does not line up in time. In fact, this type of

---

experimental set-up allows one to determine whether the single realization of the observations is contained within the "envelope" of possible climate solutions that the CGCM simulates.

[6]  Residual effects of these modes of variability will remain in the data.

[7]  See, for example, the Ocean A and Ocean E results in Figure 3 of Sun and Hansen (2003).

[8]  The computer models constructed by different research groups can have quite different "structures" in terms of their horizontal and vertical resolution, atmospheric dynamics (so-called "dynamical cores"), numerical implementation (*e.g.,* spectral versus grid-point), and physical parameterizations. They do, however, share many common assumptions.

---

**BOX 5.2: Uncertainties in Simulated Temperature Changes**

In discussing the major sources of uncertainty in observational estimates of temperature change, Chapter 2 partitioned uncertainties into three distinct categories: "structural," "parametric," and "statistical." Uncertainties in simulated temperature changes fall into similar categories. In the modeling context, "structural" uncertainties can be thought of as the uncertainties resulting from the choice of a particular climate model, model configuration (Section 2), or forcing data set (Section 3).

Within a given model, there are small-scale physical processes (such as convection, cloud formation, precipitation, *etc.*) that cannot be simulated explicitly. Instead, so-called "parameterizations" represent the large-scale effects of these unresolved processes. Each of these processes has uncertainties in the values of one or more key parameters.[a] Varying these parameters within plausible ranges introduces "parametric" uncertainty in climate change simulations (Allen, 1999; Forest *et al.*, 2002; Murphy *et al.*, 2004). Finally (analogous to the observational case), there is statistical uncertainty that arises from the unpredictable "noise" of internal climate variability, from the choice of a particular statistical metric to describe climate change, or from the application of a selected metric to noisy data.

[a] Note that some of these parameters influence not only the climate response, but also the portrayal of the forcing itself. Examples include parameters related to the size of sulfate aerosols, and how aerosol particles scatter incoming sunlight.

---

Section 5 considers results from a range of state-of-the-art CGCMs, and thus samples some of the "structural uncertainty" in model simulations of 20th century climate change (Table 5.1). A further component of the spread in simulations of 20th century climate is introduced by uncertainties in the climate forcings with which models are run (Table 5.2). These are discussed in the following Section.

## 3. FORCINGS IN SIMULATIONS OF RECENT CLIMATE CHANGE

In an ideal world, there would be reliable quantitative estimates of all climate forcings – both natural and human-induced – that have made significant contributions to surface and tropospheric temperature changes. We would have detailed knowledge of how these forcings had changed over space and time. Finally, we would have used standard sets of forcings to perform climate-change experiments with a whole suite of numerical models, thus isolating uncertainties arising from structural differences in the models themselves (see Box 5.2).

Unfortunately, this ideal situation does not exist. As part of the IPCC Third Assessment Report, Ramaswamy *et al.* (2001b) assigned subjective confidence levels to our current "level of scientific understanding" (LOSU) of the changes in a dozen different climate forcings. Only in the case of well-mixed greenhouse gases ("GHGs"; carbon dioxide [$CO_2$], methane, nitrous oxide, and halocarbons) was the LOSU characterized as "high." The LOSU of changes

in stratospheric and tropospheric ozone was judged to be "medium." For all other forcings (various aerosols, mineral dust, land use-induced albedo changes, solar, *etc.*), the LOSU was estimated to be "low" or "very low" (see Chapter 1, Table 1.1 and Section 1.2)[9].

In selecting the forcings for simulating the climate of the 20th century, there are at least three strategies that modeling groups can adopt. The first strategy is to incorporate only those forcings whose changes and effects are thought to be better understood, and for which time- and space-resolved data sets suitable for performing 20CEN experiments are readily available. The second strategy is to include a large number of different forcings, even those for which the LOSU is "very low." A third strategy is to vary the size of poorly known 20CEN forcings. This yields a range of simulated climate responses, which are then used to estimate the levels of the forcings that are consistent with observations (*e.g.,* Forest *et al.,* 2002).

The pragmatic focus of Chapter 5 is on climate forcings that have been incorporated in many CGCM simulations of 20th century climate. The primary forcings that we consider are changes in well-mixed GHGs, the direct effects of sulfate aerosol particles, tropospheric and stratospheric ozone, volcanic aerosols, and solar

---

[9] We note that there is no direct relationship between the LOSU of a given forcing and the contribution of that forcing to 20th century climate change. Forcings with "low" or "very low" LOSU may have had significant climatic impacts at regional and even global scales.

## BOX 5.3:  Example of a Spatially-Heterogeneous Forcing: Black Carbon Aerosols

Carbon-containing aerosols (also known as "carbonaceous" aerosols) exist in a variety of chemical forms (Penner *et al.*, 2001). Two main classes of carbonaceous aerosol are generally distinguished: "black carbon" (BC) and "organic carbon" (OC). Both types of aerosol are emitted during fossil fuel and biomass burning. Most previous modeling work has focused on BC aerosols rather than OC aerosols. Some of the new model experiments described in Section 5 have now incorporated both types of aerosol in CGCM simulations of 20th century climate changes (see Tables 5.2 and 5.3).

Black carbon aerosols absorb sunlight and augment the GHG-induced warming of the troposphere (Hansen *et al.*, 2000; Satheesh and Ramanathan, 2000; Penner *et al.*, 2001; Hansen, 2002; Penner *et al.*, 2003)[a]. Their effects on atmospheric temperature profiles are complex, and depend on such factors as the chemical composition, particle size, and height distribution of the aerosols (*e.g.*, Penner *et al.*, 2003).

Menon *et al.* (2002) showed that the inclusion of fossil fuel and biomass aerosols over China and India[b] directly affected simulated vertical temperature profiles by heating the lower troposphere and cooling the surface. In turn, this change in atmospheric heating influenced regional circulation patterns and the hydrological cycle. Krishnan and Ramanathan (2002) found that an increase in black carbon aerosols has reduced the surface solar insolation (exposure to sunlight) over the Indian subcontinent. Model experiments performed by Penner *et al.* (2003) suggest that the net effect of carbonaceous aerosols on global-scale surface temperature changes depends critically on how aerosols affect the vertical distribution of clouds. On regional scales, the surface temperature effects of these aerosols are complex, and vary in sign (Penner *et al.*, 2006).

[a]  Note that soot particles are sometimes transported long distances by winds, and can also have a "far field" effect on climate by reducing the reflectivity of snow in areas remote from pollution sources (Hansen and Nazarenko, 2003; Jacobson, 2004).

[b]  During winter and spring, black carbon aerosols contribute to a persistent haze over large areas of Southern Asian and the Northern Indian Ocean (Ramanathan *et al.*, 2001).

ed in many of the new CGCM simulations of 20th century climate described in Section 5 (see Tables 5.1 and 5.2).

Clearly, we will never have complete and reliable information on all forcings that are thought to have influenced climate over the late 20th century. A key question is whether those forcings most important for understanding the differential warming problem are reliably represented. This is currently difficult to answer. What we can say, with some certainty, is that the expected atmospheric temperature signal due to forcing by well-mixed GHGs alone is distinctly different from the signal due to the combined effects of multiple natural and human forcing factors (Chapter 1; Santer *et al.*, 1996a; Tett *et al.*, 1996; Hansen *et al.*, 1997, 2002; Bengtsson *et al.*, 1999; Santer *et al.*, 2003a).

irradiance. These are forcings whose effects on surface and atmospheric temperatures have been quantified in rigorous fingerprint studies (see Section 4.4). This does not diminish the importance of other climate forcings, whose global-scale contributions to "differential warming" have not been reliably quantified to date.

Examples of these "other forcings" include carbon-containing aerosols produced during fossil fuel or biomass combustion, human-induced changes in land surface properties, and the indirect effects of tropospheric aerosols on cloud properties. There is emerging scientific evidence that such spatially variable forcings may have had important impacts on regional and even on global climate (NRC, 2005). Some of this evidence is summarized in Box 5.3 and Box 5.4 for the specific cases of carbonaceous aerosols and land use change. These and other previously neglected forcings have been includ-

This is illustrated by the 20CEN and "single forcing" experiments performed with the Parallel Climate Model (PCM; Washington *et al.*, 2000). In PCM, changes in the vertical profile of atmospheric temperature over 1979 to 1999 are primarily forced by changes in well-mixed GHGs, ozone, and volcanic aerosols (Figure 5.1). Changes in solar irradiance and the scattering effects of sulfate aerosols are of secondary importance over this period. Even without performing formal statistical tests, it is visually obvious from Figure 5.1 that radiosonde-based estimates of observed stratospheric and tropospheric temperature changes are in better agreement with the PCM 20CEN experiment than with the PCM "GHG only" run.

This illustrates the need for caution in comparisons of modeled and observed atmospheric temperature change. The differences evident in such comparisons have multiple interpretations. They may be due to real errors in the

---

**BOX 5.4: Example of a Spatially-Heterogeneous Forcing: Land Use Change**

Humans have transformed the surface of the planet through such activities as conversion of forest to cropland, urbanization, irrigation, and large water diversion projects (see Chapter 4). These changes can affect a variety of physical properties of the land surface, such as the albedo (reflectivity), the release of water by plants (transpiration), the moisture-holding capacity of soil, and the surface "roughness." Alterations in these physical properties may in turn affect runoff, heat and moisture exchanges between the land surface and atmospheric boundary layer, wind patterns, and even rainfall (*e.g.,* Pitman *et al.,* 2004). Depending on the nature of the change, either warming or cooling of the land surface may occur (Myhre and Myhre, 2003).

At the regional level, modeling studies of the Florida peninsula (Marshall *et al.,* 2004) and southwest Western Australia (Pitman *et al.,* 2004) have linked regional-scale changes in atmospheric circulation and rainfall to human transformation of the natural vegetation. Modeling work focusing on North America suggests that the conversion of natural forest and grassland to agricultural production has led to a cooling in summertime (Oleson *et al.,* 2004). The global-scale signal of land use/land cover (LULC) changes from pre-industrial times to the present is estimated to be a small net cooling of surface temperature (Matthews *et al.,* 2003, 2004; Brovkin *et al.,* 2004; Hansen *et al.,* 2005a; Feddema *et al.,* 2005). Larger regional trends of either sign are likely to be evident (*e.g.,* Hansen *et al.,* 2005a)[a].

> [a] Larger regional trends do not necessarily translate to enhanced detectability. Although the signals of LULC and other spatially-heterogeneous forcings are likely to be larger regionally than globally, the "noise" of natural climate variability is also larger at smaller spatial scales. It is not obvious *a priori,* therefore, how signal-to-noise relationships (and detectability of a given forcing's climate effects) behave as one moves from global to continental to regional scales.

---

models,[10] errors in the forcings used to drive the models, the neglect of important forcings, and residual inhomogeneities in the observations themselves. They may also be due to different manifestations of natural variability noise in the observations and a given CGCM realization. All of these factors can be important in model evaluation work.

## 4. PUBLISHED COMPARISONS OF MODELED AND OBSERVED TEMPERATURE CHANGES

A number of observational and modeling studies have attempted to shed light on the possible causes of "differential warming"[11]. We have attempted to organize the discussion of results so that investigations with similar analysis methods are grouped together[12]. Our discussion proceeds from simple to more complex and statistically rigorous analyses.

### 4.1 Regression Studies Using Observed Global-mean Temperature Data

One class of study that has attempted to address the causes of recent tropospheric temperature change relies on global-mean observational data only (Jones, 1994; Christy and McNider, 1994; Michaels and Knappenberger, 2000; Douglass and Clader, 2002). Such work uses a multiple regression model to quantify the statistical relationships between various "predictor variables" (typically time series of ENSO variability,

---

[10] These may lie in the physics, parameterizations, inadequate horizontal or vertical resolution, *etc.*

[11] We do not discuss studies which provide empirical estimates of "equilibrium climate sensitivity" – the steady-state warming of the Earth's surface that would eventually be reached after the climate system equilibrated to a doubling of pre-industrial $CO_2$ levels. This is often referred to as $\Delta T_{2xCO_2}$. Estimates of $\Delta T_{2xCO_2}$ have been obtained by studying Earth's temperature response to "fast," "intermediate," and "slow" forcing of the climate system. Examples include the "fast" (<10-year) response of surface and tropospheric temperatures to massive volcanic eruptions (Hansen *et al.,* 1993; Lindzen and Giannitsis, 1998; Douglass and Knox, 2005; Wigley *et al.,* 2005a,b; Robock, 2005); the "intermediate" (100- to 150-year) response of surface temperatures to natural and human-caused forcing changes over the 19th and 20th centuries (Andronova and Schlesinger, 2001; Forest *et al.,* 2002; Gregory *et al.,* 2002; Harvey and Kaufmann, 2002) or

to solar and volcanic forcing changes over the past 1-2 millennia (Crowley, 2000), and the "slow" (100,000-year) response of Earth's temperature to orbital changes between glacial and interglacial conditions (Hoffert and Covey, 1992; Hansen *et al.,* 1993). These investigations are not directly relevant to elucidation of the causes of changes in the vertical structure of atmospheric temperatures, which is the focus of this Chapter.

[12] It is useful to mention one technical issue relevant to model-data comparisons. As noted in Chapter 2, the satellite-based Microwave Sounding Unit (MSU) monitors the temperature of very broad atmospheric layers. To facilitate comparisons with observed MSU data sets, many of the studies reported on here calculate "synthetic" MSU temperatures from climate model experiments. Technical aspects of these calculations are discussed in Chapter 2, Box 2.1.

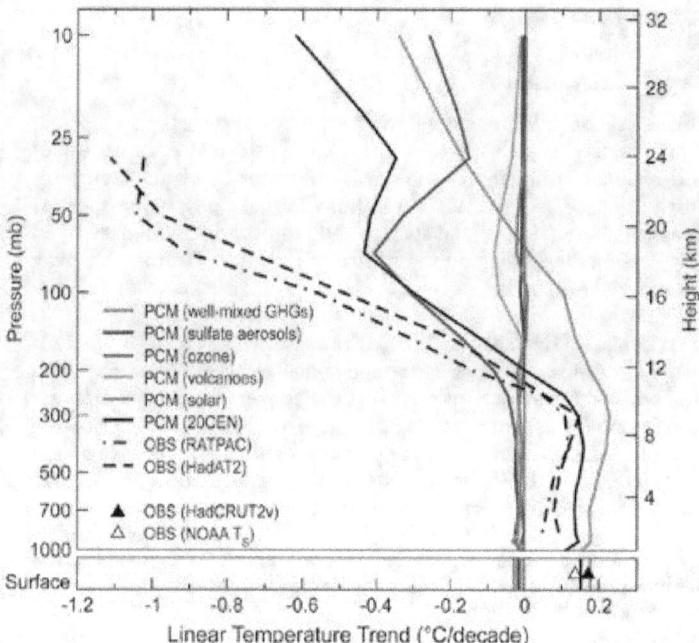

**Figure 5.1:** Vertical profiles of global-mean atmospheric temperature change over 1979 to 1999. Surface temperature changes are also shown. Results are from two different radiosonde data sets (HadAT2 and RATPAC; see Chapter 3) and from single forcing and combined forcing experiments performed with the Parallel Climate Model (PCM; Washington *et al.*, 2000). PCM results for each forcing experiment are averages over four different rea izations of that experiment. All trends were calculated with monthly mean anomaly data.

as predictors rather than the climate responses to those forcings. Distinctions between forcing and response are important (Wigley *et al.*, 2005a). Accounting for uncertainties in predictor variables (and use of responses rather than forcings as predictors) expands the range of uncertainties in estimates of residual $T_{2LT}$ trends (Santer *et al.*, 2001)[14].

Regression methods have also been used to estimate the net effects of ENSO and volcanoes on trends in global-mean surface and tropospheric temperatures. For $T_{2LT}$, both Jones (1994) and Christy and McNider (1994) found that ENSO effects induced a small net warming of 0.03 to 0.05°C/decade over 1979 to 1993, while volcanoes caused a cooling of 0.18°C/decade over the same period. Michaels and Knappenberger (2000) also reported a relatively small ENSO influence on $T_{2LT}$ trends[15]. Santer *et al.* (2001) noted that over 1979 to 1997, volcanoes had likely cooled the troposphere by more than the surface. Removing the combined volcano and ENSO effects from surface and UAH $T_{2LT}$ data helped to explain some of the observed differential warming: the "raw" $T_S$-minus-$T_{2LT}$ trend over 1979 to 1997 decreased from roughly 0.15°C/decade to 0.05-0.13°C/decade.[16] Removal of volcano and ENSO influences also brought observed lapse rate trends closer to model results, but could not fully reconcile modeled and observed lapse rate trends[17].

volcanic aerosol loadings, and solar irradiance) and a single "predictand" (typically $T_{2LT}$). The aim is to remove the effects of the selected predictors on tropospheric temperature, and to estimate the residual trend that may arise from human-induced forcings. The quoted values for this residual trend in $T_{2LT}$ range from 0.04 to 0.09°C/decade[13].

These studies often make the unrealistic assumption that the uncertainties inherent in such statistical signal separation exercises are very small. They do not explore the sensitivity of regression results to uncertainties in the predictor variables or the observational record, and generally use solar and volcanic forcings

---

[13] The studies by Jones (1994) and Christy and McNider (1994) remove volcano and ENSO effects from $T_{2LT}$, and estimate residual trends of 0.093 and 0.090 C/decade over 1979 to 1993. A similar investigation by Michaels and Knappenberger (2000) obtained a residual trend of 0.041 C/decade over 1979 to 1999. The error bars on these residual trend estimates are either not given, or claimed to be very small (*e.g.*, ± 0.005 C/decade in Christy and McNider). A fourth study removed combined ENSO, volcano, and solar effects from $T_{2LT}$, and estimated a residual trend of 0.065 ± 0.012 C/decade over 1979 to 2000 (Douglass and Clader, 2002).

---

[14] Santer *et al.* (2001) obtain residual $T_{2LT}$ trends ranging from 0.06 to 0 16 C/decade over 1979 to 1999. Their regression model is iterative, and involves removal of ENSO and volcano effects only.

[15] The ENSO components of their $T_{2LT}$ trends were 0.04 C/decade over 1979 to 1998 and 0.01 C/decade over 1979 to 1999. This difference in the net ENSO influence on $T_{2LT}$ (with the addition of only a single year of record) arises from the El Niño event in 1997/98, and illustrates the sensitivity of this kind of analysis to so-called "end effects."

[16] The latter results were obtained with the HadCRUTv surface data (Jones *et al.*, 2001) and version d03 of the UAH $T_{2LT}$ data. The range of residual lapse-rate trends arises from parametric uncertainty, *i.e.*, from the different choices of ENSO predictor variables and volcano parameters.

[17] Santer *et al.* (2001) analyzed model experiments performed with the ECHAM4/OPYC model developed at the Max-Planck Institute for Meteorology in Hamburg (Roeckner *et al.*, 1999). The experiments included forcing by well-mixed greenhouse gases, direct and indirect sulfate aerosol effects, tropospheric and stratospheric ozone, and volcanic aerosols (Pinatubo only).

## 4.2 Regression Studies Using Spatially Resolved Temperature Data

Other regression studies have attempted to remove natural variability influences using spatially resolved temperature data. Regression is performed "locally" at individual grid-points and/or atmospheric levels. To obtain a clearer picture of volcanic effects on atmospheric temperatures, Free and Angell (2002) removed the effects of variability in ENSO and the Quasi-Biennial Oscillation (QBO) from Hadley Centre radiosonde data[18]. Their work clearly shows that the cooling effect of massive volcanic eruptions has been larger in the upper troposphere than in the lower troposphere. The implication is that volcanic effects probably contribute to slow changes in observed lapse rates.

Hegerl and Wallace (2002) used regression methods to identify and remove different components of natural climate variability from gridded fields of surface temperature data, UAH $T_{2LT}$, and "synthetic" $T_{2LT}$ calculated from radiosonde data. They focused on the variability associated with ENSO and the so-called "cold ocean warm land" (COWL) pattern (Wallace *et al.*, 1995). While ENSO and COWL variability made significant contributions to the month-to-month and year-to-year variability of temperature differences between the surface and $T_{2LT}$, their analysis indicated that it had very little impact on decadal fluctuations in lapse rate. The authors concluded that natural variability alone was unlikely to explain these slow lapse-rate changes. However, the removal of ENSO and COWL effects more clearly revealed a volcanic contribution, consistent with the findings of Santer *et al.* (2001) and Free and Angell (2002). A climate model control run (with no changes in forcings) and a 20CEN experiment were unable to adequately reproduce the observed decadal changes in lapse rate[19].

## 4.3 Other Studies of Global and Tropical Lapse-rate Trends

Several studies have investigated lapse-rate trends without attempting to remove volcano effects or natural climate noise. Brown *et al.* (2000) used surface, radiosonde, and satellite data to identify slow, tropic-wide changes in the lower tropospheric lapse rate[20]. In their analysis, the surface warmed relative to the troposphere between the early 1960s and mid-1970s and after the early 1990s. Between these two periods, the tropical troposphere warmed relative to the surface. The spatial coherence of these variations (and independent evidence of concurrent variations in the tropical general circulation) led Brown *et al.* (2000) to conclude that tropical lapse rate changes were unlikely to be an artifact of residual errors in the observations.

Very similar decadal changes in lower tropospheric lapse rate were reported by Gaffen *et al.* (2000)[21]. Their study analyzed radiosonde-derived temperature and lapse rate changes over two periods: 1960 to 1997 and 1979 to 1997. Tropical lapse rates decreased over the longer period[22] and increased over the satellite era[23]. To evaluate whether natural climate variability could explain these slow variations, Gaffen *et al.* (2000) computed lapse rates from the control runs performed with three different CGCMs. Each control run was 300 years in length. These long runs provided estimates of the "sampling variability" of modeled lapse rate changes on timescales relevant to the two observational periods (38 and 19 years)[24]. Model-based esti-

The cooling effect of massive volcanic eruptions has been larger in the upper troposphere than in the lower troposphere.

[18] The HadRT2 1 data set of Parker *et al.* (1997). Like Santer *et al.* (2001), Free and Angell (2002) also found some sensitivity of the estimated volcanic signals to "parametric" uncertainty.
[19] The model was the ECHAM4/OPYC CGCM used by Bengtsson *et al.* (1999). The 20CEN experiment analyzed by Hegerl and Wallace (2002) involved combined changes in well-mixed greenhouse gases, the direct and indirect effects of sulfate aerosols, and tropospheric ozone. Forcing by volcanoes and stratospheric ozone depletion was not included.

[20] The Brown *et al.* (2000) study employed UKMO surface data (HadCRUT), version d of the UAH $T_{2LT}$, and an early version of the Hadley Centre radiosonde data set (HadRT2.0) that was uncorrected for instrumental biases.
[21] Gaffen *et al.* (2000) used a different radiosonde data set from that employed by Brown *et al.* (2000). The two groups also analyzed different surface temperature data sets.
[22] Corresponding to a tendency towards a more stable atmosphere.
[23] These lapse-rate changes were accompanied by increases and decreases in tropical freezing heights (which were inferred from the same radiosonde data).
[24] Each control run was used to generate distributions of 38-year and 19-year lapse rate trends. For example, a 300-year control run can be split up into 15 different "segments" that are each of length 19 years (assuming there is no overlap between segments). From these segments, one obtains 15 different estimates of how the lapse rate might vary in the absence of any forcing

**Different climate forcings have different characteristic patterns of temperature response ("fingerprints").**

mates of natural climate variability could not adequately explain the observed tropical lapse rate changes over 1979 to 1997. Similar conclusions were reached by Hansen *et al.* (1995) and Santer *et al.* (2000). Including natural and anthropogenic forcings in the latter study narrowed the gap between modeled and observed estimates of recent lapse-rate changes, although a significant discrepancy between the two still remained.

It should be emphasized that all of the studies reported on to date in Section 4 relied on satellite data from one group only (UAH), on early versions of the radiosonde data[25], and on experiments performed with earlier model "vintages." It is likely, therefore, that this work may have underestimated the structural uncertainties in observed and simulated estimates of lapse rate changes. We will consider in Section 5 whether modeled and observed lapse rate changes can be better reconciled by the availability of more recent 20CEN runs and more comprehensive estimates of structural uncertainties in observations.

### 4.4 Pattern-based "Fingerprint" Detection Studies

Fingerprint detection studies rely on patterns of temperature change (Box 5.5). The patterns are typically either latitude-longitude "maps" (*e.g.*, for $T_4$, $T_2$, $T_S$, *etc.*) or latitude-height cross-sections through the atmosphere[26]. The basic premise in fingerprinting is that different climate forcings have different characteristic patterns of temperature response ("fingerprints"), particularly in the free atmosphere (Chapter 1, Figure 1.3; Hansen *et al.*, 1997, 2002, 2005a; Bengtsson *et al.*, 1999; Santer *et al.*, 1996a; Tett *et al.*, 1996).

Most analysts rely on a climate model to provide physically based estimates of each fingerprint's structure, size, and evolution. The model fingerprints are searched for in observational climate records, using rigorous statistical methods to quantify the degree of correspondence with observed patterns of climate change[27]. Fingerprints are also compared with patterns of climate change in model control runs. This helps to determine whether the correspondence between the fingerprint and observations is truly significant, or could arise through internal variability alone (Box 5.5). Model errors in internal variability[28] can bias detection results, although most detection work tries to guard against this possibility by performing "consistency checks" on modeled and observed variability (Allen and Tett, 1999), and by using variability estimates from multiple models (Hegerl *et al.*, 1997; Santer *et al.*, 2003a,b).

The application of fingerprint methods involves a variety of decisions, which introduce uncertainty in detection results (Box 5.5). Our confidence in fingerprint detection results is increased if they are shown to be consistent across a range of plausible choices of statistical methods, processing options, and model and observational data sets.

#### SURFACE TEMPERATURE CHANGES

Most fingerprint detection studies have focused on surface temperature changes. The common denominator in this work is that the model fingerprints resulting from forcing by well-mixed GHGs and sulfate aerosols[29] are statistically identifiable in observed surface temperature records (Hegerl *et al.*, 1996, 1997; North and Stevens, 1998; Tett *et al.*, 1999, 2002; Stott *et*

---

changes. The observed lapse rate change over 1979 to 1997 is then compared with the model trend distribution to determine whether the observed result could be explained by natural variability alone.

[25]  These radiosonde data sets were either unadjusted for inhomogeneities, or had not been subjected to the rigorous adjustment procedures used in more recent work (Lanzante *et al.*, 2003; Thorne *et al.*, 2005).

[26]  In constructing these cross-sections, the temperature changes are generally averaged along individual bands of latitude. Zonal averages are then displayed at individual pressure levels, starting at the lowest model or radiosonde level and ending at the top of the model atmosphere or highest reported radiosonde level (see, *e.g.*, Chapter 1, Figure 3).

[27]  The fingerprint can be either the response to an individual forcing or a combination of forcings. One strategy, for example, is to search for the climate fingerprint in response to combined changes in a suite of different human-caused forcings.

[28]  For example, current CGCMs fail to simulate the stratospheric temperature variability associated with the QBO or with solar-induced changes in stratospheric ozone (Haigh, 1994). Such errors may help to explain why one particular CGCM underestimated observed temperature variability in the equatorial stratosphere (Gillett *et al.*, 2000). In the same model, however, the variability of temperatures and lapse rates in the tropical troposphere was in reasonable agreement with observations.

[29]  Most of this work considers only the direct scattering effects of sulfate aerosols on incoming sunlight, and not indirect aerosol effects on clouds.

## BOX 5.5: Fingerprint Studies

Detection and attribution ("D&A") studies attempt to represent an observed climate data set as a linear combination of the climate signals ("fingerprints") arising from different forcing factors and the noise of natural internal climate variability (Section 4.4). A number of different fingerprint methods have been applied to the problem of identifying human-induced climate change. Initial studies used relatively simple pattern correlation methods (Barnett and Schlesinger, 1987; Santer et al., 1996a,b; Tett et al., 1996). Later work involved variants of the "optimal detection" approach suggested by Hasselmann (1979, 1993, 1997)[a]. These are essentially regression-based techniques that seek to estimate the strength of a given fingerprint pattern in observational data (i.e., how much a given fingerprint pattern has to be scaled up or down in order to best match observations). For example, if the regression coefficient for a GHG-induced $T_S$ fingerprint is significantly different from zero, GHG effects are deemed to be "detected" in observed surface temperature records. Attribution tests address the question of whether these regression coefficients are also consistent with unity – in other words, whether the size of the model fingerprint is consistent with its amplitude in observations (e.g., Allen and Tett, 1999; Mitchell et al., 2001).

There are two broad classes of regression-based D&A methods (Mitchell et al., 2001). One class assumes that although the fingerprint's amplitude changes over time, its spatial pattern does not (Hegerl et al., 1996, 1997; Santer et al., 2003a,b, 2004). The second class explicitly considers both the spatial structure and time evolution of the fingerprint (Allen and Tett, 1999; Allen et al., 2006; Stott and Tett, 1998; Stott et al., 2000; Tett et al., 1999, 2002; Barnett et al., 2001, 2005). This is particularly useful if the time evolution of the fingerprint contains specific information (such as a periodic 11-year solar cycle) that may help to distinguish it from natural internal climate variability (North et al., 1995; North and Stevens, 1998).

A number of choices must be made in applying D&A methods to real-world problems. One of the most important decisions relates to "reduction of dimensionality". D&A methods require some knowledge of the correlation structure of natural climate variability[b]. This structure is difficult to estimate reliably, even from long model control runs, because the number of time samples available to estimate correlation behavior is typically much smaller than the number of spatial points in the field. In practice, the total amount of spatial information (the "dimensionality") must be reduced. This is often done by using a mathematical tool (Empirical Orthogonal Functions) to reduce a complex space-time data set to a very small number of spatial patterns ("EOFs") that capture most of the information content of the data set[c]. Different analysts use different procedures to determine the number of patterns to retain. Further decisions relate to the choice of data used for estimating fingerprint and noise, the number of fingerprints considered, the selection of observational data, the treatment of missing data, etc.[d].

D&A methods have some limitations. They do not work well if fingerprints are highly uncertain, or if the fingerprints arising from two different forcings are similar[e]. They make at least two important assumptions: that model-based estimates of natural climate variability are a reliable representation of "real-world" variability, and that the sum of climate responses to individual forcing mechanisms is equivalent to the response obtained when these factors are varied in concert. Testing the validity of both assumptions remains an important research activity (Allen and Tett, 1999; Santer et al., 2003a; Gillett et al., 2004a).

[a] Hasselmann (1979) noted that the engineering field had extensive familiarity with the problem of identifying coherent signals embedded in noisy data, and that many of the techniques routinely used in signal processing were transferable to the problem of detecting a human-induced climate change signal.

[b] The relationship between variabiity at different points in a spatial field.

[c] The number of patterns retained is often referred to as the "truncation dimension." How the truncation dimension should be determined is a key decision in optimal detection studies (Hegerl et al., 1996; Allen and Tett, 1999).

[d] Another important choice determines whether global-mean changes are included or removed from the detection analysis. Removal of global means focuses attention on smaller-scale features of modeled and observed climate-change patterns, and provides a more stringent test of model performance.

[e] This problem is known as "degeneracy." Formal tests of fingerprint degeneracy are sometimes applied (e.g., Tett et al., 2002).

al., 2000). These results are robust to a wide range of uncertainties (Allen et al., 2006)[30]. In summarizing this body of work, the IPCC concluded that "There is new and stronger evidence that most of the warming observed over the last 50 years is attributable to human activities" (Houghton et al., 2001, page 4). The causes of surface temperature change over the first half of the 20th century are more ambiguous (IDAG, 2005).

---

[30] For example, to uncertainties in the applied greenhouse-gas and sulfate aerosol forcings, the model responses to those forcings, and model-based estimates of natural internal climate variability.

Most of the early fingerprint detection work dealt with global-scale patterns of surface temperature change. The positive detection results obtained for "GHG-only" fingerprints were driven by model-data pattern similarities at very large spatial scales (*e.g.*, at the scale of individual hemispheres, or land-versus-ocean behavior). Fingerprint detection of GHG effects becomes more challenging at continental or sub-continental scales[31]. It is at these smaller scales that spatially heterogeneous forcings, such as those arising from changes in aerosol loadings and land use patterns, may have large impacts on regional climate (see Box 5.3 and 5.4). This is illustrated by the work of Stott and Tett (1998), who found that a combined GHG and sulfate aerosol signal was identifiable at smaller spatial scales than a "GHG-only" signal.

Recently, Stott (2003) and Zwiers and Zhang (2003) have reported positive identification of the continental- or even sub-continental features of combined GHG and sulfate aerosol fingerprints in observed surface temperature records.[32] Using a variant of "classical" fingerprint methods,[33] Min *et al.* (2005) identified a GHG signal in observed records of surface temperature change over East Asia. Karoly and Wu (2005) suggest that GHG and sulfate aerosol effects are identifiable at even smaller spatial scales ("of order 500 km in many regions of the globe"). These preliminary investigations raise the intriguing possibility of formal detection of anthropogenic effects at regional scales that are of direct relevance to policymakers.

### CHANGES IN LATITUDE/LONGITUDE PATTERNS OF ATMOSPHERIC TEMPERATURE OR LAPSE RATE

Fingerprint methods have also been applied to spatial "maps" of changes in layer-averaged

atmospheric temperatures (Santer *et al.*, 2003b; Thorne *et al.*, 2003) and lapse rate (Thorne *et al.*, 2003). The study by Santer *et al.* (2003b) compared modeled and observed changes in $T_2$ and $T_4$. Model fingerprints were estimated from 20CEN experiments performed with PCM (see Table 5.1), while observations were taken from two different satellite data sets (UAH and RSS; see Christy *et al.*, 2003, and Mears *et al.*, 2003). The aim of this work was to assess the sensitivity of detection results to structural uncertainties in observed MSU data.

For the $T_4$ layer, the model fingerprint of combined human and natural effects was consistently detectable in both satellite data sets. In contrast, PCM's $T_2$ fingerprint was identifiable in RSS data (which show net warming over the satellite era), but not in UAH data (which show little overall change in $T_2$; see Chapter 3). Encouragingly, once the global-mean differences between RSS and UAH data were removed, the PCM $T_2$ fingerprint was detectable in both observed data sets. This suggests that the structural uncertainties in RSS and UAH $T_2$ data are most prominent at the global-mean level, and that this global-mean difference masks underlying similarities in smaller-scale pattern structure (Chapter 4; Santer *et al.*, 2004).

Thorne *et al.* (2003) applied a "space-time" fingerprint method to six individual climate variables. These variables contained information on patterns[34] of temperature change at the surface, in broad atmospheric layers (the upper and lower troposphere), and in the lapse rates between these layers[35]. Thorne *et al.* explicitly considered uncertainties in the searched-for fingerprints, the observed radiosonde data[36], and in various data processing/fingerprinting options. They also assessed the detectability of fingerprints arising from multiple forcings[37].

> Preliminary investigations raise the intriguing possibility of formal detection of anthropogenic effects at regional scales that are of direct relevance to policymakers.

---

[31] This is partly due to the fact that natural climate noise is larger (and models are less skillful) on smaller spatial scales.

[32] Another relevant "sub-global" detection study is that by Karoly *et al.* (2003). This showed that observed trends in a variety of area-averaged "indices" of North American climate (*e.g.*, surface temperature, daily temperature range, and the amplitude of the seasonal cycle) were consistent with model-predicted trends in response to anthropogenic forcing, but were inconsistent with model estimates of natural climate variability.

[33] Involving Bayesian statistics.

[34] The "patterns" are in the form of temperature averages calculated over large areas rather than temperatures on a regular latitude/longitude grid.

[35] Thorne *et al.* (2003) calculated the lapse rate changes between the surface and lower troposphere, the surface and upper troposphere, and the lower and upper troposphere.

[36] The model fingerprint was estimated from 20CEN runs performed with two different versions of the Hadley Centre CGCM (HadCM2 and HadCM3). Observational data were taken from two early compilations of the Hadley Centre radiosonde data (HadRT2.1 and HadRT2.1s).

[37] Well-mixed greenhouse gases, the direct effects of

The "bottom-line" conclusion of Thorne *et al.* is that two human-caused fingerprints – one arising from changes in well-mixed GHGs alone, and the other due to combined GHG and sulfate aerosol effects – were robustly identifiable in the observed surface, lower tropospheric, and upper tropospheric temperatures. Evidence for the existence of a detectable volcanic signal was more equivocal. Volcanic and human-caused fingerprints were not consistently identifiable in observed patterns of lapse rate change[38].

### CHANGES IN LATITUDE/HEIGHT PROFILES OF ATMOSPHERIC TEMPERATURE

Initial detection work with zonal-mean profiles of atmospheric temperature change used pattern correlations to compare model fingerprints with radiosonde data (Karoly *et al.*, 1994; Santer *et al.*, 1996a; Tett *et al.*, 1996; Folland *et al.*, 1998; Sexton *et al.*, 2001). These early investigations found that model fingerprints of the stratospheric cooling and tropospheric warming in response to increases in atmospheric $CO_2$ were identifiable in observations (Chapter 1, Figure 1.3A). The pattern similarity between modeled and observed changes generally increased over the period of the radiosonde record.

The inclusion of other human-induced forcings in 20CEN experiments – particularly the effects of stratospheric ozone depletion and sulfate aerosols – tended to improve agreement with observations (Santer *et al.*, 1996a; Tett *et al.*, 1996; Sexton *et al.*, 2001). The addition of ozone depletion cooled the lower stratosphere and upper troposphere. This brought the height of the "transition level" between stratospheric cooling and tropospheric warming lower down in the atmosphere, and in better accord with observations (Chapter 1, Figure 1.3F). It also improved the agreement between simulated and observed patterns of $T_4$ (Ramaswamy *et al.*, 1996), and decreased the size of the "warming maximum" in the upper tropical troposphere, a prominent feature of $CO_2$-only experiments

(compare Figures 1.3A and 1.3F in Chapter 1).

Early work on the direct scattering effects of sulfate aerosols suggested that this forcing was generally stronger in the Northern Hemisphere (NH) than in the Southern Hemisphere (SH), due to the larger emissions of sulfur dioxide in industrialized regions of the NH. This asymmetry in the distribution of anthropogenic sulfur dioxide sources should yield greater aerosol-induced tropospheric cooling in the NH (Santer *et al.*, 1996a,b). Other forcings can lead to different hemispheric temperature responses. Increases in atmospheric $CO_2$, for example, tend to warm land more rapidly than ocean (Chapter 1). Since there is more land in the NH than in the SH, the expected signal due to $CO_2$ increases is greater warming in the NH than in the SH. Because the relative importance of $CO_2$ and sulfate aerosol forcings evolves in a complex way over time (Tett *et al.*, 2002; Hansen *et al.*, 2002),[39] the "imprints" of these two forcings on NH and SH temperatures must also vary with time (Santer *et al.*, 1996b; Stott *et al.*, 2006).

Initial attempts to detect sulfate aerosol effects on atmospheric temperatures did not account for such slow changes in the hemispheric-scale features of the aerosol fingerprint. They searched for a time-invariant fingerprint pattern in observed radiosonde data (Santer *et al.*, 1996a). This yielded periods of agreement and periods of disagreement between the (fixed) aerosol fingerprint and the time-varying effect of aerosols on atmospheric temperatures. Some have interpreted the periods of disagreement as "evidence of absence" of a sulfate aerosol signal (Michaels and Knappenberger, 1996). However, subsequent studies (see below) illustrate that such behavior is expected if one uses a fixed sulfate aerosol fingerprint, and that it is important for detection studies to account for large temporal changes in the fingerprint.

---

sulfate aerosols, combined greenhouse-gas and sulfate aerosol effects, volcanic aerosols, and solar irradiance changes.

[38] The failure to detect volcanic signals is probably due to the coarse time resolution of the input data (five-year averages) and the masking effects of ENSO variability in the radiosonde observations. Note that the two models employed in this work yielded different estimates of the size of the natural and human-caused fingerprints.

---

[39] See, for example, Figure 1a in Tett *et al.* (2002) and Figure 8b in Hansen *et al.* (2002).

**Table 5.1: Acronyms of climate models referenced in this Chapter. All 19 models performed simulations of 20th century climate change ("20CEN") in support of the IPCC Fourth Assessment Report. The ensemble size "ES" is the number of independent realizations of the 20CEN experiment that were analyzed here.**

|   | Model Acronym | Country | Institution | ES |
|---|---|---|---|---|
| 1 | CCCma-CGCM3.1(T47) | Canada | Canadian Centre for Climate Modelling and Analysis | 1 |
| 2 | CCSM3 | United States | National Center for Atmospheric Research | 5 |
| 3 | CNRM-CM3 | France | Météo-France/Centre National de Recherches Météorologiques | 1 |
| 4 | CSIRO-Mk3.0 | Australia | CSIRO[a.] Marine and Atmospheric Research | 1 |
| 5 | ECHAM5/MPI-OM | Germany | Max-Planck Institute for Meteorology | 3 |
| 6 | FGOALS-g1.0 | China | Institute for Atmospheric Physics | 3 |
| 7 | GFDL-CM2.0 | United States | Geophysical Fluid Dynamics Laboratory | 3 |
| 8 | GFDL-CM2.1 | United States | Geophysical Fluid Dynamics Laboratory | 3 |
| 9 | GISS-AOM | United States | Goddard Institute for Space Studies | 2 |
| 10 | GISS-EH | United States | Goddard Institute for Space Studies | 5 |
| 11 | GISS-ER | United States | Goddard Institute for Space Studies | 5 |
| 12 | INM-CM3.0 | Russia | Institute for Numerical Mathematics | 1 |
| 13 | IPSL-CM4 | France | Institute Pierre Simon Laplace | 1 |
| 14 | MIROC3.2(medres) | Japan | Center for Climate System Research / NIES[b.] / JAMSTEC[c.] | 3 |
| 15 | MIROC3.2(hires) | Japan | Center for Climate System Research / NIES[b.] / JAMSTEC[c.] | 1 |
| 16 | MRI-CGCM2.3.2 | Japan | Meteorological Research Institute | 5 |
| 17 | PCM | United States | National Center for Atmospheric Research | 4 |
| 18 | UKMO-HadCM3 | United Kingdom | Hadley Centre for Climate Prediction and Research | 1 |
| 19 | UKMO-HadGEM1 | United Kingdom | Hadley Centre for Climate Prediction and Research | 1 |

[a.] CSIRO is the Commonwealth Scientific and Industrial Research Organization.

[b.] NIES is the National Institute for Environmental Studies.

[c.] JAMSTEC is the Frontier Research Center for Global Change in Japan.

Fingerprint detection studies provide consistent evidence that human-induced changes in greenhouse gases and sulfate aerosols are identifiable in radiosonde records of free atmospheric temperature change.

"Space-time" optimal detection schemes explicitly account for time variations in the signal pattern and in observational data (Box 5.5). Results from recent space-time detection studies support previous claims of an identifiable sulfate aerosol effect on surface temperature (Stott *et al.*, 2006) and on zonal-mean profiles of atmospheric temperature (Allen and Tett, 1999; Forest *et al.*, 2001, 2002; Thorne *et al.*, 2002; Tett *et al.*, 2002; Jones *et al.*, 2003). This work also illustrates that the identification of human effects on atmospheric temperatures can be achieved using tropospheric temperatures alone (Thorne *et al.*, 2002). Positive detection results are not solely driven by the inclusion of strong stratospheric cooling in the vertical pattern of temperature change (as has been claimed by Weber, 1996).

In summary, fingerprint detection studies

provide consistent evidence that human-induced changes in greenhouse gases and sulfate aerosols are identifiable in radiosonde records of free atmospheric temperature change. The fingerprint evidence is much more equivocal in the case of solar and volcanic signals in the troposphere. These natural signals have been detected in some studies (Jones *et al.*, 2003) but not in others (Tett *et al.*, 2002), and their identification appears to be more sensitive to specific processing choices that are made in applying fingerprint methods (Leroy, 1998; Thorne *et al.*, 2002, 2003).

# 5. NEW COMPARISONS OF MODELED AND OBSERVED TEMPERATURE CHANGES

In this section, we evaluate selected results from recently completed CGCM 20CEN experiments that have been performed in support of the IPCC Fourth Assessment Report (AR4). The runs analyzed here were performed with 19 different models, and involve modeling groups in nine different countries (Table 5.1). They use new model versions, and incorporate historical changes in many (but not all) of the natural and human forcings that are thought to have influenced atmospheric temperatures over the past 50 years[40] (Table 5.2). These new experiments provide our current best estimates of the expected climate change due to combined human and natural effects.

The new 20CEN runs constitute an "ensemble of opportunity" (Allen and Stainforth, 2002). The selection and application of natural and anthropogenic forcings was not coordinated across modeling groups.[41] For example, only seven of the 19 models were run with time-varying changes in LULC (Table 5.2). Modeling groups that included LULC effects did not always use the same observational data set for specifying this forcing, or apply it in the same way (Table 5.3). Only six models included some representation of the indirect effects of anthropogenic aerosols, which are thought to have had a net cooling influence on surface temperatures through their effects on cloud properties (Ramaswamy *et al.*, 2001b).

One important implication of Tables 5.2 and 5.3 is that model-to-model differences in the applied forcings are intertwined with model-to-model differences in the climate responses to those forcings. This makes it more difficult to isolate systematic errors that are common to a number of models, or to identify problems with a specific forcing data set. Note, however, that the lack of a coordinated experimental design is also an advantage, since the "ensemble of opportunity" spans a wide range of uncertainty in current estimates of climate forcings.

In addition to model forcing and response uncertainty, the 20CEN ensemble also encompasses uncertainties arising from inherently unpredictable climate variability (Boxes 5.1, 5.2). Roughly half of the modeling groups that submitted 20CEN data performed multiple realizations of their historical forcing experiment (see Section 2 and Table 5.1). For example, the five-member ensemble of CCSM3.0 20CEN runs contains an underlying signal (which one might define as the ensemble-average climate response to the forcings varied in CCSM3.0) plus five different sequences of climate noise. Such multi-member ensembles provide valuable information on the relative sizes of signal and noise. In all, a total of 49 20CEN realizations were examined here[42].

The following Section presents preliminary results from analyses of these 20CEN runs and the new observational data sets described in Chapters 2-4. Our primary focus is on the tropics, since previous work by Gaffen *et al.* (2000) and Hegerl and Wallace (2002) suggests that this is where any differences between observations and models are most critical. We also discuss comparisons of global-mean changes in atmospheric temperatures and lapse rates. We do not discount the importance of comparing modeled and observed lapse-rate changes at much smaller scales (particularly in view of the incorporation of regional-scale forcing changes in many of the runs analyzed here), but no comprehensive regional-scale comparisons were available for us to assess.

In order to facilitate "like with like" comparisons between modeled and observed atmospheric temperature changes, we calculate synthetic MSU $T_4$, $T_2$, and $T_{2LT}$ from the model 20CEN results (see Chapter 2, Box 1). Both observed and synthetic MSU $T_2$ data include a contribution from the cooling stratosphere (Fu *et al.*, 2004a,b), and hence complicate the interpretation of slow changes in $T_2$. To provide a less ambiguous measure of "bulk" tropospheric

---

[40] This was not the case in previous model intercomparison exercises, such as AMIP (Gates *et al.*, 1999) and CMIP2 (Meehl *et al.*, 2000).

[41] In practice, experimental coordination is very difficult across a range of models of varying complexity and sophistication. Aerosols are a case in point. Some modeling groups that contributed 20CEN simulations to the IPCC AR4 do not have the technical capability to explicitly include aerosols, and instead attempt to represent their net radiative effects by adjusting the surface albedo.

[42] 49 individual realizations of the IPCC 20CEN run were available at the time this Chapter was written. An analysis of lapse-rate changes in these realizations has been published (Santer *et al.*, 2005). At present, the IPCC database contains 82 realizations of the 20CEN experiment. Relevant analyses of these additional 33 realizations are currently unpublished and unreviewed, and have not been included here.

**Table 5.2: Forcings used in IPCC simulations of 20th century climate change. This Table was compiled using information provided by the participating modeling centers (see http://www-pcmdi.llnl.gov/ipcc/model.documentation/ipcc_model_documentation.php). Eleven different forcings are listed: well-mixed greenhouse gases (G), tropospheric and stratospheric ozone (O), sulfate aerosol direct (SD) and indirect effects (SI), black carbon (BC) and organic carbon aerosols (OC), mineral dust (MD), sea salt (SS), land use/land cover (LU), solar irradiance (SO), and volcanic aerosols (V). Shading denotes inclusion of a specific forcing. As used here, "inclusion" means specification of a time-varying forcing, with changes on interannual and longer timescales. Forcings that were varied over the seasonal cycle only are not shaded.**

| | MODEL | G | O | SD | SI | BC | OC | MD | SS | LU | SO | V |
|---|---|---|---|---|---|---|---|---|---|---|---|---|
| 1 | CCCma-CGCM3.1(T47) | | | | | | | | | | | |
| 2 | CCSM3 | | | | | | | | | | | |
| 3 | CNRM-CM3 | | | | | | | | | | | |
| 4 | CSIRO-Mk3.0 | | | | | | | | | | | |
| 5 | ECHAM5/MPI-OM | | | | | | | | | | | |
| 6 | FGOALS-g1.0 | | | | | | | | | | | |
| 7 | GFDL-CM2.0 | | | | | | | | | | | |
| 8 | GFDL-CM2.1 | | | | | | | | | | | |
| 9 | GISS-AOM | | | | | | | | | | | |
| 10 | GISS-EH | | | | | | | | | | | |
| 11 | GISS-ER | | | | | | | | | | | |
| 12 | INM-CM3.0 | | | | | | | | | | | |
| 13 | IPSL-CM4 | | | | | | | | | | | |
| 14 | MIROC3.2(medres) | | | | | | | | | | | |
| 15 | MIROC3.2(hires) | | | | | | | | | | | |
| 16 | MRI-CGCM2.3.2 | | | | | | | | | | | |
| 17 | PCM | | | | | | | | | | | |
| 18 | UKMO-HadCM3 | | | | | | | | | | | |
| 19 | UKMO-HadGEM1 | | | | | | | | | | | |

temperature changes, we use the statistical approach of Fu *et al.* (2004a, 2005) to remove stratospheric influences, thereby obtaining $T^*_G$ and $T^*_T$ in addition to $T_{2LT}$[43]. As a simple measure of lapse-rate changes, we consider temperature differences between the surface and three different atmospheric layers ($T_{2LT}$, $T^*_G$, and $T^*_T$). Each of these layers samples slightly different portions of the troposphere (Chapter 2, Figure 2.2).

The trend comparisons shown in Sections 5.1 and 5.2 do not involve any formal statistical significance tests (see Appendix A). While such tests are entirely appropriate for comparisons of individual model and observational trends,[44] they are less relevant here, where we compare a 49-member ensemble of model trends with a relatively small number of observationally based estimates. The model ensemble encapsulates uncertainties in climate forcings and model responses, as well as the effects of climate noise on trends. The observational range characterizes current structural uncertainties in historical changes. We simply assess whether the observations are contained within the simu-

---

[43] There is still some debate over the reliability of $T^*_G$ trends estimated with the Fu *et al.* (2004a) statistical approach (Tett and Thorne, 2004, Gillett *et al.*, 2004; Kiehl *et al.*, 2005; Fu *et al.*, 2004b; Chapter 4). $T^*_T$ is derived mathematically (from the overlap between the $T_4$ and $T_2$ weighting functions) rather than statistically, and is now generally accepted as a reasonable measure of temperature change in the tropical troposphere.

[44] For example, such tests have been performed by Santer *et al.* (2003b) in comparisons between observed MSU trends (in RSS and UAH) and synthetic MSU trends in four PCM 20CEN realizations.

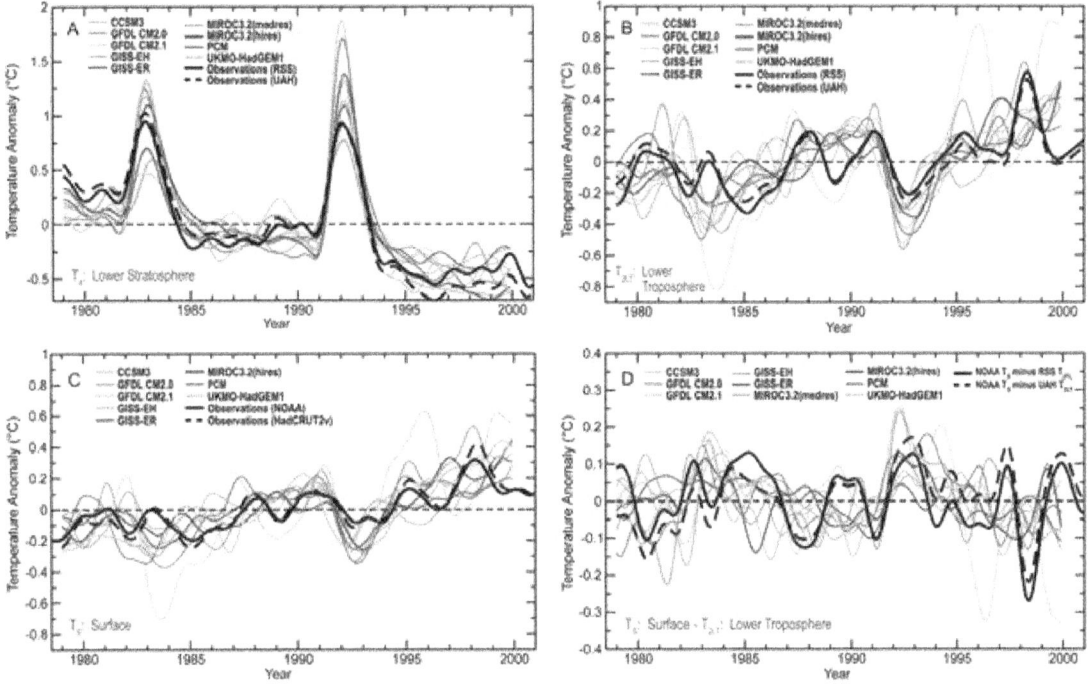

**Figure 5.2:** Modeled and observed changes in global-mean monthly-mean temperature of the lower stratosphere ($T_4$; A), the lower troposphere ($T_{2LT}$; B), the surface ($T_S$; C), and the surface minus the lower troposphere ($T_S - T_{2LT}$; D). A simple weighting function approach (Box 2.1, Chapter 2) was used to calculate "synthetic" $T_4$ and $T_{2LT}$ temperatures (equivalent to the MSU $T_4$ and $T_{2LT}$ monitored by satellites) from model temperature data. Simulated surface and atmospheric temperatures are from 20CEN experiments performed with nine different models (see Table 5.1). These models were chosen because they satisfy certain minimum requirements in terms of the forcings applied in the 20CEN run: all nine were driven by changes in well-mixed GHGs, sulfate aerosol direct effects, tropospheric and stratospheric ozone, volcanic aerosols, and solar irradiance (in addition to other forcings; see Table 5.2). Observed satellite-based estimates of $T_4$ and $T_{2LT}$ changes were obtained from both RSS and UAH (see Chapter 3). Observed $T_S$ results in C are from NOAA and HadCRUT2v, while observed $T_S - T_{2LT}$ differences in D use both observed $T_{2LT}$ datasets, but NOAA $T_S$ data only. All anoma ies are expressed as departures from a 1979 to 1999 reference period average, and were smoothed with the same filter. To make it easier to compare temperature variability in models with different ensemble sizes (see Table 5.1), only the first 20CEN realization is plotted from each model. This also facilitates comparisons of modeled and observed variability.

lated trend distributions.[45] Our goal here is to determine where model results are qualitatively consistent with observations, and where serious inconsistencies are likely to exist. This does not obviate the need for the more rigorous statistical

comparisons described in Box 5.5, which should be a high priority (see Recommendations).

## 5.1 Global-Mean Temperature and Lapse-Rate Trends

In all but two of the 49 20CEN realizations, the global-mean temperature of the lower strato-sphere experiences a net cooling over 1979 to 1999 (Figures 5.2A, 5.3A)[46]. The model average $T_4$ trend is –0.25°C/decade (Table 5.4A). Most of this cooling is due to the combined effects of stratospheric ozone depletion and increases in well-mixed GHGs (Ramaswamy *et al.*, 2001a,b), with the former the dominant influence on $T_4$ changes over the satellite era (Ramaswamy *et al.*, 1996; Santer *et al.*, 2003a). The

---

[45] The 49 20CEN realizations analyzed here are a very small sample from the large population of results that could have been generated by accounting for existing uncertainties in physics parameterizations and historical forcings (*e.g.*, Allen, 1999; Stainforth *et al.*, 2005). Likewise, the observational datasets that we consider in this report probably only capture part of the true "construction uncertainty" inherent in the development of homogeneous climate records from raw temperature measurements. We do not know *a priori* whether temperature changes inferred from these small samples are representative of the true temperature changes that would be estimated from the much larger (but unknown) populations of model and observational results. This is another reason why we are cautious about making formal assessments of the statistical significance of differences between modeled and observed temperature trends. We do, however, attempt to characterize some basic statistical properties of the model results (see Tables 5.4A,B).

[46] In the following, all inter-model and model-data comparisons are over January 1979 to December 1999. This is the longest period of overlap (at least during the satellite era) between the model experiments (which generally end in 1999) and the satellite data (which start in 1979).

**Table 5.3: Forcings used in 20CEN experiments performed with the PCM, CCSM3.0, GFDL CM2.1, and GISS-EH models. Shading indicates a forcing that was not incorporated or that did not vary over the course of the experiment.**

| | PCM | CCSM3.0 | GFDL CM2.1 | GISS-EH |
|---|---|---|---|---|
| **Well-mixed greenhouse gases** | IPCC Third Assessment Report. | IPCC Third Assessment Report. | IPCC Third Assessment Report and World Meteorological Organization (2003). | $CH_4$, $N_2O$ and CFC spatial distributions are fit to Minschwaner et al. (1998). |
| **Sulfate aerosols (direct effects)** | Spatial patterns of sulfur dioxide [$SO_2$] emissions prescribed over seasonal cycle. Year-to-year changes scaled by estimates of historical changes in $SO_2$ emissions.[a.] | Sulfur cycle model using time and space-varying $SO_2$ emissions (Smith et al., 2001, 2005).[b.] | Computed from an atmospheric chemistry transport model.[c.] | Based on simulations of Koch et al. (1999) and Koch (2001).[d.] |
| **Sulfate aerosols (indirect effects)** | Not included. | Not included. | Not included. | Parameterization of aerosol indirect effects on cloud albedo and cloud cover.[d.] |
| **Stratospheric ozone** | Assumed to be constant up to 1970. After 1970 prescribed from a NOAA dataset.[a.] | Assumed to be constant up to 1970. After 1970 prescribed from a NOAA dataset.[b.] | Specified using data from Randel and Wu (1999). | Specified using data from Randel and Wu (1999).[d.] |
| **Tropospheric ozone** | Computed from an atmospheric chemistry transport model. Held constant after 1990.[a.] | Computed from an atmospheric chemistry transport model. Held constant after 1990.[b.] | Computed from an atmospheric chemistry transport model.[c.] | Computed from an atmospheric chemistry transport model (Shindell et al., 2003).[d.] |
| **Black carbon aerosols** | Not included. | Present-day estimate of distribution and amount of black carbon, scaled by population changes over 20th century.[b.] | Computed from an atmospheric chemistry transport model.[c.] | Based on simulations of Koch et al. (1999) and Koch (2001).[d.] |
| **Organic aerosols** | Not included. | Not included. | Computed from an atmospheric chemistry transport model.[c.] | Based on simulations of Koch et al. (1999) and Koch (2001).[d.] |
| **Sea salt** | Not included. | Distributions held fixed in 20th century at year 2000 values.[b.] | Distributions held fixed at 1990 values. | Decadally varying. |
| **Dust** | Not included. | Distributions held fixed in 20th century at year 2000 values.[b.] | Distributions held fixed at 1990 values. | Decadally varying. |
| **Land use change** | Distributions held fixed at present-day values. | Distributions held fixed at present-day values. | Knutson et al. (2006) global land use reconstruction history. Includes effect on surface albedo, surface roughness, stomatal resistance, and effective water capacity. | Uses Ramankutty and Foley (1999) and Klein Goldewijk (2001) time-dependent datasets. Effects on albedo and evapotranspiration included, but no irrigation effects.[d.] |
| **Volcanic stratospheric aerosols** | Ammann et al. (2003). | Ammann et al. (2003). | "Blend" between Sato et al. (1993) and Ramachandran et al. (2000). | Update of Sato et al. (1993). |
| **Solar irradiance** | Hoyt and Schatten (1993). | Lean et al. (1995). | Lean et al. (1995). | Uses solar spectral changes of Lean (2000). |

a. See Dai et al. (2001) for further details.
b. See Meehl et al.(2005) for further details.
c. The chemistry transport model (MOZART; see Horowitz et al., 2003; Tie et al., 2005) was driven by meteorology from the

Middle Atmosphere version of the Community Climate Model ("MACCM"; version 3). "1990" weather from MACCM3 was used for all years between 1860 and 2000.
d. See Hansen et al. (2005a) for further details.

model average cooling is larger (–0.35°C/decade) and closer to the satellite-based estimates if it is calculated from the subset of 20CEN realizations that include forcing by ozone depletion. The range of model $T_4$ trends encompasses the trends derived from satellites, but not the larger trends estimated from radiosondes. The most likely explanation for this discrepancy is a residual cooling trend in the radiosonde data (Chapter 4)[47]. The neglect of stratospheric water vapor increases

**Table 5.4A: Summary statistics for global-mean temperature trends calculated from 49 different realizations of 20CEN experiments performed with 19 different coupled models. Results are for four different atmospheric layers ($T_4$, $T_2$, $T^*_G$, and $T_{2LT}$), the surface ($T_S$), and differences between the surface and the troposphere ($T_S$ minus $T^*_G$ and $T_S$ minus $T_{2LT}$). All trends were calculated over the 252-month period from January 1979 to December 1999 using global-mean monthly-mean anomaly data. Results are in °C/decade. The values in the "Mean" column correspond to the locations of the red lines in the seven panels of Figure 5.3. For each layer, means, medians and standard deviations were calculated from a sample size of n = 19, i.e., from ensemble means (if available) and individual realizations (if ensembles were not performed). This avoids placing too much weight on results from a single model with a large number of realizations. Maximum and minimum values were calculated from all available realizations (i.e., from a sample size of n = 49).**

| Layer | Mean | Median | Std. Dev. (1σ) | Minimum | Maximum |
|---|---|---|---|---|---|
| $T_4$ | -0.25 | -0.28 | 0.19 | -0.70 | 0.08 |
| $T_2$ | 0.14 | 0.12 | 0.08 | 0.02 | 0.35 |
| $T^*_G$ | 0.18 | 0.17 | 0.08 | 0.05 | 0.38 |
| $T_{2LT}$ | 0.20 | 0.19 | 0.07 | 0.06 | 0.39 |
| $T_S$ | 0.16 | 0.16 | 0.06 | 0.05 | 0.33 |
| $T_S - T^*_G$ | -0.02 | -0.02 | 0.05 | -0.11 | 0.08 |
| $T_S - T_{2LT}$ | -0.03 | -0.03 | 0.03 | -0.10 | 0.05 |

in most of the 20CEN runs considered here (Shine *et al.*, 2003) may be another contributory factor[48].

Superimposed on the overall cooling of $T_4$ are the large stratospheric warming signals in response to the eruptions of El Chichón (in April 1982) and Mt. Pinatubo (in June 1991)[49]. Nine of the 19 IPCC models explicitly included volcanic aerosols (Figure 5.2A and Table 5.2)[50]. Seven of these nine models overestimate the observed stratospheric warming after Pinatubo. GFDL CM2.1 simulates the Pinatubo response reason-

ably well, but underestimates the response to El Chichón. Differences in the magnitude of the applied volcanic aerosol forcings must account for some of the inter-model differences in the $T_4$ warming signals (Table 5.3)[51].

Over 1979 to 1999, the global-mean troposphere warms in all 49 20CEN simulations considered here (Figures 5.2B, 5.3B-D). The shorter-term cooling signals of the El Chichón and Mt. Pinatubo eruptions are superimposed on this gradual warming[52]. Because of the influence of stratospheric cooling on $T_2$, the model average

---

[47] Recent work suggests that this residual trend is largest in the lower stratosphere and upper troposphere, and is primarily related to temporal changes in the solar heating of the temperature sensors carried by radiosondes (and failure to properly correct for this effect; see Sherwood *et al.*, 2005; Randel and Wu, 2006).

[48] Recent stratospheric water vapor increases are thought to be partly due to the oxidation of methane, and are expected to have a net cooling effect on $T_4$. To our knowledge, $CH_4$-induced stratospheric water vapor increases were explicitly incorporated in only two of the 19 models considered here (GISS-EH and GISS-ER; Hansen *et al.*, 2005a).

[49] These warming signals occur because volcanic aerosols absorb both incoming solar radiation and outgoing thermal radiation (Ramaswamy *et al.*, 2001a).

[50] The MRI-CGCM2.3.2 model incorporated volcanic effects indirectly rather than explicitly, using estimated volcanic forcing data from Sato *et al.* (1993) to adjust the solar irradiance at the top of the model atmosphere. This procedure would not yield volcanically-induced stratospheric warming signals.

[51] More subtle details of the forcing are also relevant to interpretation of inter-model $T_4$ differences, such as different assumptions regarding the aerosol size distribution, the vertical distribution of the volcanic aerosol relative to the model tropopause, *etc.* Note that observed $T_4$ changes over the satellite era are not well-described by a simple linear trend, and show evidence of a step-like decline in stratospheric temperatures after the El Chichón and Mt. Pinatubo eruptions (Pawson *et al.*, 1998; Seidel and Lanzante, 2004). Inter-model differences in the applied ozone forcings and solar forcings may help to explain why the GFDL, GISS, and HadGEM1 models appear to reproduce some of this step-like behavior, particularly after El Chichón, while $T_4$ decreases in PCM are much more linear (Dameris *et al.*, 2005; Ramaswamy *et al.*, 2006).

[52] Because of differences in the timing of modeled and observed ENSO events (Section 5.2), the tropospheric and surface cooling caused by El Chichón is more noticeable in all models than in observations (where it was partially masked by the large 1982/83 El Niño; Figures 5.2B,C).

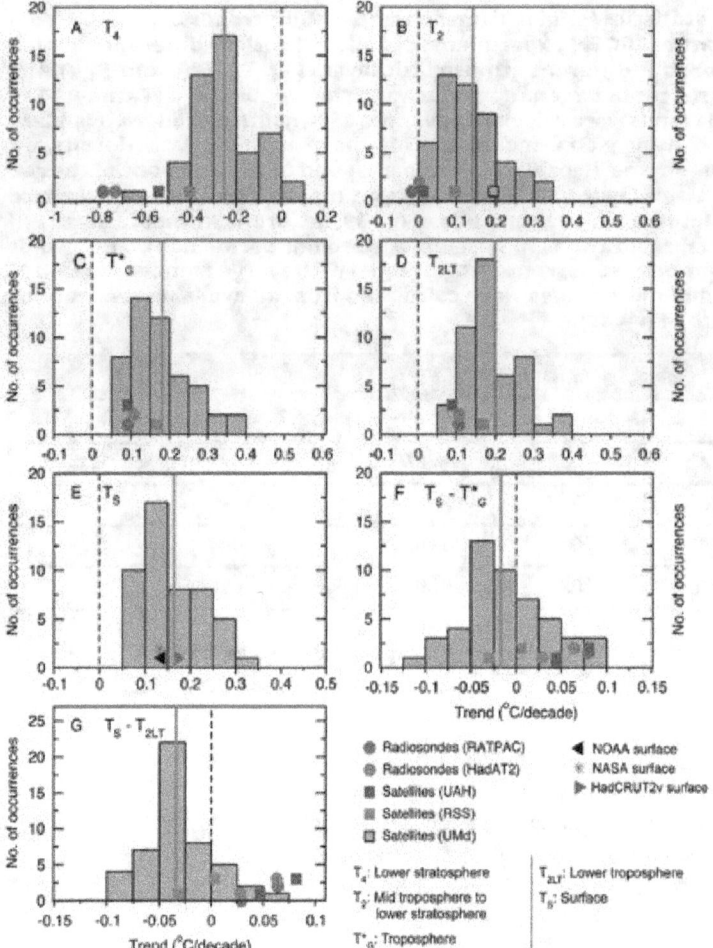

**Figure 5.3:** Modeled and observed trends in time series of global-mean $T_4$ (panel A), $T_2$ (panel B), $T^*_G$ (panel C), $T_{2LT}$ (panel D), $T_S$ (panel E), $T_S$ minus $T^*_G$ (panel F), and $T_S$ minus $T_{2LT}$ (panel G). All trends were calculated using monthly-mean anomaly data. The analysis period is 1979 to 1999. Model results are displayed in the form of histograms. Each histogram is based on results from 49 individual realizations of the 20CEN experiment, performed with 19 different models (Table 5.1). The applied forcings are listed in Table 5.2. The vertical red  ine in each panel is the mean of the model trends, calculated with a sample size of n = 19 (see Table 5.4A). Observed trends are estimated from two radiosonde and three satellite datasets ($T_2$), two radiosonde and two satellite datasets ($T_4$, $T^*_G$ and $T_{2LT}$), and three different surface datasets ($T_S$) (see Chapter 3). The bottom "rows" of the observed difference trends in panels F and G were calculated with NOAA $T_S$ data. The top "rows" of observed results in panels F and G were computed with HadCRUT2v $T_S$ data. The vertical offsetting of observed results in these panels (and also in panels B-E) is purely for the purpose of simplifying the visual display – observed trends bear no relation to the y-axis scale. To simplify the display, the Figure does not show the statistical uncertainties arising from the fitting of  inear trends to noisy data. GISS observed $T_S$ trends (not shown) are very close to those estimated with NOAA $T_S$ data (see Chapter 3).

trend is smaller for this layer than for either $T_{2LT}$ or $T^*_G$, which are more representative of temperature changes in the bulk of the troposphere (Table 5.4A)[53]. All of the satellite- and

radiosonde-based trends in $T_{2LT}$ and $T^*_G$ are contained within the spread of model results. This illustrates that there is no fundamental discrepancy between modeled and observed trends in global-mean tropospheric temperature.

In contrast, the $T_2$ trends in both radiosonde data sets are either slightly negative or close to zero, and are smaller than all of the model results. This difference is most likely due to contamination from residual stratospheric and upper-tropospheric cooling biases in the radiosonde data (Chapter 4; Sherwood *et al.*, 2005; Randel and Wu, 2006). The satellite-based $T_2$ trends are either close to the model average (RSS and VG) or just within the model range (UAH; Fig. 5.3B). Even without formal statistical tests, it is clear that observational uncertainty is an important factor in assessing the consistency between modeled and observed changes in mid- to upper tropospheric temperature (Santer *et al.*, 2003b).

Observed $T_S$ trends closely bracket the model average (Figure 5.3E). There is no inconsistency between modeled and observed surface temperature changes. Structural uncertainties in observed $T_S$ trends are much smaller than for trends in $T_4$ or tropospheric layer-average temperatures (see Chapter 4).

The model-simulated ranges of lapse-rate trends also encompass virtually all observational results (Figures 5.3F,G)[54]. Closer inspection reveals that the model-average trends in tropospheric lapse rate are slightly negative,[55] indicating larger warming aloft than at the surface. Most combinations of observed $T_S$, $T^*_G$, and $T_{2LT}$ data sets yield the converse result, and show smaller warming aloft than at the surface. As in the case of global-mean $T^*_G$ and $T_{2LT}$ trends, RSS-based lapse-rate trends are invariably closest to the model average results. Both models and observations show a tendency towards positive values of $T_S$ minus $T_{2LT}$ for several years after the El Chichón and Mt. Pinatubo eruptions, indicative of larger

---

[53] Due to ozone-induced cooling of the lower stratosphere, the model-average $T_2$ trend is slightly smaller (0.12 C/decade) and closer to the RSS result if it is estimated from the subset of 20CEN runs that include stratospheric ozone depletion. Subsetting in this way has little impact on the model-average $T_{2LT}$ and $T^*_G$

trends.

[54] Note that the subtraction of temperature variability common to surface and troposphere decreases (by about a factor of two) the large range of model trends in $T_S$, $T^*_G$, and $T_{2LT}$ (Table 5.4A).

[55] Values are –0.02 C/decade in the case of $T_S$ minus $T^*_G$ and –0.03 C/decade for $T_S$ minus $T_{2LT}$.

cooling aloft than at the surface (Figure 5.2D; Section 5.4).

## 5.2 Tropical Temperature and Lapse-Rate Trends

The previous section examined whether simulated global-mean temperature trends were contained within current estimates of structural uncertainty in observations. Since ENSO is primarily a tropical phenomenon, its influence on surface and tropospheric temperature is more pronounced in the tropics than in global averages. Observations contain only one specific sequence of ENSO fluctuations from 1979 to present, and only one sequence of ENSO effects on tropical temperatures. The model 20CEN runs examined here provide many different sequences of ENSO variability. We therefore expect – and find – that these runs yield a wide range of trends in tropical surface and tropospheric temperature (Figure 5.4)[56]. It is of interest whether this large model range encompasses the observed trends.

At the surface, results from the multi-model ensemble include all observational estimates of tropical temperature trends (Figure 5.4E; Table 5.4B). Observed results are close to the model average $T_S$ trend of +0.16°C/decade. There is no evidence that the models significantly over- or underestimate the observed surface warming. In the troposphere, all observational results are still within the range of possible model solutions, but the majority of model results show tropospheric warming that is larger than observed (Figures 5.4B-D). As in the case of the global-mean $T_4$ trends, the cooling of the tropical stratosphere in both radiosonde data sets is larger than in any of the satellite data sets or model results (Figure 5.4A)[57]. The UAH and RSS $T_4$ trends are close to the model average[58].

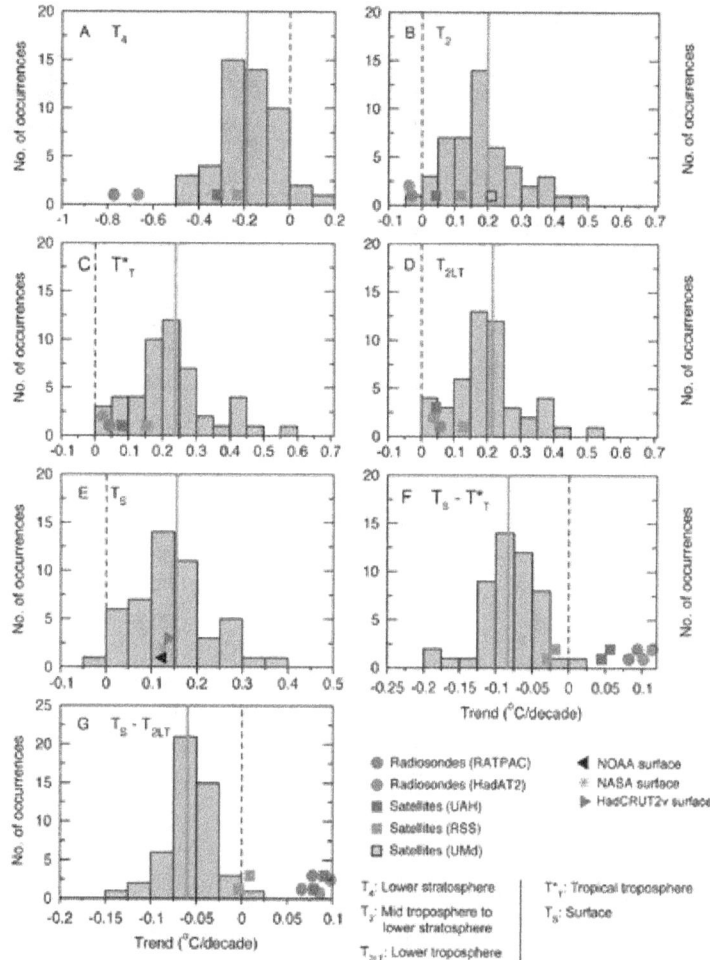

**Figure 5.4:** As for Figure 5.3, but for trends in the tropics (20°N-20°S).

In the model results, trends in the two measures of tropical lapse rate ($T_S$ minus $T_{2LT}$ and $T_S$ minus $T^*_T$) are almost invariably negative, indicating larger warming aloft than at the surface (Figure 5.4F,G). Similar behavior is evident in only one of the four upper-air data sets examined here (RSS)[59]. The RSS trends are just within the range of model solutions[60].

---

[56] This would be true even for a hypothetical "perfect" climate model run with "perfect" forcings. The large model range of tropical temperature trends is not solely due to the effects of ENSO and other modes of internal variability. It also arises from uncertainties in the models and forcings (see Boxes 5 1 and 5.2 and Table 5.2).

[57] This supports recent findings of a residual cooling bias in tropical radiosonde data (Sherwood *et al.*, 2005; Randel and Wu, 2006).

[58] The model average is –0.27 C/decade when estimated from the subset of 20CEN runs that include stratospheric ozone depletion.

---

[59] The UMd group does not provide either a stratospheric or lower-tropospheric temperature retrieval, and so could not be included in the comparison of modeled and observed trends in $T_S$ minus $T^*_T$ or $T_S$ minus $T_{2LT}$. Assuming that the relationships between the UMd $T_2$, $T_{2LT}$ and $T^*_T$ trends were similar to those for the UAH and RSS data, the UMd data would yield $T_{2LT}$ and $T^*_T$ trends that were larger than in RSS. This would expand the range of observational uncertainty shown in Figures 5.4F,G.

[60] Three of the four RSS-based results in Figures 5.4 F and G are within two standard deviations of the model average values (see Table 5.4B). Note also that for their tropical $T_{2LT}$ trend, RSS claims a 2σ uncertainty of ± 0.09 C/decade (Mears and Wentz 2005; Mears personal communication). This uncertainty is not included here.

**Table 5.4B: As for Table 5.4A, but for tropical temperature trends (calculated from spatial averages over 20°N-20°S).**

| Layer | Mean | Median | Std. Dev. (1σ) | Minimum | Maximum |
|---|---|---|---|---|---|
| $T_4$ | -0.19 | -0.19 | 0.15 | -0.49 | 0.13 |
| $T_2$ | 0.20 | 0.19 | 0.10 | -0.01 | 0.48 |
| $T^*_T$ | 0.24 | 0.21 | 0.11 | 0.01 | 0.56 |
| $T_{2LT}$ | 0.22 | 0.19 | 0.09 | 0.01 | 0.51 |
| $T_S$ | 0.16 | 0.14 | 0.07 | -0.02 | 0.37 |
| $T_S - T^*_T$ | -0.08 | -0.08 | 0.04 | -0.19 | 0.02 |
| $T_S - T_{2LT}$ | -0.06 | -0.05 | 0.03 | -0.15 | 0.01 |

high latitudes in both hemispheres.

To illustrate structural uncertainties in the observed data, we show two different patterns of trends in $T_S$ minus $T_{2LT}$. Both rely on the same NOAA surface data, but use either UAH (Figure 5.5E) or RSS (Figure 5.5F) as their source of $T_{2LT}$ results. The "NOAA minus UAH" combination provides a picture that is very different from the model results, with coherent warming of the surface relative to the troposphere over much of the world's tropical oceans. While "NOAA minus RSS" also has relative warming of the surface in the Western and tropical Pacific, it shows relative warming of the troposphere in the eastern tropical Pacific and Atlantic Oceans. This helps to clarify why simulated lapse-rate trends in Figures 5.4F and 5.4G are closer to NOAA minus RSS results than to NOAA minus UAH results.

As pointed out by Santer *et al.* (2003b) and Christy and Spencer (2003), we cannot use such model-data comparisons alone to determine whether the UAH or RSS $T_{2LT}$ data set is closer to (an unknown) "reality." As the next section will show, however, models and basic theory can be used to identify aspects of observational behavior that require further investigation, and may help to constrain observational uncertainty.

The model results that overlap with the RSS-derived tropical lapse-rate trends exhibit less surface warming than the observations. This analysis is revisited in Section 5.4 using a metric that more directly addresses the relationship between surface and tropospheric temperature changes. Tropical lapse-rate trends in both radiosonde data sets and in the UAH satellite data are always positive (larger warming at the surface than aloft), and lie outside the range of model results.

This comparison suggests that discrepancies between our current best estimates of simulated and observed lapse-rate changes may be larger and more serious in the tropics than in globally averaged data. Large structural uncertainties in the observations (even in the sign of the trend in tropical lapse-rate changes) make it difficult to reach more definitive conclusions regarding the significance and importance of model-data discrepancies (see Section 5.4).

**5.3 Spatial Patterns of Lapse-Rate Trends**

Maps of the trends in lower tropospheric lapse rate help to identify geographical regions where the model-data discrepancies in Figures 5.4F and 5.4G are most pronounced. We focus on four U.S. models run with the most complete set of forcings: CCSM3.0, PCM, GFDL CM2.1, and GISS-EH (Table 5.3). These show qualitatively similar patterns of trends in $T_S$ minus $T_{2LT}$ (Figures 5.5A-D). Over most of the tropical ocean, the simulated warming is larger in the troposphere than at the surface. All models have some tropical land areas where the surface warms relative to the troposphere. The largest relative warming of the surface occurs at

**5.4 Tropospheric Amplification of Tropical Surface Temperature Changes**

When surface and lower tropospheric temperature changes are spatially averaged over the deep tropics, and when day-to-day tropical temperature changes are averaged over months, seasons, or years, it is evident that temperature changes aloft are larger than at the surface. This "amplification" behavior has been described in many observational and modeling studies, and is a consequence of the release of latent heat by moist convecting air (*e.g.,* Manabe and Stouffer, 1980; Horel and Wallace, 1981; Pan and

Oort, 1983; Yulaeva and Wallace, 1994; Hurrell and Trenberth, 1998; Soden, 2000; Wentz and Schabel, 2000; Hegerl and Wallace, 2002; Knutson and Tuleya, 2004)[61].

A recent study by Santer *et al.* (2005) examined this amplification behavior in the same 20CEN runs and observational data sets considered in the present report. The sole difference (relative to the data used here) was that Santer *et al.* analyzed a version of the UAH $T_{2LT}$ data that had not yet been adjusted for a recently discovered error (Mears and Wentz, 2005)[62]. The amplification of tropical surface temperature changes was assessed on different timescales (monthly, annual, and multi-decadal) and in different atmospheric layers ($T^*_T$ and $T_{2LT}$).

On short timescales (month-to-month and year-to-year variations in temperature), the estimated tropospheric amplification of surface temperature changes was in good agreement in all model and observational data sets considered, and was in accord with basic theory. This is illustrated in Figure 5.6, which shows the standard deviations of monthly-mean $T_S$ anomalies plotted against the standard deviations of monthly-mean anomalies of $T_{2LT}$ (panel A) and $T^*_T$ (panel B). All model and observational results lie above the black line indicating equal temperature variability aloft and at the surface. All have similar "amplification factors" between their surface

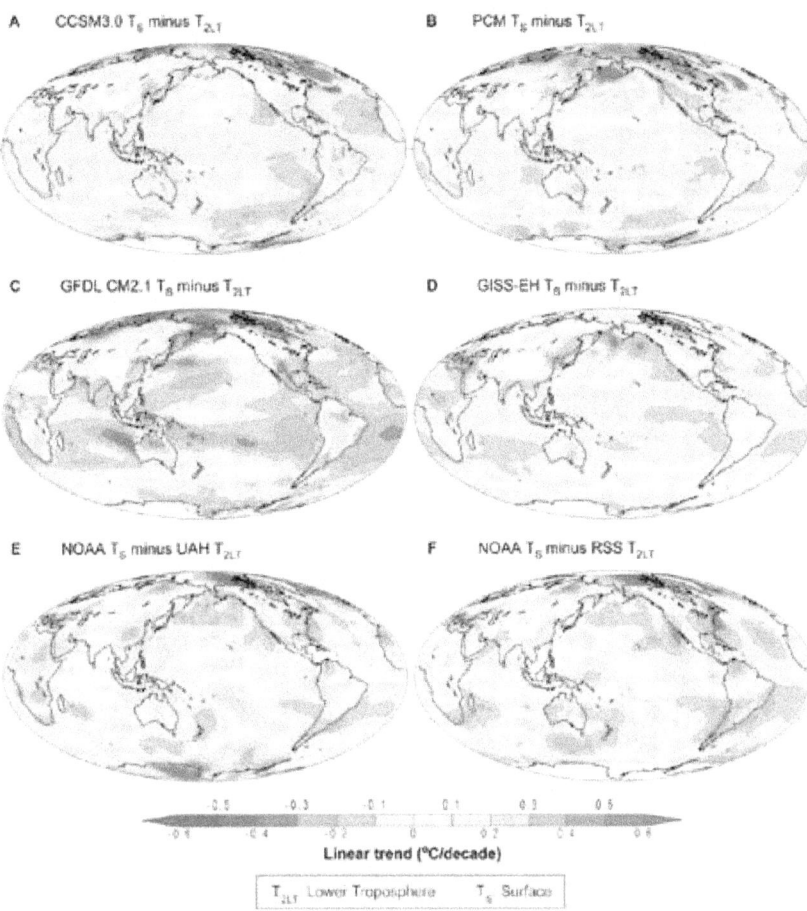

**Figure 5.5:** Modeled and observed maps of the differences between trends in $T_S$ and $T_{2LT}$. All trends in $T_S$ and $T_{2LT}$ were calculated over the 252-month period from January 1979 to December 1999. Model results are ensemble means from 20CEN experiments performed with CCSM3.0 (panel A), PCM (panel B), GFDL CM2.1 (panel C), and GISS-EH (panel D). Observed results rely on NOAA $T_S$ trends and on two different satellite estimates of trends in $T_{2LT}$, obtained from UAH (panel E) and RSS (panel F). White denotes high elevation areas where it is not meaningful to calculate synthetic $T_{2LT}$ (panels A-D). Note that RSS mask $T_{2LT}$ values in such regions, while UAH do not (compare panels E, F).

and tropospheric variability[63]. In the models, these similarities occur despite differences in physics, resolution, and forcings, and despite a large range (roughly a factor of 5) in the size of simulated temperature variability. In observations, the scaling ratios estimated from monthly temperature variability are relatively unaffected by the structural uncertainties discussed in Chapter 4.

---

[61] The essence of tropical atmospheric dynamics is that the tropics cannot support large temperature gradients, so waves (Kelvin, Rossby, gravity) even out the temperature field between convecting and non-convective regions. The temperature field throughout the tropical troposphere is more or less on the moist adiabatic lapse rate set by convection over the warmest waters. This is why there is a trade wind inversion where this profile finds itself inconsistent with boundary layer temperatures in the colder regions.

[62] The error was related to the UAH group's treatment of systematic drifts in the time of day at which satellites sample Earth's diurnal temperature cycle (see Chapter 4).

---

[63] Note that the slope of the red regression lines that has been fitted to the model results is slightly steeper for $T^*_T$ than for $T_{2LT}$ (c.f. panels 5.6A and 5.6B). This is because $T^*_T$ samples more of the mid-troposphere than $T_{2LT}$ (see Prospectus). Amplification is expected to be larger in the mid-troposphere than in the lower troposphere.

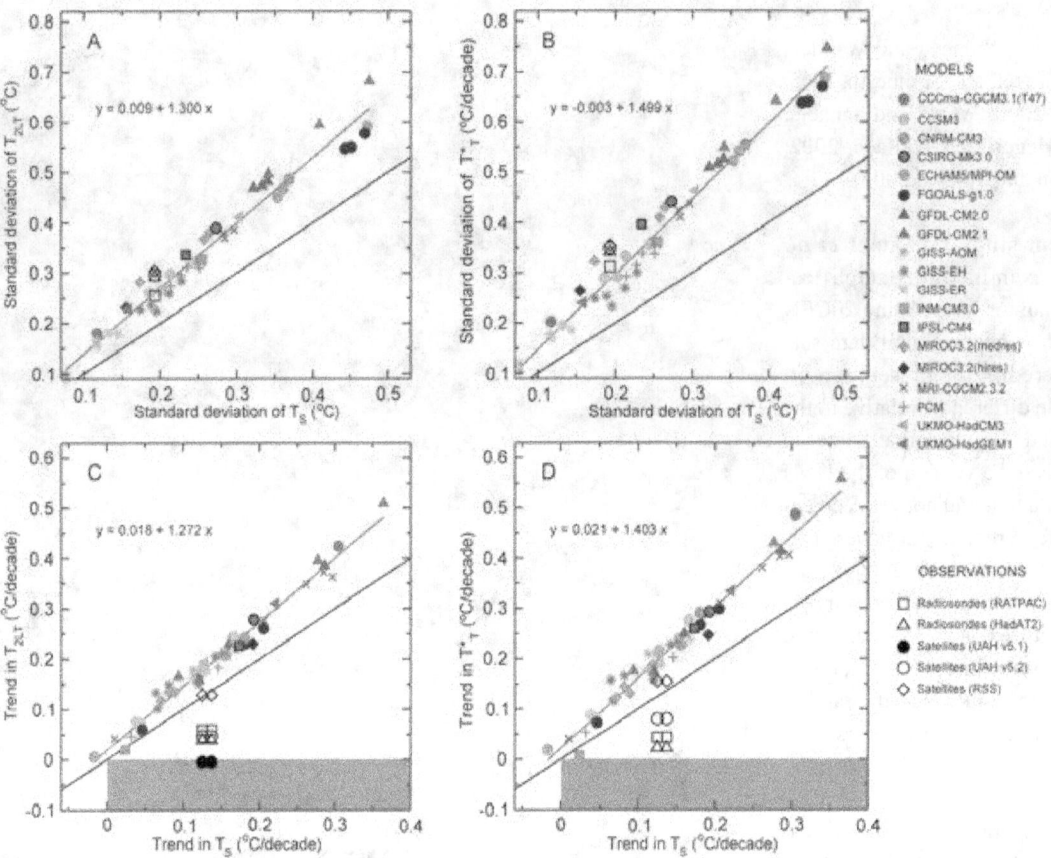

**Figure 5.6:** Scatter plots showing the relationships between tropical temperature changes at Earth's surface and in two different layers of the troposphere. All results rely on temperature data that have been spatially-averaged over the deep tropics (20°N-20°S). Model data are from 49 realizations of 20CEN runs performed with 19 different models (Table 5.1). Observational results were taken from four different upper-air datasets (two from satellites, and two from radiosondes) and two different surface temperature datasets (see Chapter 3). The two upper panels provide information on the month-to-month variability in $T_S$ and $T_{2LT}$ (panel A) and in $T_S$ and $T^*_T$ (panel B). The two bottom panels consider temperature changes on multi-decadal timescales, and show the trends (over 1979 to 1999) in $T_S$ and $T_{2LT}$ (panel C) and in $T_S$ and $T^*_T$ (panel D). The red line in each panel is the regression line through the model points. Its slope provides information on the amplification of surface temperature variability and trends in the free troposphere. The black line in each panel is given for reference purposes, and has a slope of 1. Values above (below) the black lines indicate tropospheric amplification (damping) of surface temperature changes. There are two columns of observational results in C and D. These are based on the NOAA and HadCRUT2v $T_S$ trends (0.12 and 0.14°C/decade, respectively). Note that panel C show results from published and recently-revised versions of the UAH $T_{2LT}$ data (versions 5.1 and 5.2). Since the standard deviations calculated from NOAA and HadCRUT2v monthly $T_S$ anomalies are very similar, observed results in A and B use NOAA standard deviations only. The blue shading in the bottom two panels defines the region of simultaneous surface warming and tropospheric cooling.

A different picture emerges if amplification behavior is estimated from decadal changes in tropical temperatures. Figures 5.6C and 5.6D show multi-decadal trends in $T_S$ plotted against trends in $T_{2LT}$ and $T^*_T$. The 20CEN runs exhibit amplification factors that are consistent with those estimated from month-to-month and year-to-year temperature variability[64]. Only

one observational upper-air data set (RSS) shows amplified warming aloft, and similar amplification relationships on short and on

---

[64] As in the case of amplification factors inferred from short-timescale variability, the factors estimated from multi-decadal temperature changes are relatively insensitive to inter-model differences in physics and the applied forcings (see Table 5.3). At first glance, this appears to be a somewhat surprising result in view of the large spatial and temporal heterogeneity of certain

forcings (see Section 3). Black carbon aerosols, for example, are thought to cause localized heating of the troposphere relative to the surface (Box 5.3), a potential mechanism for altering amplification behavior. The fact that amplification factors are similar in experiments that include and exclude black carbon aerosols suggests that aerosol-induced tropospheric heating is not destroying the connection of large areas of the tropical ocean to a moist adiabatic lapse rate. Single-forcing experiments (see Recommendations) will be required to improve our understanding of the physical effects of black carbon aerosols and other spatially-heterogeneous forcings on tropical temperature-change profiles.

long timescales. The other observational data sets have scaling ratios less than 1, indicating tropospheric damping of surface warming (Fu and Johanson, 2005; Santer *et al.*, 2005)[65].

These analyses shed further light on the differences between modeled and observed changes in tropical lapse rates described in Section 5.2. They illustrate the usefulness of comparing models and data on different timescales. On short timescales, it is evident that models successfully capture the basic physics that controls "real world" amplification behavior. On long timescales, model-data consistency is sensitive to structural uncertainties in the observations. One possible interpretation of these results is that in the real world, different physical mechanisms govern amplification processes on short and on long timescales, and models have some common deficiency in simulating such behavior. If so, these "different physical mechanisms" need to be identified and understood.

Another interpretation is that the same physical mechanisms control short- and long-term amplification behavior. Under this interpretation, residual errors in one or more of the observed data sets must affect their representation of long-term trends, and must lead to different scaling ratios on short and long timescales. This explanation appears to be the more likely one in view of the large structural uncertainties in observed upper-air data sets (Chapter 4) and the complementary physical evidence supporting recent tropospheric warming (see Section 6).

"Model error" and "observational error" are not mutually exclusive explanations for the amplification results shown in Figures 5.6C and D. Although a definitive resolution of this issue has not yet been achieved, the path towards such resolution is now more obvious. We have learned that models show considerable

consistency in terms of what they tell us about tropospheric amplification of surface warming. This consistency holds on a range of different timescales. Observations display consistent amplification behavior on short timescales, but radically different behavior on long timescales. Clearly, not all of the observed lapse-rate trends can be equally probable. Intelligent use of "complementary evidence" – from the behavior of other climate variables, from remote sensing systems other than MSU, and from more systematic exploration of the impacts of different data adjustment choices – should ultimately help us to constrain observational uncertainty, and reach more definitive conclusions regarding the true significance of modeled and observed lapse-rate differences.

### 5.5 Vertical Profiles of Atmospheric Temperature Change

Although formal fingerprint studies have not yet been completed with atmospheric temperature-change patterns estimated from the new 20CEN runs, it is instructive to make a brief qualitative comparison of these patterns. This helps to address the question of whether the inclusion of previously neglected forcings (like carbonaceous aerosols and land use/land cover changes; see Section 2) has fundamentally modified the "fingerprint" of human-induced atmospheric temperature changes searched for in previous detection studies.

We examine the zonal-mean profiles of atmospheric temperature change in 20CEN runs performed with four U.S. models (CCSM3, PCM, GFDL CM2.1, and GISS-EH). All four show a common large-scale fingerprint of stratospheric cooling and tropospheric warming over 1979 to 1999 (Figures 5.7A-D). The pattern of temperature change estimated from HadAT2 radiosonde data is broadly similar, although the transition height between stratospheric cooling and tropospheric warming is noticeably lower than in the model simulations (Figure 5.7E). Another noticeable difference is that the HadAT2 data show a relative lack of warming in the tropical troposphere,[66] where all four models simulate maximum warming. This particular aspect of the observed temperature-change pattern is very sensitive to data adjustments

*Models show considerable consistency in terms of what they tell us about tropospheric amplification of surface warming.*

---

[65] The previous version of the UAH $T_{2LT}$ data yielded a negative amplification factor for multi-decadal changes in tropical temperatures. The UMd data set, which exhibits greatest warming in $T_2$, has not to date produced a $T_{2LT}$ or $T^*_T$ product, and so could not be included in Figure 5.6. However, assuming an internally consistent set of channel records, the UMd data would show larger $T_{2LT}$ and $T^*_T$ trends than RSS, and would therefore have amplification factors consistent with or greater than those inferred from the models.

---

[66] Despite the "end point" effect of the large El Niño event in 1997-1998 (see Chapter 3).

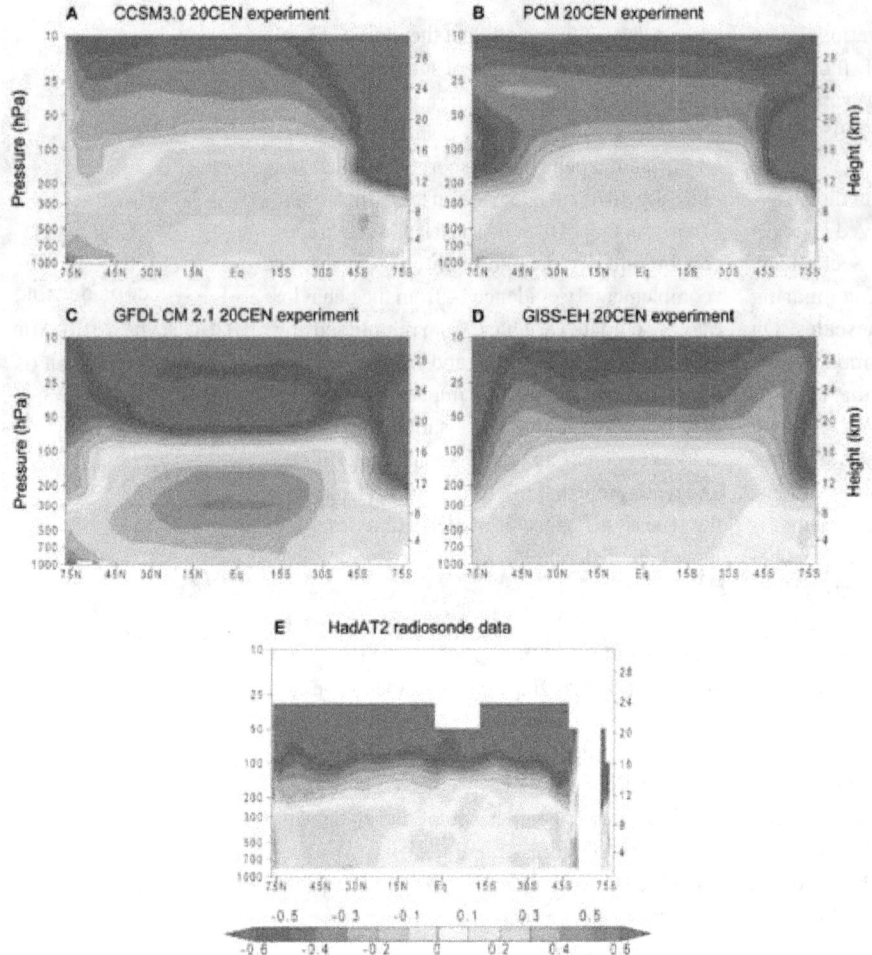

**Figure 5.7:** Zonal-mean patterns of atmospheric temperature change in "20CEN" experiments performed with four different climate models and in observational radiosonde data. Model results are for CCSM3.0 (panel A), PCM (panel B), GFDL CM 2.1 (panel C), and GISS-EH (panel D). The model experiments are ensemble means. There are differences between the sets of climate forcings that the four models used in their 20CEN runs (Table 5.3). Observed changes (panel E) were estimated with HadAT2 radiosonde data (Thorne et al., 2005, and Chapter 3). The HadAT2 temperature data do not extend above 30 hPa, and have inadequate coverage at high latitudes in the Southern Hemisphere. All temperature changes were calculated from monthly-mean data and are expressed as inear trends (in °C/decade) over 1979 to 1999.

cooling patterns, with largest cooling at high latitudes in the SH,[67] this asymmetry is less apparent in PCM and GISS-EH.

Future work should consider whether the conclusions of detection studies are robust to such fingerprint differences. This preliminary analysis suggests that the large-scale "fingerprint" of stratospheric cooling and tropospheric warming over the satellite era – a robust feature of previous detection work – has not been fundamentally altered by the inclusion of hitherto-neglected forcings like carbonaceous aerosols and LULC changes (see Table 5.3). This does not diminish the need to quantify the individual contributions of these forcings in appropriate "single forcing" experiments.

# 6. CHANGES IN "COMPLEMENTARY" CLIMATE VARIABLES

Body temperature is a simple metric of our physical well-being. A temperature of 40°C (104°F) is indicative of an illness, but does not by itself identify the cause of the illness. In medicine, investigation of causality typically requires the analysis of many different lines of evidence. Similarly, analyses of temperature alone provide incomplete information on the causes of climate change. For example, there is evidence that major volcanic eruptions affect not only the Earth's radiation budget (Wielicki et al., 2002; Soden et al., 2002) and atmospheric temperatures (Hansen et al., 1997, 2002; Free and Angell, 2002; Wigley et al., 2005a), but also water vapor (Soden et al., 2002), precipitation (Gillett et al., 2004c), atmospheric circulation patterns

(Sherwood et al., 2005; Randel and Wu, 2006). Tropospheric warming in the observations is most obvious in the NH extra-tropics, where our confidence in the reliability of radiosonde records is greatest.

Note that some of the details of the model fingerprint pattern are quite different. For example, GFDL's cooling maximum immediately above the tropical tropopause is not evident in any of the other models. Its maximum warming in the upper tropical troposphere is noticeably larger than in CCSM3.0, PCM, or GISS-EH. While CCSM and GFDL CM2.1 have pronounced hemispheric asymmetry in their stratospheric

---

[67] This may be related to an asymmetry in the pattern of stratospheric ozone depletion: the largest ozone decreases over the past two to three decades have occurred at high latitudes in the SH.

(see, *e.g.,* Robock, 2000, and Ramaswamy *et al.,* 2001a; Robock and Oppenheimer, 2003), ocean heat content and sea level (Church *et al.,* 2005), and even global-mean surface pressure (Trenberth and Smith, 2005). These responses are physically interpretable and internally consistent[68]. The combined evidence from changes in all of these variables makes a stronger case for an identifiable volcanic effect on climate than evidence from a single variable only.

A "multi-variable" perspective may also be beneficial in understanding the possible causes of differential warming. The value of "complementary" climate data sets for studying this specific problem has been recognized by Wentz and Schabel (2000) and by Pielke (2004). The former found internally consistent increases in SST, $T_{2LT}$, and marine total column water vapor over the 12-year period from 1987 to 1998[69]. Multi-decadal increases in surface and lower tropospheric water vapor were also reported in the IPCC Second Assessment Report (Folland *et al.,* 2001).[70] More recently, Trenberth

*et al.* (2005) found significant increases in total column water vapor over the global ocean[71]. At constant relative humidity, water vapor increases non-linearly with increasing temperature (Hess, 1959). Slow increases in tropospheric water vapor therefore provide circumstantial evidence in support of tropospheric warming. However, water vapor measurements are affected by many of the same data quality and temporal homogeneity problems that influence temperature measurements (Elliott, 1995; Trenberth *et al.,* 2005), so the strength of this circumstantial evidence is still questionable[72].

Other climate variables also corroborate the warming of Earth's surface over the second half of the 20th century. Examples include increases in ocean heat content (Levitus *et al.,* 2000, 2005; Willis *et al.,* 2004), sea-level rise (Cabanes *et al.,* 2001), thinning of major ice sheets and ice shelves (Krabill *et al.,* 1999; Rignot and Thomas, 2002; Domack *et al.,* 2005), and widespread glacial retreat, with accelerated rates of glacial retreat over the last several decades (Arendt *et al.,* 2002; Paul *et al.,* 2004)[73].

Changes in some of these "complementary" variables have been used in detection and attribution studies. Much of this work has focused on ocean heat content. When driven

A "multi-variable" perspective may also be beneficial in understanding the possible causes of differential warming. At constant relative humidity, water vapor increases non-linearly with increasing temperature.

---

[68] The physical consistency between the temperature and water vapor changes after the Pinatubo eruption has been clearly demonstrated by Soden *et al.* (2002). The surface and tropospheric cooling induced by Pinatubo caused a global-scale reduction in total column water vapor. Since water vapor is a strong GHG, the reduction in water vapor led to less trapping of outgoing thermal radiation by Earth's atmosphere, thus amplifying the volcanic cooling. This is referred to as a "positive feedback." Soden *et al.* "disabled" this feedback in a climate model experiment, and found that the "no water vapor feedback" model was incapable of simulating the observed tropospheric cooling after Pinatubo. Inclusion of the water vapor feedback yielded close agreement between the simulated and observed $T_{2LT}$ responses to Pinatubo. This suggests that the model used by Soden *et al.* captures important aspects of the physics linking the real world's temperature and moisture changes.

[69] The Wentz and Schabel study used NOAA optimally interpolated SST data, a version of the UAH $T_{2LT}$ data that had been corrected for orbital decay effects, and information on total column water vapor from the satellite-based Special Sensor Microwave Imager (SSM/I).

[70] More specifically, Folland *et al.* (2001) concluded, "Changes in water vapor mixing ratio have been analyzed for selected regions using *in situ* surface observations as well as lower-tropospheric measurements based on satellites and weather balloons. A pattern of overall surface and lower-tropospheric water vapor mixing ratio increases over the past few decades is emerging, although there are likely to be some time-dependent biases in these data and regional variations in trends. The more reliable data sets show that it is likely that total atmospheric water vapor has increased several percent per decade over

many regions of the Northern Hemisphere since the early 1970s. Changes over the Southern Hemisphere cannot yet be assessed."

[71] Trenberth *et al.* (2005) reported an increase in total column water vapor over 1988 to 2001 of "1.3 ± 0.3% per decade for the ocean as a whole, where the error bars are 95% confidence intervals." This estimate was obtained with an updated version of the SSM/I data set analyzed by Wentz and Schabel (2000).

[72] Note, however, that SSM/I-derived water vapor measurements may have some advantages relative to temperature measurements obtained from MSU. Wentz and Schabel (2000) point out that (under a constant relative humidity assumption), the 22 GHz water vapor radiance observed by SSM/I is three times more sensitive to changes in air temperature than the MSU $T_2$ 54 GHz radiance. Furthermore, while drift in sampling the diurnal cycle influences MSU-derived tropospheric temperatures (Chapter 4), it has a much smaller impact on SSM/I water vapor measurements.

[73] Folland *et al.* (2001) note that "Long-term monitoring of glacier extent provides abundant evidence that tropical glaciers are receding at an increasing rate in all tropical mountain areas". Accelerated retreat of high-elevation tropical glaciers is occurring within the tropical lower tropospheric layer that is a primary focus of this report, and provides circumstantial support for warming of this layer over the satellite era.

by anthropogenic forc-
ing, a number of dif-
ferent CGCMs capture
the overall increase in
observed ocean heat
content estimated by
Levitus *et al.* (2000;
2005), but not the large
decadal variability in
heat content (Barnett
*et al.,* 2001; Levitus *et
al.,* 2001; Reichert *et
al.,* 2002; Sun and Han-
sen, 2003; Pielke, 2003;
Gregory *et al.,* 2004;
Hansen *et al.,* 2005b)[74].
It is still unclear whether this discrepancy be-
tween simulated and
observed variability is
primarily due to model
deficiencies or is an ar-
tifact of how Levitus *et
al.* (2000; 2005) "infilled" data-sparse ocean
regions (Gregory *et al.,* 2004; AchutaRao *et
al.,* 2006).

In summary, the behavior of complementary
variables enhances our confidence in the reality
of large-scale warming of the Earth's surface,
and tells us that the signature of this warming
is manifest in many different aspects of the
climate system. Pattern-based fingerprint detection work performed with ocean heat content
(Barnett *et al.,* 2001; Reichert *et al.,* 2002; Barnett *et al.,* 2005; Pierce *et al.,* 2006), sea-level
pressure (Gillett *et al.,* 2003), and tropopause
height (Santer *et al.,* 2003a, 2004)[75] suggests

that anthropogenic forcing is necessary in order
to explain observed changes in these variables.
This supports the findings of the surface- and
atmospheric temperature studies described in
Section 4.4. To date, however, investigations
of complementary variables have not enabled
us to narrow uncertainties in satellite- and
radiosonde-based estimates of tropospheric
temperature change over the past 2-3 decades.
Formal detection and attribution studies involving water vapor changes may be helpful in this
regard, since observations suggest a recent
moistening of the troposphere, consistent with
tropospheric warming.

## 7. SUMMARY

This chapter has evaluated a wide range of
scientific literature dealing with the possible
causes of recent temperature changes, both at
the Earth's surface and in the free atmosphere.
It shows that many factors – both natural and
human-related – have probably contributed to
these changes. Quantifying the relative importance of these different climate forcings is a
difficult task. Analyses of observations alone
cannot provide us with definitive answers. This
is because there are important uncertainties in
the observations and in the climate forcings
that have affected them. Although computer
models of the climate system are useful in
studying cause-effect relationships, they, too,
have limitations. Advancing our understanding
of the causes of recent lapse-rate changes will
best be achieved by comprehensive comparisons of observations, models, and theory – it is
unlikely to arise from analysis of a single model
or observational data set.

---

[74] Model control runs cannot generate such large
multi-decadal increases in the heat content of the
global ocean.

[75] The tropopause is the transition zone between the
turbulently-mixed troposphere, where most weather
occurs, and the more stably-stratified stratosphere (see
Preface and Chapter 1). Increases in tropopause height
over the past 3-4 decades represent an integrated
response to temperature changes above and below
the tropopause (Highwood *et al.,* 2000; Santer *et al.,*
2004), and are evident in both radiosonde data (Highwood *et al.,* 2000; Seidel *et al.,* 2001) and reanalyses
(Randel *et al.,* 2000). In model 20CEN simulations,
recent increases in tropopause height are driven by
the combined effects of GHG-induced tropospheric
warming and ozone-induced stratospheric cooling
(Santer *et al.,* 2003a). Available reanalysis products do
not provide a consistent picture of the relative contributions of stratospheric and tropospheric temperature

changes to recent tropopause height increases (Pielke
and Chase, 2004; Santer *et al.,* 2004).

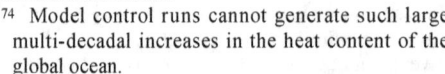

# CHAPTER 6

# What measures can be taken to improve our understanding of observed changes?

***Convening Lead Author:*** Chris K. Folland, U.K. Met. Office

***Lead Authors:*** D.E. Parker, U.K. Met. Office; R.W. Reynolds, NOAA; S.C. Sherwood, Yale Univ.; P.W. Thorne, U.K. Met. Office

## BACKGROUND

There remain differences between independently estimated temperature trends for the surface, troposphere and lower stratosphere, and differences between the observed changes and model simulations, that are, as yet, not fully understood, although recent progress is reported in previous chapters. This Chapter makes recommendations that address these specific problems rather than more general climate research aims, building on the discussions, key findings, and recommendations of the previous chapters. Because the previous chapters fully discuss the many issues, we only provide a summary here. Furthermore, we only list key references to the peer reviewed literature. To ensure traceability and to enable easy cross-referencing we refer to the chapters by e.g., (C5) for Chapter 5. We do not specifically refer to sub-sections of chapters.

Much previous work has been done to address, or plan to address, most of the problems discussed in this Report. Rather than invent brand new proposals and recommendations, we have tried to expand and build upon existing ideas emphasizing those we believe to be of highest utility. Key documents in this regard are: the Global Climate Observing System (GCOS) Implementation Plan for the Global Observing System (GCOS, 2004), the wider Global Earth System of Systems (GEOSS) 10 year Implementation Plan Reference Document (GEOSS, 2005) which explicitly includes the GCOS Implementation Plan as its climate component; and the over-arching Climate Change Science Program plan (CCSP, 2004).

The remainder of this Chapter is split into six sections. Each section discusses requirements under a particular theme, aiming to encapsulate the key findings and recommendations of the earlier chapters and culminating in one main recommendation in each of Sections 1 to 5 and two recommendations in Section 6. Sections 1 to 5 focus on key actions that should be carried out in the near future, making use of existing historical data and current climate models. Section 6 discusses future climate monitoring in relation to the vertical profile of temperature trends in the atmosphere. Figure 6.1 summarizes the recommendations and links them to the overarching aim of a better understanding of the vertical profile of temperature trends and their variations on all important space and time scales.

## I. CONSTRAINING OBSERVATIONAL UNCERTAINTY

An important advance since recent in-depth reviews of the subject of this Report (NRC, 2000a; IPCC, 2001) has been a better appreciation of the uncertainties in our estimates of recent temperature changes, particularly above the surface (C2, C3, C4). Many observations that are used in climate studies are taken primarily for the purposes of operational weather forecasting (C2). Not surprisingly, there have been numerous changes in instrumentation, observing practices, and the processing of data over time. While these changes have undoubtedly led to improved forecasts of weather, they add significant complexity to attempts to reconstruct past climate trends, (C2, C4). The

main problem is that such an evolution tends to introduce artificial (non-climatic) changes into the data (C2).

Above the Earth's surface, the spread in independently-derived estimates of climate change, representing what is referred to in this report as "construction" uncertainty (C2, C4, Appendix) (Thorne *et al.,* 2005), is of similar magnitude to the expected climate signal itself (C3, C4, C5). Changes in observing practices have been particularly pervasive aloft, where the technical challenges in maintaining robust, consistent measurements of climate variables are considerably greater than at the surface (C2, C4, C5). This does not imply that there are no problems in estimating temperature trends at the surface. Such problems include remaining uncertainties in adjustments that must be made to sea

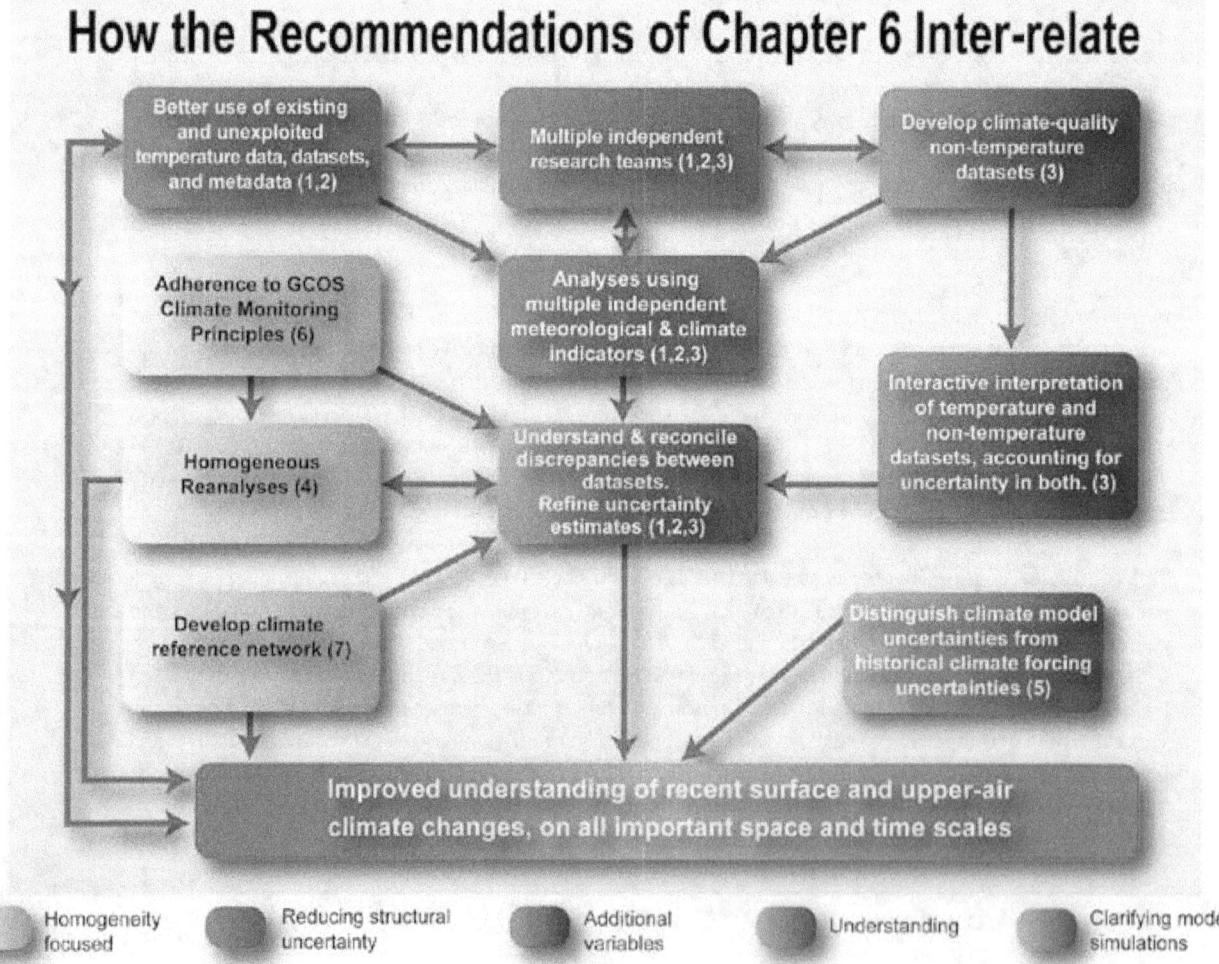

**Figure 6.1** Schematic showing how recommendations inter-relate. Recommendations relating to each box are indicated in parentheses.

surface temperatures (SSTs) in recent decades (C2, C4), and uncertainties in accounting for changes in micro-climate exposure for some individual land stations (C2, C4) or simply allowing for genuinely bad stations (Davey and Pielke, 2005). Differences between surface data sets purporting to measure the same variable become larger as the spatial resolution being considered decreases. This implies that many problems tend to have random effects on climate analyses at the large spatial scales, that are the focus of this Report, but can be systematic at much smaller scales (C2, C3, C4).

The climate system has evolved in a unique way, and, by definition the best analysis is that which most closely approaches this actual evolution. However, because we do not know the evolution of the climate system exactly, we have generally had to treat apparently well constructed but divergent data sets, of atmospheric temperature changes in particular, as equally valid (C3, C4, C5). Clearly, this approach is untenable in the longer-term. Thus, it is imperative that we reduce the uncertainty in our knowledge of how the three-dimensional structure of atmospheric temperature has evolved (C4).

To ascertain unambiguously the causes of differences in data sets generally requires extensive metadata[1] for each data set (C4; NRC, 2000b). Appropriate metadata, whether obtained from the peer-reviewed literature or from data made available on-line, should include, for data on all relevant spatial and temporal scales:

*   Documentation of the raw data and the data sources used in the data set construction to enable quantification of the extent to which the raw data overlap with other similar data sets;
*   Details of instrumentation used, the observing practices and environments and their changes over time to help assessments of, or adjustments for, the changing accuracy of the data;
*   Supporting information such as any adjustments made to the data and the numbers and locations of the data through time;
*   An audit trail of decisions about the adjustments made, including supporting evidence

that identifies non-climatic influences on the data and justifies any consequent adjustments to the data that have been made; and
*   Uncertainty estimates and their derivation.

This information should be made openly available to the research community.

There is evidence, discussed in earlier chapters, for a number of unresolved issues in existing data sets that should be addressed:

*   Systematic, historically varying biases in day-time relative to night-time radiosonde temperature data are important, particularly in the tropics (C4). These are likely to have been poorly accounted for by present approaches to quality controlling such data (Sherwood *et al.*, 2005) and may seriously affect trends.
*   Radiosonde stratospheric records are strongly suspected of retaining a spurious long-term cooling bias, especially in the tropics (C4).
*   Diurnal adjustment techniques for satellite temperature data are uncertain (C2, C4). This effect is particularly important for the $2_{LT}$ retrieval (C4). Further efforts are required to refine our quantification of the diurnal cycle, perhaps through use of reanalyses, in-situ observations, or measurements from non-sun-synchronous orbiters (C4).
*   Different methods of making inter-satellite bias adjustments, particularly for satellites with short periods of overlap, can lead to large discrepancies in trends (C4) (see also Section 6).
*   Variable biases in modern SST data remain that have not been adequately addressed (C4). Some historical metadata are now available for the first time, but are yet to be fully exploited (Rayner *et al.*, 2006). Better metadata, better use of existing metadata, and use of recently bias-adjusted day-time marine air temperature data are needed to assess remaining artifacts (C4).
*   Land stations may have had undocumented changes in the local environment that could lead to their records being unrepresentative of regional- or larger-scale changes (C2, C4).

Much previous work has been done to address, or plan to address, most of the problems discussed in this Report. Rather than invent brand new proposals and recommendations, we have tried to expand and build upon existing ideas emphasizing those we believe to be of highest utility.

In addition to making data sets and associated metadata openly available and addressing the

---

[1]   Metadata are literally "data about data" and are typically records of instrumentation used, observing practices, the environmental context of observations, and data-processing procedures.

issues discussed above, it would be useful to develop a set of guidelines that can be used to help assess the quality of data sets (C4). It is important that numerous tests be applied to reduce ambiguity. There are three types of check that may be used:

**1. Internal consistency checks**

For example, we expect only relatively small real changes in the diurnal cycle of temperature above the atmospheric boundary layer (C1) (Sherwood *et al.,* 2005), so an apparently homogenized data set that shows large changes in the diurnal cycle in these regions should be closely scrutinized.

**2. Inter-data set comparisons**

For example, comparisons are needed between radiosonde and MSU temperature measures representing the same regions (Christy and Norris, 2004).

**3. Consistency with changes in other climate variables and parameters**

This is a potentially powerful but much under-utilized approach and is discussed further in Section 3.

---

### Recommendation I

The independent development of data sets and analyses by several independent scientists or teams will serve to quantify structural uncertainty and to provide objective corroboration of the results. In order to encourage further independent scrutiny, data sets and their full metadata (footnote I) should be made openly available. Comprehensive analyses should be carried out to ascertain the causes of remaining differences between data sets and to refine uncertainty estimates.

---

## 2. MAKING BETTER USE OF EXISTING OBSERVATIONAL DATA

There is a considerable body of observational data that have either been under-utilized or not used at all when constructing the data sets of historical temperature changes discussed

in this Report (C2, Table 2.1). Estimates of temperature changes can potentially be made from several satellite instruments beside the (Advanced) Microwave Sounding Unit data considered here (C2, C3). In particular, largely overlooked satellite data sets should be re-examined to try to extend, fortify or corroborate existing microwave-based temperature records for climate research, *e.g.,* microwave data from other instruments such as the Nimbus 5 (Nimbus E) Microwave Spectrometer (NEMS) (1972) and the Nimbus 6 Scanning Microwave Spectrometer (SCAMS) (1975), infra-red data from the High Resolution Infrared Radiation Sounder (HIRS) suite, and radio occultation data from Global Positioning System (GPS) satellites (C2). Some of these instruments may allow us to extend the records back to the early 1970s. Many unused radiosonde measurements of a relatively short length exist in regions of relatively sparse coverage and, with some effort, could be advantageously used to fill gaps. Many additional surface temperature data exist, mainly over land over the period considered in this Report, but are either not digitized or not openly available. This latter problem is particularly common in many tropical regions where much of the interest in this Report resides. Given the needed level of international cooperation, we could significantly improve our current estimates of tropical temperature changes over land and derive better estimates of the changing temperature structure of the lower atmosphere (C2).

In addition to the recovery and use of such existing data, we need to improve the access to metadata for existing raw observations (C2). Additional information on when and how changes occurred in observing practices, the local environment, etc., is potentially available in national meteorological and hydrometeorological services. Such metadata would help reduce current uncertainties in estimates of observed climate change. In the absence of comprehensive metadata, investigators have to make decisions regarding the presence of heterogeneities (non-climatic jumps or trends) using statistical methods alone. Statistical methods of adjusting data for inhomogeneities have a very useful role, but are much more valuable in the presence of good and frequent metadata that can be used to confirm the presence, type, and timing of non-

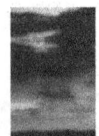

climatic influences. Metadata requirements will vary according to observing system, but, if in doubt, all potentially important information should be included. For example, surface temperature metadata may include:

- Current and historical photographs and site sketches to ascertain changes in micro-climate exposure and their timing, collected during the routine site inspections made by most meteorological services;
- The history of instrumentation changes;
- Changes in the way stations are maintained and in their immediate environment;
- Changes in observers; and
- Changes in observing and reporting practices.

For other instrument types, *e.g.*, for humidity measurements, the detailed metadata requirements will vary. A further discussion on the challenges of collecting climate data can be found in Folland *et al.* (2000).

---

**Recommendation 2**

Efforts should be made to archive and make openly available for independent analysis surface, balloon-based, and satellite data and metadata that have not previously been exploited. Emphasis should be placed on the tropics, and on recovery and inclusion of satellite data before 1979, which may allow better characterization of the climate regime shift in the mid-1970s.

---

## 3. MULTIVARIATE ANALYSES

Temperature changes alone are a necessary, but insufficient, constraint on understanding the evolution of the climate system. Even with a perfect knowledge of temperature changes, knowledge about changes in the climate system would be incomplete. Consequently, understanding temperature trends also requires knowledge about changes in other measures of the climate system. For example, changes in atmospheric circulation and accompanying dynamical effects, and also in latent heat transport, have significant implications for vertical profiles of temperature trends (C1).

Changes in variables other than temperature may be used to confirm the attribution of climate change to given causes (C5) and to test the physical plausibility of reported temperature changes (C3, C4). It is likely that to fully understand changes in atmospheric temperature, it will be necessary to consider changes in at least some of the following physical parameters and properties of the climate system beside its temperature:

- Water vapor content (C1, C5)
- Ocean heat content (C5)
- The height of the tropopause (C5)
- Wind fields
- Cloud cover and the characteristics of clouds
- Radiative fluxes
- Aerosols and trace gases
- Changes in glacial mass, sea ice volume, permafrost and snow cover (C5)

Our current ability to undertake such multivariate analyses of climate changes is constrained by the relative paucity of accurate climate data sets for variables other than temperature. Furthermore, since our analysis of temperature data sets has highlighted the importance of construction uncertainty in determining trends (C2, C4, Appendix A), it is very likely that similar considerations will pertain to these other data types. It is therefore necessary to construct further independent estimates of the changes in these variables even where data sets already exist. Similar considerations to those discussed in Section 1 are also important for these additional data.

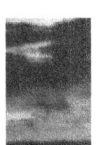

---

**Recommendation 3**

Efforts should be made to develop or reprocess[a] data sets for a range of variables other than temperature, creating climate quality[b] analyses. These should subsequently be compared with each other and with temperature data to determine whether they are consistent with our physical understanding. It is important to create several independent estimates for each parameter in order to assess the magnitude of construction uncertainties.

[a] See http://copes.ipsl.jussieu.fr/organization/COPESStructure/WGOA.html
[b] "Climate quality" refers to a record for which the best possible efforts have been made to identify and remove non-climatic effects that produce spurious changes over time. (NRC, 2004)

## 4. CLIMATE QUALITY REANALYSES

Reanalyses are derived from Numerical Weather Prediction (NWP) (forecast) models run retrospectively with historical observations to produce physically consistent, fully global fields with high temporal and spatial resolution. As in NWP, reanalyses employ all available observations to produce their analysis and minimize the instantaneous differences between the available observations and a background forecast field initiated a number of hours earlier. Reanalyses also use the same NWP model throughout the reanalysis period. However, as for observed climate data sets, pervasive changes in the raw observations lead to discontinuities and spurious drifts (C2). Because such discontinuities and drifts have been identified in the temperature fields of the current generation of reanalyses, these have been deemed inappropriate for the purpose of long-term temperature trend characterisation by this Report's authors (C2, C3). However, it is recognised that some progress has been made (*e.g.,* Simmons *et al.,* 2004, C2). This does not preclude the usefulness of reanalyses for characterizing seasonal to interannual timescale variability and processes, or trends in other, related, variables such as tropopause height (C5). Indeed, they have proven to be a very important tool for the climate research community.

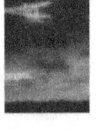

A more homogeneous reanalysis that minimized time-dependent biases arising from changes in the observational network would be of enormous benefit for multivariate analyses of climate change (C2, C3). Advances in NWP systems, which will continue to happen regardless of climate requirements, will in the future inevitably lead to better reanalyses of interannual climate variability. Some advances, such as so-called 'feedback files'[2] from the data assimilation of reanalyses, could be uniquely helpful for climate reanalysis and should be encouraged for this reason if no other. However, to determine trends accurately from reanalyses will also require intensive efforts by the reanalysis community to understand which observations are critical for trend characterization and to homogenize these data insofar as possible to

eliminate non-climatic changes before input to the reanalysis system. This in turn requires observing system experiments where the impact on trends of new or different observation types from land, radiosonde, and space-based observations are assessed. A few possible examples (far from an exhaustive list) are:

- Successively include or remove specific satellite retrievals (*e.g.,* MSU Channel 2).
- Carry out test reanalyses for one or more decades with different adjustments to the observed data for inhomogeneities within their construction uncertainty estimates.
- Run a short period (*e.g.,* a year) of reanalysis with and without radiosondes.

Progress would depend on reanalyses and data construction experts from all the key groups working closely together.

### Recommendation 4

Consistent with Key Action 24 of GCOS (2004)[a] and a 10 Year Climate Target of GEOSS (2005), efforts should be made to create several homogeneous atmospheric reanalyses. Particular care needs to be taken to identify and homogenize critical input climate data, and to more effectively manage large-scale changes in the global observing system to avoid non-climatic influences[b].

[a] Parties are urged to give high priority to establishing a sustained capacity for global climate reanalysis, to develop improved methods for such reanalysis, and to ensure coordination and collaboration among Centers conducting reanalyses.

[b] A focal point for planning of future U.S. reanalysis efforts is the CCSP Synthesis and Assessment Product 1.3: "Re-analyses of historical climate data for key atmospheric features. Implications for attribution of causes of observed change." Ongoing progress in the planning of future U.S. reanalysis efforts can be found at: http://www.joss.ucar.edu/joss_psg/meetings/climatesystem/

---

[2] "Feedback files" are diagnostic summaries of adjustments applied to data during their assimilation.

## 5. BETTER UNDERSTANDING OF UNCERTAINTIES IN MODEL ESTIMATES

New state-of-the-art global climate models have simulated the influences of natural and anthropogenic climate forcings on tropospheric and surface temperature. The simulations generally cover the period since the late nineteenth century, but results are only reported over the period of primary interest to this Report, 1979-1999 (the satellite era), in Chapter 5. Taken together, these models, for the first time, consider most of the recognized first-order climate forcings and feedbacks as identified in IPCC (2001), NRC (2003), and NRC (2005). This is an important step forward (C5).

However, most individual models considered in this Report still do not make use of all likely important climatic forcings (C5, Table 5.2). In addition, many of the forcings are not yet well quantified. Models that appear to include the same forcings often differ in both the way the forcings are quantified and how these forcings are applied to the model. This makes it difficult to separate intrinsic differences between models from the effects of different forcings on predicted temperature trends. Thus, within the "ensemble of opportunity" considered in this Report (C5), it is difficult to separate differences in:

- Model physics and resolution;
- The details of the way the forcings are applied in the experiments;
- The chosen history of the changes in the forcing.

To better quantify the impacts of the various forcings on vertical temperature trends, a further suite of experiments is needed along the following lines:

- Runs with one forcing applied in a single experiment with a given model; these are already required in some detection and attribution studies (C5). They have been performed for a small number of models already. This approach is particularly important for the recently developed and spatially heterogeneous land use / land cover change and black carbon aerosol forcings (C5).
- Apply the same forcing in exactly the same manner to a suite of models so that the dif-

ferences that result are due unambiguously to model differences (C5).
- Apply the full range of important forcings, with their uncertainties explicitly sampled to a small subset of the most advanced models to gain an overall estimate of the effects on temperature trends of the uncertainties in these forcings.

It is recognized that there are many problems in achieving this, so a considerable effort will be needed over a number of years. In addition, these model runs should be compared to the full range of observational estimates to avoid ambiguity (C5). Finally, detection and attribution studies should be undertaken using this new range of observations and model-based estimates to refine our understanding of human-induced influences on climate (C5).

> ### Recommendation 5
>
> Models that appear to include the same forcings often differ in both the way the forcings are quantified and how these forcings are applied to the model. Hence, efforts are required to more formally separate uncertainties arising from model structure from the effects of forcing uncertainties. This requires running multiple models with standardized forcings, and running the same models individually under a range of plausible scenarios for each forcing.

## 6. FUTURE MONITORING OF CLIMATE

Much of this Report hitherto has concerned historical climate measurements. However, over the coming decades new, mainly space-based, observations will yield very large increases in the volume and types of data available. These will come from many different instruments making measurements with greater accuracy and detail, especially in the vertical direction, and with greater precision (C2, C3). In fact, new types of more accurate data such as temperature and moisture profiles from GPS radio-occultation measurements are already available, although, as yet, few efforts have been made

to analyse them (C2, C3). Current and planned multi-spectral infra-red satellite sounders such as the Atmospheric InfraRed Sounder (AIRS) and the Infrared Atmospheric Sounding Interferometer (IASI) have much finer vertical resolution than earlier satellite sounders used in the Report. They have the potential to resolve quite fine vertical and horizontal details of temperature and humidity through the depth of much of the atmosphere. These higher spectral resolution data should also permit a continuation of records equivalent to earlier coarser infrared satellite data (*e.g.,* from the HIRS satellite instruments). The new suite of satellite data will not only prove useful for sensing changes aloft. For example, satellite data to remotely sense sea-surface temperatures now include microwave products that can sense surface temperatures even in cloudy conditions (C4). The Global Ocean Data Assimilation Experiment (GODAE) High-resolution Sea Surface Temperature (SST) Pilot Project (GHRSST-PP) has been established to give international focus and coordination to the development of a new generation of global, multi-sensor, high-resolution SST products (Donlon *et al.,* 2005).

Many other agencies and bodies (*e.g.,* NRC, 2000b; GCOS, 2004; GEOSS, 2005; CCSP, 2004) have already made recommendations for managing such new data developments. These include such subjects as:

- Adherence to the GCOS Climate Monitoring Principles, needed to create and maintain homogenous data sets of climate quality and for which there is a special set for satellites (GCOS, 2004, Appendix 3)
- Continuation of records equivalent to current monitoring abilities: *e.g.,* use new and more detailed satellite data to create equivalent MSU measures of temperature to allow the indefinite extension of the historical records used in this Report.
- Full implementation of national and international climate monitoring networks such as the GCOS Upper-Air Network and the GCOS Surface Network.
- Overlap of measurement systems as they evolve in time.

This last point is of primary importance. It was given prominence by NRC (2000b) and is emphasized in the GCOS Climate Monitoring

Principles and leads to the following recommendation. If this recommendation had been followed in the past, one of the major problems in producing a homogeneous record of MSU temperatures would have been largely removed (C4):

## Recommendation 6

The GCOS Climate Monitoring Principles should be fully adopted. In particular when any type of instrument for measuring climate is changed or re-sited, the period of overlap between the old and new instruments or configurations should be sufficient to allow analysts to adjust for the change with small uncertainties that do not prejudice the analysis of climate trends. The minimum period is a full annual cycle of the climate. Thus, replacement satellite launches should be planned to take place at least a year prior to the expected time of failure of a key instrument.

Finally, we expand on a recommendation made in GCOS (2004) that is imperative for successful future monitoring of temperatures at and above the Earth's surface. The main lesson learned from this Report is that great difficulties in identifying and removing non-climatic influences from upper-air observations have led to a very large spread in trend estimates (C2, C3, C4). These differences can lead to fundamentally different interpretations both of the extent of any discrepancies in trends between the surface and the troposphere (C3,C4); and of the skill of climate models (C5). The problem has arisen because there has been no high quality reference or "ground truth" data, however restricted in scope, against which routine observations can be compared to facilitate rigorous removal of non-climatic influences.

Our key recommendation in this regard is a set of widely distributed (perhaps about 5% of the operational radiosonde network) reference sites that will provide high quality data for anchoring more globally-extensive monitoring efforts (satellites, reanalyses, etc.). At such reference

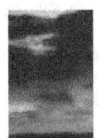

sites (which could coincide with selected GCOS Upper Air Network [GUAN], GCOS Surface Network [GSN] or Global Atmospheric Watch [GAW] sites) there would be full, high-quality measurements of atmospheric column properties, both physical and chemical. This requires a large suite of instrumentation and redundancy in measurements[3]. These globally distributed reference sites should incorporate upward looking instruments (radar, lidar, GPS-related data, microwave sensors, wind profilers, etc.) along with high-quality temperature, relative humidity and wind measurements on balloons regularly penetrating well into the stratosphere[4] A key requirement is an end-to-end management system including archiving of coincident observations made from over-flying satellites. The data need to be made openly available. The development of such a reference network is recommended in outline by GCOS (2004). The ideas are currently being discussed in more detail as part of an on-going process led by NOAA and WMO. Further details can be found at http://www.oco.noaa.gov/workshop/.

---

### Recommendation 7

Following Key Action 12 of the GCOS Implementation Plan[a] (GCOS, 2004), develop and implement a subset of about 5% of the operational radiosonde network as reference network sites for all kinds of climate data from the surface to the stratosphere.

[a] Parties need to: ... establish a high-quality reference network of about 30 precision radiosonde stations and other collocated observations.

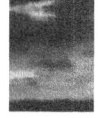

---

[3] Measurement of the same parameter by two or more independent instruments

[4] Recent inter-comparisons under the auspices of WMO suggest that new operational sondes are as accurate as proposed reference sondes (C4; Nash *et al.*, 2005), which may reduce costs.

APPENDIX A

# Statistical Issues Regarding Trends

**Author:** Tom M.L. Wigley, NSF NCAR

*With contributions by:* Benjamin D. Santer, DOE LLNL; John R. Lanzante, NOAA

## Abstract

The purpose of this Appendix is to explain the statistical terms and methods used in this Report. We begin by introducing a number of terms: mean, standard deviation, variance, linear trend, sample, population, signal, and noise. Examples are given of linear trends in surface, tropospheric, and stratospheric temperatures. The least squares method for calculating a best- fit linear trend is described. The method for quantifying the statistical uncertainty in a linear trend is explained, introducing the concepts of standard error, confidence intervals, and significance testing. A method to account for the effects of temporal autocorrelation on confidence intervals and significance tests is described. The issue of comparing two data sets to decide whether differences in their trends could have occurred by chance is discussed. The analysis of trends in state-of-the-art climate model results is a special case because we frequently have an ensemble of simulations for a particular forcing case. The effect of ensemble averaging on confidence intervals is illustrated. Finally, the issue of practical versus statistical significance is discussed. In practice, it is important to consider construction uncertainties as well as statistical uncertainties. An example is given showing that these two sources of trend uncertainty can be of comparable magnitude.

## (1) WHY DO WE NEED STATISTICS?

Statistical methods are required to ensure that data are interpreted correctly and that apparent relationships are meaningful (or "significant") and not simply chance occurrences.

A "statistic" is a numerical value that describes some property of a data set. The most commonly used statistics are the average (or "mean") value, and the "standard deviation," which is a measure of the variability within a data set around the mean value. The "variance" is the square of the standard deviation. The linear trend is another example of a data "statistic."

Two important concepts in statistics are the "population" and the "sample." The population is a theoretical concept, an idealized representation of the set of all possible values of some measured quantity. An example would be if we were able to measure temperatures continuously at a single site for all time – the set of all values (which would be infinite in size in this case) would be the population of temperatures for that site. A sample is what we actually see and can measure: *i.e.,* what we have available for statistical analysis, and a necessarily limited subset of the population. In the real world, all we ever have is limited samples, from which we try to estimate the properties of the population.

As an analogy, the population might be an infinite jar of marbles, a certain proportion of which (say 60%) is blue and the rest (40%) are red. We can only draw off a finite number of these marbles (a sample) at a time; and, when we measure the numbers of blue and red marbles in the sample, they need not be in the precise ratio 60:40. The ratio we measure is called a "sample statistic." It is an estimate of some hypothetical underlying population value (the corresponding "population parameter"). The techniques of statistical science allow us to make optimum use of the sample statistic and obtain a best estimate of the population parameter. Statistical science also allows us to quantify the uncertainty in this estimate.

## (2) DEFINITION OF A LINEAR TREND

If data show underlying smooth changes with time, we refer to these changes as a trend. The simplest type of change is a linear (or straight line) trend, a continuous increase or decrease over time. For example, the net effect of increasing greenhouse-gas concentrations and other human-induced factors is expected to cause warming at the surface and in the troposphere and cooling in the stratosphere (see Figure 1). Warming corresponds to a positive (or increasing) linear trend, while cooling corresponds to a negative (or decreasing)

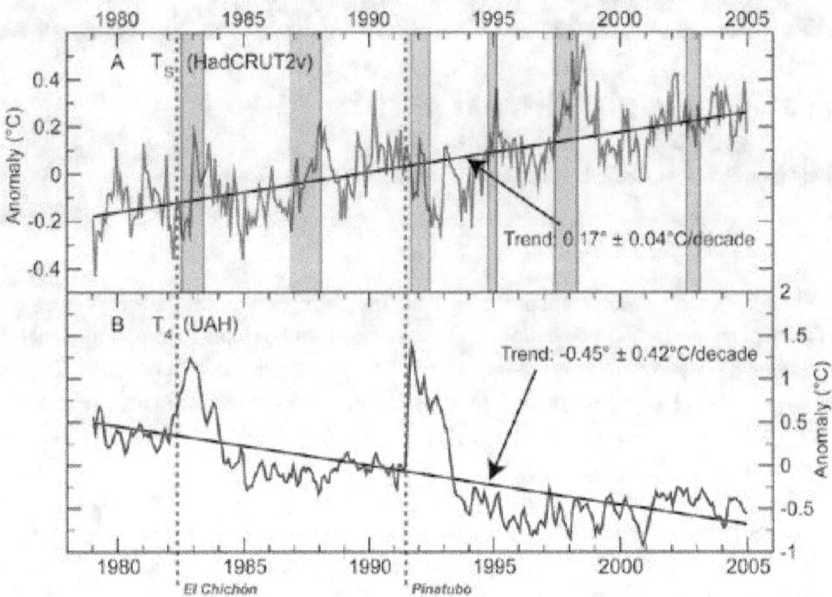

**Figure 1:** Examples of temperature time series with best-fit (least squares) linear trends: A, global-mean surface temperature from the UKMO Hadley Centre/Climatic Research Unit data set (HadCRUT2v); and B, global-mean MSU channel 4 data ($T_4$) for the lower stratosphere from the University of Alabama at Huntsville (UAH). Note the much larger temperature scale on the lower panel. Temperature changes are expressed as anomalies relative to the 1979 to 1999 mean (252 months). Dates for the eruptions of El Chichón and Mt. Pinatubo are shown by vertical lines. El Niños are shown by the shaded areas. The trend values are as given in Chapter 3, Table 3.3. The ± values define the 95% confidence intervals for the trends, also from Chapter 3, Table 3.3. The smaller confidence interval for the surface data shows that the straight line fit in this case is better than the straight ine fit to the stratospheric data.

represent the true underlying behavior.

A linear trend may therefore be deceptive if the trend number is given in isolation, removed from the original data. Nevertheless, used appropriately, linear trends provide the simplest and most convenient way to describe the overall change over time in a data set, and are widely used.

Linear temperature trends are usually quantified as the temperature change per year or per decade (even when the data are available on a month by month basis). For example, the trend for the surface temperature data shown in Figure 1 is 0.169°C per decade. (Note that 3 decimals are given here purely for mathematical convenience. The accuracy of these trends is much less, as is shown by the confidence intervals given in the Figure and in the Tables in Chapter 3. Precision should not be confused with accuracy.) Giving trends per decade is a more convenient representation than the trend per month, which, in this case, would be 0.169/120 = 0.00141°C per month, a very small number. An alternative method is to use the "total change" over the full data period – *i.e.,* the total change for the fitted linear trend line from the start to the end of the record (see Figure 2 in the Executive Summary). In Figure 1 here, the data shown span January 1979 through December 2004 (312 months or 2.6 decades). The total change is therefore 0.169x2.6 = 0.439°C.

trend. Over the present study period (1958 onwards), the expected changes due to "anthropogenic" (human-induced) effects are expected to be approximately linear. In some cases, natural factors have caused substantial deviations from linearity (see, *e.g.,* the lower stratospheric changes in Figure 1B), but the linear trend still provides a simple way of characterizing the overall change and of quantifying its magnitude.

Alternatively, there may be some physical process that causes a rapid switch or change from one mode of behavior to another. In such a case the overall behavior might best be described as a linear trend to the change-point, a step change at this point, followed by a second linear trend portion. Tropospheric temperatures from radiosondes show this type of behavior, with an apparent step increase in temperature occurring around 1976 (see Chapter 3, Figure 3.2a, or Figure 1 in the Executive Summary).

Step changes can lead to apparently contradictory results. For example, a data set that shows an initial cooling trend, followed by a large upward step, followed by a renewed cooling trend could have an overall warming trend. To state simply that the data showed overall warming would mis-

## (3) EXPECTED TEMPERATURE CHANGES: SIGNAL AND NOISE

Different physical processes generally cause different spatial and temporal patterns of change. For example, anthropogenic emissions of halocarbons at the surface have led to a reduction in stratospheric ozone and a contribution to stratospheric cooling over the past three or four decades. Now that these chemicals are controlled under the Montreal Protocol, the concentrations of the controlled species are decreasing and there is a trend towards a recovery of the ozone layer. The eventual long-term effect on stratospheric temperatures is expected to be non-linear: a cooling up until the late 1990s followed by a warming as the ozone layer recovers.

This is not the only process affecting stratospheric temperatures. Increasing concentrations of greenhouse gases lead to stratospheric cooling; and explosive volcanic eruptions cause sharp, but relatively short-lived stratospheric warmings (see Figure 1)[1]. There are also natural variations, most notably those associated with the Quasi-Bienniel Oscillation (QBO)[2]. Stratospheric temperature changes (indeed, changes at all levels of the atmosphere) are therefore the combined results of a number of different processes acting across all space and time scales.

In climate science, a primary goal is to identify changes associated with specific physical processes (causal factors) or combinations of processes. Such changes are referred to as "signals." Identification of signals in the climate record is referred to as the "detection and attribution" (D&A) problem. "Detection" is the identification of an unusual change, through the use of statistical techniques like significance testing (see below). "Attribution" is the association of a specific cause or causes with the detected changes in a statistically rigorous way.

The reason why D&A is a difficult and challenging statistical problem is because climate signals do not occur in isolation. In addition to these signals, temperature fluctuations in all parts of the atmosphere occur even in the absence of external driving forces. These internally generated fluctuations represent the "noise" against which we seek to identify specific externally forced signals. All climate records, therefore, are "noisy," with the noise of this natural variability tending to obscure the externally driven changes. Figure 1 illustrates this. At the surface, a primary noise component is the variability associated with ENSO (the El Niño/Southern

Oscillation phenomenon), while, in the stratosphere, if our concern is to identify anthropogenic influences, the warmings after the eruptions of El Chichón and Mt. Pinatubo constitute noise.

If the underlying response to external forcing is small relative to the noise, then, by chance, we may see a trend in the data due to random fluctuations purely as a result of the noise. The science of statistics provides methods through which we can decide whether the trend we observe is "real" *(i.e.,* a signal associated with some causal factor) or simply a random fluctuation *(i.e.,* noise).

## (4) DERIVING TREND STATISTICS

There are a number of different ways to quantify linear trends. Before doing anything, however, we should always inspect the data visually to see whether a linear trend model is appropriate. For example, in Figure 1, the linear warming trend appears to be a reasonable description for the surface data (top panel), but it is clear that a linear cooling model for the lower stratosphere (lower panel) fails to capture some of the more complex changes that are evident in these data. Nevertheless, the cooling trend line does give a good idea of the magnitude of the overall change.

There are different ways to fit a straight line to the data. Most frequently, a "best-fit" straight line is defined by finding the particular line that minimizes the sum, over all data points, of the squares of deviations about the line (these deviations are generally referred to as "residuals" or "errors"). This is an example of a more general procedure called least squares regression.

In linear regression analysis, a predictand (Y) is expressed as a linear combination of one or more predictors ($X_i$):

$$Y_{est} = b_0 + b_1 X_1 + b_2 X_2 + \ldots \qquad \ldots.. (1)$$

Where the subscript "est" is used to indicate that this is the estimate of Y that is given by the fitted relationship. Differences between the actual and estimated values of Y, the residuals, are defined by

$$e = Y - Y_{est} \qquad \ldots.. (2)$$

For linear trend analysis of temperature data (T) there is a single predictor, time (t; t = 1,2,3, …). The time points are almost always evenly spaced, month-by-month, year-by-year, etc. – but this is not a necessary restriction. In the linear trend case, the regression equation becomes:

---

[1]  Figure 1 shows a number of interesting features. In the stratosphere, the warmings following the eruptions of El Chichón (April 1982) and Mt Pinatubo (June 1991) are pronounced. For El Chichón, the warming appears to start before the eruption, but this is just a chance natural fluctuation. The overall cooling trend is what is expected to occur due to anthropogenic influences. At the surface, on short time scales, there is a complex combination of effects. There is no clear cooling after El Chichón, primarily because this was offset by the very strong 1982/83 El Niño. Cooling after Pinatubo is more apparent, but this was also partly offset by the El Niño around 1992/93 (which was much weaker than that of 1982/83). El Niño events, characterized by warm temperatures in the tropical Pacific, have a noticeable effect on global-mean temperature, but the effect lags behind the Pacific warming by 3-7 months. This is very clear in the surface temperature changes at and immediately after the 1986/87 and 1997/98 El Niños, also very large events. The most recent El Niños were weak and have no clear signature in the surface temperatures.

[2]  The QBO is a quasi-periodic reversal in winds in the tropical stratosphere that leads to alternating warm and cold tropical stratospheric temperatures with a periodicity of 18 to 30 months.

$$T_{est} = a + b\,t \qquad \qquad \ldots\ldots (3)$$

In equation (3), "b" is the slope of the fitted line – *i.e.,* the linear trend value. This is a sample statistic, *i.e.,* it is an estimate of the corresponding underlying population parameter. To distinguish the population parameter from the sample value, the population trend value is denoted ß.

The formula for b is:

$$b = [\Sigma((t - \bar{t})T_t)]/[\Sigma((t - \bar{t})^2)] \qquad \ldots\ldots (4)$$

Where $\bar{t}$ denotes the mean value, and the summation is over t = 1,2,3, … n *(i.e.,* the sample size is n). $T_t$ denotes the value of temperature, T, at time "t". Equation (4) produces an unbiased estimate[3] of population trend, ß.

For the usual case of evenly spaced time points, $\bar{t} = (n+1)/2$, and

$$\Sigma((t - \bar{t})^2)=n(n^2-1)/12 \qquad \ldots\ldots (5)$$

When we are examining deviations from the fitted line the sign of the deviation is not important. This is why we consider the squares of the residuals in least squares regression. An important and desirable characteristic of the least squares method is that the average of the residuals is zero.

Estimates of the linear trend are sensitive to points at the start or end of the data set. For example, if the last point, by chance, happened to be unusually high, then the fitted trend might place undue weight on this single value and lead to an estimate of the trend that was too high. This is more of a problem with small sample sizes *(i.e.,* for trends over short time periods). For example, if we considered tropospheric data over 1979 through 1998, because of the unusual warmth in 1998 (associated with the strong 1997/98 El Niño; see Figure 1), the calculated trend may be an overestimate of the true underlying trend.

There are alternative ways to estimate the linear trend that are less sensitive to endpoints. Although we recognize this problem, for the data used in this Report tests using different trend estimators give results that are virtually the same as those based on the standard least-squares trend estimator.

---

[3]  An unbiased estimator is one where, if the same experiment were to be performed over and over again under identical conditions, then the long-run average of the estimator will be equal to the parameter that we are trying to estimate. In contrast, in a biased estimator, there will always be some slight difference between the long-run average and the true parameter value that does not tend to zero no matter how many times the experiment is repeated. Since our goal is to estimate population parameters, it is clear that unbiased estimators are preferred.

## (5) TREND UNCERTAINTIES

Some examples of fitted linear trend lines are shown in Figure 1. This Figure shows monthly temperature data for the surface and for the lower stratosphere (MSU channel 4) over 1979 through 2004 (312 months). In both cases there is a clear trend, but the fit is better for the surface data. The trend values *(i.e.,* the slopes of the best fit straight lines that are shown superimposed on monthly data) are +0.17°C/decade for the surface and −0.45°C/decade for the stratosphere. For the stratosphere, although there is a pronounced overall cooling trend, as noted above, describing the change simply as a linear cooling considerably oversimplifies the behavior of the data[1].

A measure of how well the straight line fits the data *(i.e.,* the "goodness of fit") is the average value of the squares of the residuals. The smaller this is, the better is the fit. The simplest way to define this average would be to divide the sum of the squares of the residuals by the sample size *(i.e.,* the number of data points, n). In fact, it is usually considered more correct to divide by n − 2 rather than n, because some information is lost as a result of the fitting process and this loss of information must be accounted for. Dividing by n − 2 is required in order to produce an unbiased estimator[3].

The population parameter we are trying to estimate here is the standard deviation of the trend estimate, or its square, the variance of the distribution of b, which we denote Var(b). The larger the value of Var(b), the more uncertain is b as an estimate of the population value, ß.

The formula for Var(b) is …

$$Var(b) = [\sigma^2]/[\Sigma((t - \bar{t})^2] \qquad \ldots\ldots (6)$$

where $\sigma^2$ is the population value for the variance of the residuals. Unfortunately, we do not in general know what $\sigma^2$ is, so we must use an unbiased sample estimate of $\sigma^2$. This estimate is known as the Mean Square Error (MSE), defined by …

$$MSE = [\Sigma(e^2)]/(n - 2) \qquad \ldots\ldots (7)$$

Hence, equation (6) becomes

$$Var(b) = (SE)^2 = MSE/[\Sigma((t - \bar{t})^2)] \qquad \ldots\ldots (8)$$

where SE, the square root of Var(b), is called the "standard error" of the trend estimate. The smaller the value of the standard error, the better the fit of the data to the linear change description and the smaller the uncertainty in the sample trend as an estimate of the underlying population trend value. The standard error is the primary measure of trend uncertainty. The standard error will be large if the MSE is large, and the MSE will be large if the data points show large scatter about the fitted line.

There are assumptions made in going from equation (6) to (8): viz. that the residuals have mean zero and common variance, that they are Normally (or "Gaussian") distributed[4], and that they are uncorrelated or statistically independent. In climatological applications, the first two assumptions are generally valid. The third assumption, however, is often not justified. We return to this below.

## (6) CONFIDENCE INTERVALS AND SIGNIFICANCE TESTING

In statistics we try to decide whether a trend is an indication of some underlying cause, or merely a chance fluctuation. Even purely random data may show periods of noticeable upward or downward trends, so how do we identify these cases?

There are two common approaches to this problem, through significance testing and by defining confidence intervals. The basis of both methods is the determination of the "sampling distribution" of the trend, *i.e.,* the distribution of trend estimates that would occur if we analyzed data that were randomly scattered about a given straight line with slope ß. This distribution is approximately Gaussian with a mean value equal to ß and a variance (standard deviation squared) given by equation (8). More correctly, the distribution to use is Student's "t" distribution, named after the pseudonym "Student" used by the statistician William Gosset. For large samples, however (n more than about 30), the distribution is very nearly Gaussian.

### Confidence Intervals

The larger the standard error of the trend, the more uncertain is the slope of the fitted line. We express this uncertainty probabilistically by defining confidence intervals for the trend associated with different probabilities. If the distribution of trend values were strictly Gaussian, then the range b − SE to b + SE would represent the 68% confidence interval (C.I.) because the probability of a value lying in that range for a Gaussian distribution is 0.68. The range b − 1.645(SE) to b + 1.645(SE) would give the 90% C.I.; the range b − 1.96(SE) to b + 1.96(SE) would give the 95% C.I.; and so on. Quite often, for simplicity, we use b − 2(SE) to b + 2(SE) to repre-

sent (to a good approximation) the 95% confidence interval. (This is often called the "two-sigma" confidence interval.) Examples of 95% confidence intervals are given in Figure 1. Here, the smaller value for the surface data compared with the stratospheric data shows that a straight line fits the surface data better than it does the stratospheric data.

Because of the way C.I.s are usually represented graphically, as a bar centered on the best-fit estimate, they are often referred to as "error bars." Confidence intervals may be expressed in two ways, either (as above) as a range, or as a signed error magnitude. The approximate 95% confidence interval, therefore, may be expressed as b ± 2(SE), with appropriate numerical values inserted for b and SE.

As will be explained further below, showing confidence interval for linear temperature trends may be deceptive, because the purely statistical uncertainties that they represent are not the only sources of uncertainty. Such confidence intervals quantify only one aspect of trend uncertainty, that arising from statistical noise in the data set. There are many other sources of uncertainty within any given temperature data set and these may be as or more important than statistical uncertainty. Showing just the statistical uncertainty may therefore provide a false sense of accuracy in the calculated trend.

### Significance Testing

An alternative method for assessing trends is hypothesis testing. In practice, it is much easier to disprove rather than prove a hypothesis. Thus, the standard statistical procedure in significance testing is to set up a hypothesis that we would like to disprove; we call this a "null hypothesis." In the linear trend case, we are often interested in trying to decide whether an observed data trend that is noticeably different from zero is sufficiently different that it could not have occurred by chance − or, at least, that the probability that it could have occurred by chance is very small. The appropriate null hypothesis in this case would be that there was no underlying trend (ß = 0). If we disprove *(i.e.,* "reject") the null hypothesis, then we say that the observed trend is "statistically significant" at some level of confidence and we must accept some alternate hypothesis. The usual alternate hypothesis in temperature analyses is that the data show a real, externally forced warming (or cooling) trend. (In cases like this, the statistical analysis is predicated on the assumption that the observed data are reliable, which is not always the case. If a trend were found to be statistically significant, then an alternative possibility might be that the observed data were flawed.)

---

4   The "Gaussian" distribution (often called the "Normal" distribution) is the most well-known probability distribution. This has a characteristic symmetrical "bell" shape, and has the property that values near the center (or mean value) of the distribution are much more likely than values far from the center.

An alternative null hypothesis that often arises is when we are comparing an observed trend with some model expectation. Here, the null hypothesis is that the observed trend is equal to the model value. If our results led us to reject this null hypothesis, then (assuming again that the observed data are reliable) we would have to infer that the model result was flawed – either because the external forcing applied to the model was incorrect and/or because of deficiencies in the model itself.

An important factor in significance testing is whether we are concerned about deviations from some hypothesized value in any direction or only in one direction. This leads to two types of significance test, referred to as "one-tailed" (or "one-sided") and "two-tailed" tests. A one-tailed test arises when we expect a trend in a specific direction (such as warming in the troposphere due to increasing greenhouse-gas concentrations). Two-tailed tests arise when we are concerned only with whether the trend is different from zero, with no specification of whether the trend should be positive or negative. In temperature trend analyses we generally know the sign of the expected trend, so one-tailed tests are more common.

The approach we use in significance testing is to determine the probability that the observed trend could have occurred by chance. As with the calculation of confidence intervals, this involves calculating the uncertainty in the fitted trend arising from the scatter of points about the trend line, determined by the standard error of the trend estimate (equation [8]). It is the ratio of the trend to the standard error (b/SE) that determines the probability that a null hypothesis is true or false. A large ratio (greater than 2, for example) would mean that (except for very small samples) the 95% C.I. did not include the zero trend value. In this case, the null hypothesis is unlikely to be true, because the zero trend value, the value assumed under the null hypothesis, lies outside the range of trend values that are likely to have occurred purely by chance.

If the probability that the null hypothesis is true is small, and less than a predetermined threshold level such as 0.05 (5%) or 0.01 (1%), then the null hypothesis is unlikely to be correct. Such a low probability would mean that the observed trend could only have occurred by chance one time in 20 (or one time in 100), a highly unusual and therefore "significant" result. In technical terms we would say that "the null hypothesis is rejected at the prescribed significance level", and declare the result "significant at the 5% (or 1%) level." We would then accept the alternate hypothesis that there was a real deterministic trend and, hence, some underlying causal factor.

Even with rigorous statistical testing, there is always a small probability that we might be wrong in rejecting a null hypothesis. The reverse is also true – we might accept a null hypothesis of no trend even when there is a real trend in the data. This is more likely to happen when the sample size is small. If the real trend is small and the magnitude of variability about the trend is large, it may require a very large sample in order to identify the trend above the background noise.

For the null hypothesis of zero trend, the distribution of trend values has mean zero and standard deviation equal to the standard error. Knowing this, we can calculate the probability that the actual trend value could have exceeded the observed value by chance if the null hypotheses were true (or, if we were using a two-tailed test, the probability that the magnitude of the actual trend value exceeded the magnitude of the observed value). This probability is called the "p-value." For example, a p-value of 0.03 would be judged significant at the 5% level (since 0.03<0.05), but not at the 1% level (since 0.03>0.01).

Since both the calculation of confidence intervals and significance testing employ information about the distribution of trend values, there is a clear link between confidence intervals and significance testing.

## A Complication:
## The Effect of Autocorrelation

The significance of a trend, and its confidence intervals, depend on the standard error of the trend estimate. The formula given above for this standard error (equation [8]) is, however, only correct if the individual data points are unrelated, or statistically independent. This is not the case for most temperature data, where a value at a particular time usually depends on values at previous times; *i.e.,* if it is warm today, then, on average, it is more likely to be warm tomorrow than cold. This dependence is referred to as "temporal autocorrelation" or "serial correlation." When data are auto-correlated *(i.e.,* when successive values are not independent of each other), many statistics behave as if the sample size was less than the number of data points, n.

One way to deal with this is to determine an "effective sample size," which is less than n, and use it instead of n in statistical formulae and calculations. The extent of this reduction from n to an effective sample size depends on how strong the autocorrelation is. Strong autocorrelation means that individual values in the sample are far from being independent, so the effective number of independent values must be much smaller than the sample size. Strong autocorrelation is common in temperature time series. This is accounted for by reducing the divisor "n – 2" in the mean square error term (equation [7]) that is crucial in determining the standard error of the trend (equation [8]).

There are a number of ways that this autocorrelation effect may be quantified. A common and relatively simple method is described in Santer *et al.* (2000). This method makes the assumption that the autocorrelation structure of the temperature data may be adequately described by a "first-order autoregressive" process, an assumption that is a good approximation for most climate data. The lag-1 autocorrelation coefficient $_{(r1)}$ is calculated from the observed data[5], and the effective sample size is determined by

$$n_{eff} = n \ (1 - r_1)/(1 + r_1) \qquad \qquad ..... (9)$$

There are more sophisticated methods than this, but testing on observed data shows that this method gives results that are very similar to those obtained by more sophisticated methods.

If the effective sample size is noticeably smaller than n, then, from equations (7) and (8) it can be seen that the standard error of the trend estimate may be much larger than one would otherwise expect. Since the width of any confidence interval depends directly on this standard error (larger SE leading to wider confidence intervals), then the effect of autocorrelation is to produce wider confidence intervals and greater uncertainty in the trend estimate. A corollary of this is that results that may show a significant trend if autocorrelation is ignored are frequently found to be non-significant when autocorrelation is accounted for.

## (7) COMPARING TRENDS IN TWO DATA SETS

Assessing the magnitude and confidence interval for the linear trend in a given data set is standard procedure in climate data analysis. Frequently, however, we want to compare two data

sets and decide whether differences in their trends could have occurred by chance. Some examples are:

(a) comparing data sets that purport to represent the same variable (such as two versions of a satellite data set) – an example is given in Figure 2;

(b) comparing the same variable at different levels in the atmosphere (such as surface and tropospheric data); or

(c) comparing models and observations.

In the first case (Figure 2), we know that the data sets being compared are attempts to measure precisely the same thing, so that differences can arise only as a result of differences in the methods used to create the final data sets from the same "raw" original data. Here, there is a pitfall that some practitioners fall prey to by using what, at first thought, seems to be a reasonable approach. In this naive method, one would first construct C.I.s for the individual trend estimates by applying the single sample methods described above. If the two C.I.s overlapped, then we would conclude that there was no significant difference between the two trends. This approach, however, is seriously flawed.

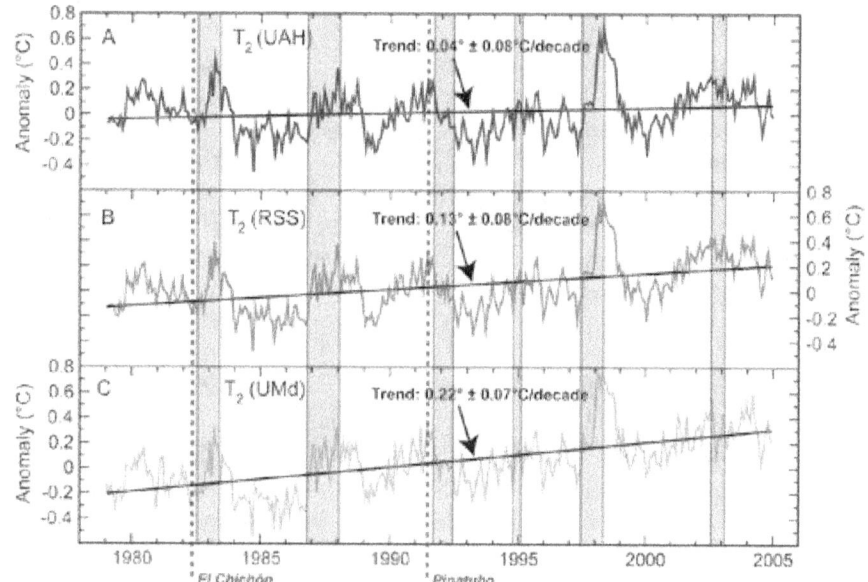

**Figure 2:** Three estimates of global-mean temperature changes for MSU channel 2 ($T_2$), expressed as anomalies relative to the 1979 to 1999 mean. Data are from: A, the University of Alabama in Huntsville (UAH); B, Remote Sensing Systems (RSS); and C, the University of Maryland (UMd) The estimates employ the same "raw" satellite data, but make different choices for the adjustments required to merge the various satelite records and to correct for instrument biases. The statistical uncertainty is virtually the same for all three series. Differences between the series give some idea of the magnitude of structural uncertainties. Volcano eruption and El Niño information are as in Figure 1. The trend values are as given in Chapter 3, Table 3.3. The ± values define the 95% confidence intervals for the trends, also from Chapter 3, Table 3.3.

---

[5] From the time series of residuals about the fitted line.

An analogous problem, comparing two means rather than two trends, discussed by Lanzante (2005), gives some insights. In this case, it is necessary to determine the standard error for the difference between two means. If this standard error is denoted "s", and the individual standard errors are $s_1$ and $s_2$, then

$$s^2 = (s_1)^2 + (s_2)^2 \quad .....(10)$$

The new standard error is often called the pooled standard error, and the pooling method is sometimes called "combining standard errors in quadrature." In some cases, when the trends come from data series that are unrelated (as when model and observed data are compared; case (c) above) a similar method may be applied to trends. If the data series are correlated with each other, however (cases (a) and (b)), this procedure is not correct. Here, the correct method is to produce a difference time series by subtracting the first data point in series 1 from the first data point in series 2, the second data points, the third data points, etc. The result of doing this with the microwave sounding unit channel 2 (MSU $T_2$) data shown in Figure 2 is shown in Figure 3. To assess the significance of trend differences we then apply the same methods used for trend assessment in a single data series to the difference series.

Analyzing differences removes the variability that is common to both data sets and isolates those differences that may be due to differences in data set production methods, temperature measurement methods (as in comparing satellite and radiosonde data), differences in spatial coverage, etc.

Figures 2 and 3 provide a striking example of this. Here, the three series in Figure 2 have very similar volcanic and ENSO signatures. In the individual series, these aspects are noise that obscures the underlying linear trend and inflates the standard error and the trend uncertainty. Since this noise is common to each series, differencing has the effect of canceling out a large fraction of the noise. This is clear from Figure 3, where the variability about the trend lines is substantially reduced. Figure 4 shows the effects on the trend confidence intervals (taking due account of autocorrelation effects). Even though the individual series look very

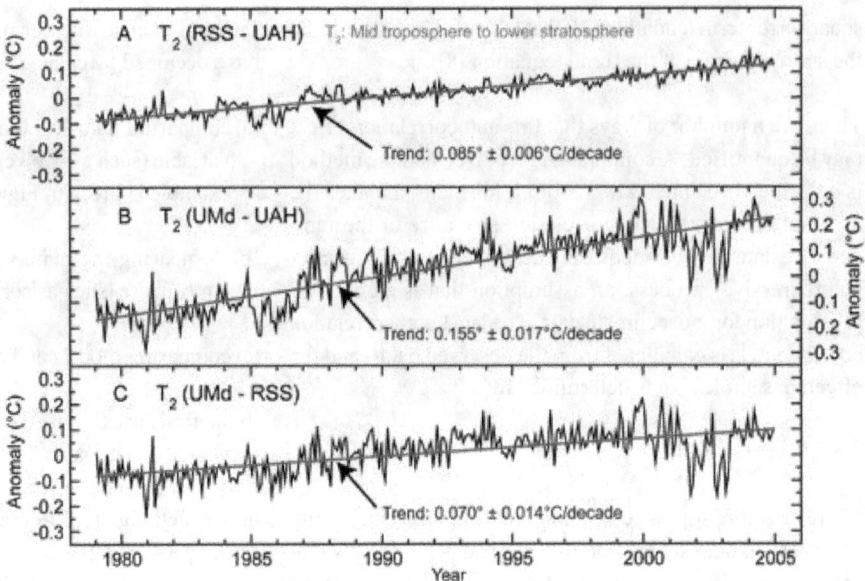

**Figure 3:** Difference series for the global-mean MSU $T_2$ series shown in Figure 2. Variability about the trend line is least for the UAH minus RSS series indicating closer correspondence between these two series than between UMd and either UAH or RSS. The trend values are consistent with results given in Chapter 3, Table 3.3, with greater precision given purely for mathematical convenience. The ± values define the 95% confidence intervals for the trends (see also Figure 4).

similar in Figure 2, this is largely an artifact of similarities in the noise. It is clear from Figures 3 and 4 that there are, in fact, very significant differences in the trends, reflecting differences in the methods of construction used for the three MSU $T_2$ data sets.

Comparing model and observed data for a single variable, such as surface temperature, tropospheric temperature, etc., is a different problem. Here, when using data from a state-of-the-art climate model (a coupled Atmosphere/Ocean General Circulation Model[6], or "AOGCM"), there is no reason to expect the background variability to be common to both the model and observations. AOGCMs generate their own internal variability entirely independently of what is going on in the real world. In this case, standard errors for the individual trends can be combined in quadrature (equation [10]). (There are some model/observed data comparison cases where an examination of the difference series may still be appropriate, such as in experiments where an atmospheric GCM is forced by observed sea surface temperature varia-

---

[6]   An AOGCM interactively couples together a three-dimensional Ocean General Circulation Model (OGCM) and an Atmospheric GCM (AGCM). The components are free to interact with one another and they are able to generate their own internal variability in much the same way that the real-world climate system generates its internal variability (internal variability is variability that is unrelated to external forcing). This differs from some other types of model (*e.g.*, an AGCM) where there can be no component of variability arising from the ocean. An AGCM, therefore, cannot generate variability arising from ENSO, which depends on interactions between the atmosphere and ocean.

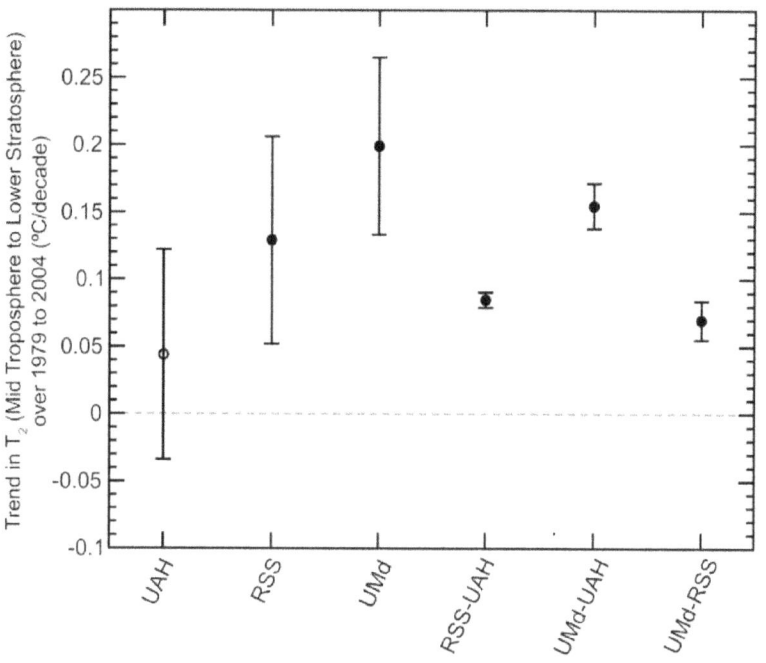

**Figure 4:** 95% confidence intervals for the three global-mean MSU $T_2$ series shown in Figure 2 (see Table 3.3 in Chapter 3), and for the three difference series shown in Figure 3.

tions so that ocean-related variability should be common to both the observations and the model.)

For other comparisons, the appropriate test will depend on the degree of similarity between the data sets expected for perfect data. For example, a comparison between MSU $T_2$ and MSU $T_{2LT}$ produced by a single group should use the difference test – although interpretation of the results may be tricky because differences may arise either from construction methods or may represent real physical differences arising from the different vertical weighting profiles, or both.

There is an important implication of this comparison issue. While it may be common practice to use error bars to illustrate C.I.s for trends of individual time series, when the primary concern (as it is in many parts of this Report) is the comparison of trends, individual C.I.s can be misleading. A clear example of this is given in Figure 4 (based on information in Figures 2 and 3). Individual C.I.s for the three MSU $T_2$ series overlap, but the C.I.s for the difference series show that there are highly significant differences between the three data sets. Because of this, in some cases in this Report, where it might seem that error bars should be given, we consider the disadvantage of their possible misinterpretation to outweigh their potential usefulness. Individual C.I.s for all trends are, however, given in Tables 3.2, 3.3, 3.4 and 3.5 of Chapter 3; and we also express individual trend uncertainties through the use of significance levels. As noted in Section

(9) below, there are other reasons why error bars can be misleading.

## (8) MULTIPLE AOGCM SIMULATIONS

Both models and the real world show weather variability and other sources of internal variability that are manifest on all time scales, from daily up to multi-decadal. With AOGCM simulations driven by historical forcing spanning the late-19th and 20th centuries, therefore, a single run with a particular model will show not only the externally forced signal, but also, superimposed on this, underlying internally generated variability that is similar to the variability we see in the real world. In contrast to the real world, however, in the model world we can perturb the model's initial conditions and re-run the same forcing experiment. This will give an entirely different realization of the model's internal variability. In each case, the output from the model is a combination of signal (the response to the forcing) and noise (the internally generated component). Since the noise parts of each run are unrelated, averaging over a number of realizations will tend to cancel out the noise and, hence, enhance the visibility of the signal. It is common practice, therefore, for any particular forcing experiment with an AOGCM, to run multiple realizations of the experiment (*i.e.*, an ensemble of realizations). An example is given in Figure 5, which shows four separate realizations and their ensemble average for a simulation using realistic 20th century forcing (both natural and anthropogenic).

This provides us with two different ways to assess the uncertainties in model results, such as in the model-simulated temperature trend over recent decades. One method is to express uncertainties using the spread of trends across the ensemble members (see, *e.g.*, Figures 3 and 4 in the Executive Summary). Alternatively, the temperature series from the individual ensemble members may be averaged and the trend and its uncertainty calculated using these average data.

Ensemble averaging, however, need not reduce the width of the trend confidence interval compared with an individual realization. This is because of compensating factors: the time series variability will be reduced by the averaging process (as is clear in Figure 5), but, because averaging can inflate

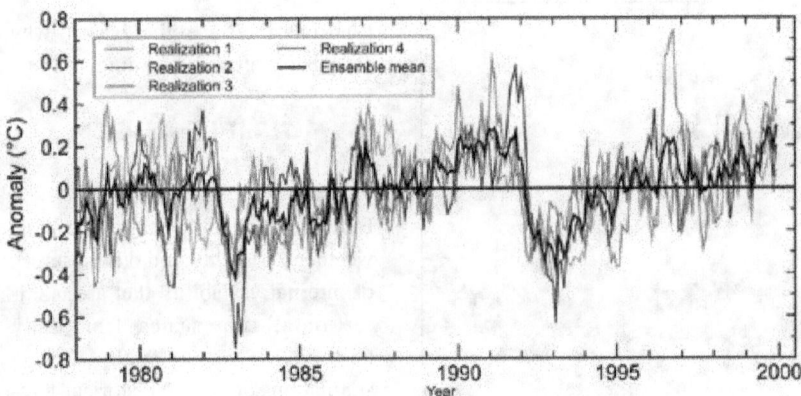

**Figure 5:** Four separate realizations of model realizations of global-mean MSU channel 2 (T₂) temperature changes, and their ensemble average, for a simulation using realistic 20th Century forcing (both natural and anthropogenic) carried out with one of the National Centre for Atmospheric Research's AOGCMs, the Parallel Climate Model (PCM). The cooling events around 1982/3 and 1991/2 are the result of imposed forcing from the eruptions of El Chichón (1982) and Mt. Pinatubo (1991). Note that the El Chichón cooling in these model simulations is more obvious than in the observed data shown in Figure 1. In the real world, a strong El Niño warming event occurred at the same time as the volcanic cooling, largely masking this cooling. In the four model worlds shown here, the sequences of El Niño events, which necessarily occurred at different times in each simulation, never overlapped with the El Chichón cooling.

records that are going to be used for trend (or other statistical) analyses, we attempt to minimize construction uncertainties by removing, as far as possible, non-climatic biases that might vary over time and so impart a spurious trend or trend component – a process referred to as "homogenization."

The need for homogenization arises in part because most observations are made to serve the short-term needs of weather forecasting (where the long-term stability of the observing system is rarely an important consideration). Most records therefore contain the effects of changes in instrumentation, instrument exposure, and observing practices made for a variety of reasons. Such changes generally introduce spurious non-climatic changes into data records that, if not accounted for, can mask (or possibly be mistaken for) an underlying climate signal.

the level of autocorrelation, there may be a compensating increase in uncertainty due to a reduction in the effective sample size. This is illustrated in Figure 6.

Averaging across ensemble members, however, does produce a net gain. Although the width of the C.I. about the mean trend may not be reduced relative to individual trend C.I.s, averaging leaves just a single best-fit trend rather than a spread of best-fit trend values.

An added problem arises because temperatures are not always measured directly, but through some quantity related to temperature. Adjustments must therefore be made to obtain

## (9) PRACTICAL VERSUS STATISTICAL SIGNIFICANCE

The Sections above have been concerned primarily with statistical uncertainty, uncertainty arising from random noise in climatological time series – *i.e.,* the uncertainty in how well a data set fits a particular "model" (a straight line in the linear trend case). Statistical noise, however, is not the only source of uncertainty in assessing trends. Indeed, as amply illustrated in this Report, other sources of uncertainty may be more important.

The other sources of uncertainty are the influences of non-climatic factors. These are referred to in this Report as "construction uncertainties." When we construct climate data

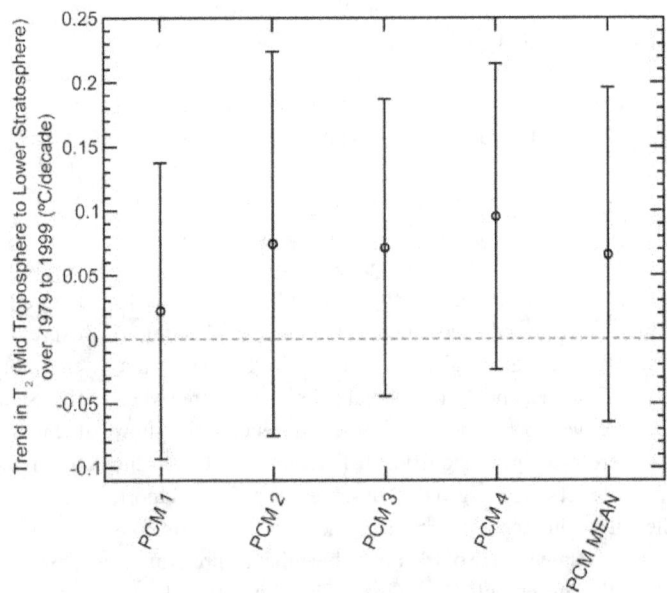

**Figure 6:** 95% confidence intervals for individual model realizations of global-mean MSU T₂ temperature changes (as shown in Figure 5), compared with the 95% confidence interval for the four-member ensemble average.

temperature information. The satellite-based microwave sounding unit (MSU) data sets provide an important example. For MSU temperature records, the quantity actually measured is the upwelling emission of microwave radiation from oxygen atoms in the atmosphere. MSU data are also affected by numerous changes in instrumentation and instrument exposure associated with the progression of satellites used to make these measurements.

Thorne *et al.* (2005) divide construction uncertainty into two components: "structural uncertainty" and "parametric uncertainty." Structural uncertainty arises because different investigators may make different plausible choices for the method (or "model") that they apply to make corrections or "adjustments" to the raw data. Differences in the choice of adjustment model and its structure lead to structural uncertainties. Parametric uncertainties arise because, once an adjustment model has been chosen, the values of the parameters in the model still have to be quantified. Since these values must be determined from a finite amount of data, they will be subject to statistical uncertainties.

Sensitivity studies using different parameter choices may allow us to quantify parametric uncertainty, but this is not always done. Quantifying structural uncertainty is very difficult because it involves consideration of a number of fundamentally different (but all plausible) approaches to data set homogenization, rather than simple parameter "tweaking." Differences between results from different investigators give us some idea of the magnitude of structural uncertainty, but this is a relatively weak constraint. There are a large number of conceivable approaches to homogenization of any particular data set, from which we are able only to consider a small sample – and this may lead to an under-estimation of structural uncertainty. Equally, if some current homogenization techniques are flawed then the resulting uncertainty estimate will be too large.

An example is given above in Figure 2, showing three different MSU $T_2$ records with trends of 0.044°C/decade, 0.129°C/decade, and 0.199°C/decade over 1979 through 2004. These differences, ranging from 0.070°C/decade to 0.155°C/decade (Figure 3), represent a considerable degree of construction uncertainty. For comparison, the statistical uncertainty in the individual data series, as quantified by the 95% confidence intervals, ranges between ±0.066 and ±0.078°C/decade; so uncertainties from these two sources are of similar magnitude.

An important implication of this comparison is that statistical and construction uncertainties may be of similar magnitude. For this reason, showing, through confidence intervals, information about statistical uncertainty alone, without giving any information about construction uncertainty, can be misleading.

# GLOSSARY

**Aerosols**
tiny particles suspended in the air

**Adjusted**
refers to time series data that have been "homogenized" to remove time dependent biases; owing to uncertainties inherent in data bias removal, the term "adjusted" is often used instead of "corrected"

**Albedo**
the fraction of incident light that is reflected from a surface

**Anthropogenic**
human-induced

**Black carbon**
soot particles primarily from fossil fuel burning

**Climate sensitivity**
the equilibrium change in global-average surface air temperature following a change in radiative forcing; in current usage, this term generally refers to the warming that would result if atmospheric carbon dioxide concentrations were to double from their pre-industrial levels

**Contrails**
condensation trails from aircraft

**Convection**
motions in a fluid or the air that are predominantly vertical and driven by buoyancy forces; a principal means of vertical energy transfer

**Diurnal**
occurring daily; varying within the course of a day

**Dewpoint**
temperature at which water vapor condenses into liquid water temperature when cooled at constant pressure

**Error**
the difference between an estimated or observed value and the true value

**Forcing**
a natural or human-induced factor that influences climate

**Greenhouse gases**
gases including water vapor, carbon dioxide, methane, nitrous oxide, and halocarbons that trap infrared heat, warming the air near the surface and in the lower levels of the atmosphere

**Homogenization**
Removing changes in time series data that might have arisen for non-climatic reasons

**Internal variability**
natural cycles and variations in climate

**Temperature inversion**
a condition in which the air temperature increases with height, in contrast to the more common situation in which temperature decreases with altitude

**Isothermal**
constant temperature; often refers to a temperature profile meaning constant temperature with height

**Lapse rate**
the rate at which temperature decreases with increasing elevation

**Latent heat**
the heat required to change the phase of a substance, *e.g.,* solid to vapor (sublimation), liquid to vapor (vaporization), or solid to liquid (melting); the temperature does not change during these processes. Heat is released for the reverse processes, *e.g.,* vapor to solid (frost), liquid to solid (freezing), or vapor to liquid (condensation)

**Metadata**
supplemental records used to interpret measurements, such as how and where measurements were collected and processed

**Parameterization**
a mathematical representation of a process that cannot be explicitly resolved in a climate model

**Radiosonde**
a balloon carrying a thermometer or other sensing device that takes measurements in the atmosphere and transmits them by radio to a data recorder on the surface

**Reanalysis**
a mathematically blended record that incorporates a variety of observational data sets (with adjustments) in an assimilation model

**Reference networks**
a small subset of sites consisting of multiple instruments that independently measure the same variable which if well coordinated could provide full characterization of instrument errors and biases, significantly reducing uncertainty in observed climate change

**Relative humidity**
the percentage of water vapor in the air relative to what is required for saturation to occur at a given temperature

**Sensible heat**
heat that can be measured by a thermometer

**Specific humidity**
the amount of water vapor in the air in units of kilograms of water vapor per kilogram of air

**Trend**
a systematic change over time

**Uncertainty**
a term used to describe the range of possible values around a best estimate, sometimes expressed in terms of probability or likelihood (see Preface Figure 1 and discussion in Appendix A)

## ACRONYMS

| | |
|---|---|
| **20CEN** | climate model simulation of the 20th century |
| **AGCM** | Atmospheric General Circulation Model |
| **AIRS** | Atmospheric InfraRed Sounder |
| **AMIP** | Atmospheric Model Intercomparison Project |
| **AMSU** | Advanced Microwave Sounding Unit |
| **AOGCM** | Atmosphere-Ocean General Circulation Model |
| **AR4** | IPCC Fourth Assessment Report |
| **ARL** | Air Resources Laboratory |
| **ATMS** | Advanced Technology Microwave Sounder |
| **ATSR** | Along-Track Scanning Radiometer |
| **AVHRR** | Advanced Very High Resolution Radiometer |
| **CCSM** | Community Climate System Model |
| **CCSP** | Climate Change Science Program |
| **CDR** | Climate Data Record |
| **CFCs** | chlorofluorocarbons |
| **CGCM** | Coupled Atmosphere-Ocean General Circulation Model |
| **CH$_4$** | Methane |
| **C.I.** | Confidence Interval |
| **CLIVAR** | Climate Variability and Prediction |
| **CMIP** | Coupled Model Intercomparison Project |
| **CMIS** | Conical scanning Microwave Imager/Sounder |
| **CO$_2$** | Carbon Dioxide |
| **COADS** | Comprehensive Ocean-Atmosphere Data Set |
| **COWL** | Cold Ocean Warm Land |
| **CrIS** | Cross-track Infrared Sounder |
| **CRN** | Climate Reference Network |
| **CRU** | Climate Research Unit |
| **DOE** | Department of Energy |
| **EBM** | Energy Balance Model |
| **ECMWF** | European Centre for Medium-range Weather Forecasts |
| **EMIC** | Earth System Models of Intermediate Complexity |
| **ENSO** | El Niño-Southern Oscillation |
| **EOF** | Empirical Orthogonal Function |
| **ERA** | ECMWF Re-Analysis |
| **ERSST** | Extended Reconstruction Sea Surface Temperature |
| **GAW** | Global Atmospheric Watch |
| **GCM** | General Circulation Model |
| **GCOS** | Global Climate Observing System |
| **GCSM** | Global Climate System Model |
| **GEOSS** | Global Earth Observation System of Systems |

| | | | |
|---|---|---|---|
| **GFDL** | Geophysical Fluid Dynamics Laboratory | **NPOESS** | National Polar-orbiting Operational Environmental Satellite System |
| **GHCN** | Global Historical Climatology Network | **NRC** | National Research Council |
| **GHG** | Greenhouse Gas | **NSF** | National Science Foundation |
| **GHRSST-PP** | GODAE High-Resolution SST Pilot Project | **NWP** | Numerical Weather Prediction |
| | | **$O_3$** | Ozone |
| **GISS** | Goddard Institute for Space Studies | **OGCM** | Ocean General Circulation Model |
| **GODAE** | Global Ocean Data Assimilation Experiment | **PCM** | Parallel Climate Model |
| | | **PDO** | Pacific Decadal Oscillation |
| **GPS** | Global Positioning System | **QBO** | Quasi-Biennial Oscillation |
| **GSN** | GCOS Surface Network | **RATPAC** | Radiosonde Atmospheric Temperature Products for Assessing Climate |
| **GUAN** | GCOS Upper Air Network | | |
| **HadCM** | Hadley Centre Climate Model | **RSS** | Remote Sensing Systems |
| **HadRT** | Hadley Centre Radiosonde Temperatures | **SAM** | Southern Hemisphere Annual Mode |
| | | **SCAMS** | SCAnning Microwave Spectrometer |
| **hPa** | hectoPascals, a measure of pressure | **SH** | Southern Hemisphere |
| **HIRS** | High-resolution Infrared Radiation Sounder | **$SO_4$** | Sulfate |
| | | **SSM/I** | Special Sensor Microwave/Imager |
| **IASI** | Infrared Atmospheric Sounding Interferometer | **SSMI/S** | Special Sensor Microwave Imager/Sounder |
| **ICOADS** | International Comprehensive Ocean-Atmosphere Data Set | **SST** | Sea Surface Temperature |
| | | **SSU** | Stratospheric Sounding Unit |
| **IGRA** | Integrated Global Radiosonde Archive | **TAO** | Tropical Atmosphere Ocean |
| **IPCC** | Intergovernmental Panel on Climate Change | **TEAP** | Technology and Economic Assessment Panel |
| **IR** | Infrared Radiation | **TIROS** | Television InfraRed Observation Satellite |
| **ITCZ** | Inter Tropical Convergence Zone | | |
| **LBNL** | Lawrence Berkeley National Laboratory | **TLT** | Temperature of the Lower Troposphere |
| **LECT** | Local Equator Crossing Time | **TOGA** | Tropical Ocean Global Atmosphere |
| **LKS** | Lanzante, Klein, Seidel | **TOVS** | TIROS Operational Vertical Sounder |
| **LLNL** | Lawrence Livermore National Laboratory | **TRMM** | Tropical Rainfall Measuring Mission |
| | | **UAH** | University of Alabama in Huntsville |
| **LOSU** | Level of Scientific Understanding | **UMd** | University of Maryland |
| **LULC** | Land Use/Land Cover | **USHCN** | United States Historical Climatology Network |
| **MAT** | Marine Air Temperatures | | |
| **MIT** | Massachusetts Institute of Technology | **UTC** | Coordinated Universal Time |
| **MSU** | Microwave Sounding Unit | **UW** | University of Washington - Seattle |
| **NAM** | Northern Hemisphere Annual Mode | **WMO** | World Meteorological Organization |
| **NAO** | North Atlantic Oscillation | | |
| **NASA** | National Aeronautics and Space Administration | | |
| **NCAR** | National Center for Atmospheric Research | | |
| **NCDC** | National Climatic Data Center | | |
| **NCEP** | National Centers for Environmental Prediction | | |
| **NEMS** | Nimbus E Microwave Spectrometer | | |
| **NESDIS** | National Environmental Satellite, Data, and Information Service | | |
| **NH** | Northern Hemisphere | | |
| **NMAT** | Night Marine Air Temperatures | | |
| **$N_2O$** | Nitrous Oxide | | |
| **NOAA** | National Oceanic and Atmospheric Administration | | |

REFERENCES

## CHAPTER I REFERENCES

**Anderson**, T.L., R.J. Charlson, S.E. Schwartz, R. Knutti, O. Boucher, H. Rodhe, and J. Heitzenberg, 2003: Climate forcing by aerosols – a hazy picture. *Science*, **300**, 1103-1104.

**Andrews**, D.G., J.R. Holton, and C.B. Leovy, 1987: *Middle Atmosphere Dynamics*. Academic Press, FL, 489 pp.

**Bradley**, R.S., 1988: The explosive volcano eruption signal in Northern Hemisphere continental temperature records. *Climatic Change*, **12**, 221-243.

**Broccoli**, A., K. Dixon, T. Delworth, T. Knutson, and R. Stouffer, 2003: Twentieth-century temperature and precipitation trends in ensemble climate simulations including natural and anthropogenic forcing. *Journal of Geophysical Research*, **108(D24)**, 4798, doi:10.1029/2003JD003812.

**Charlson**, R.J., S.E. Schwartz, J.M. Hales, R.D. Cess, J.A. Coakley, J.E. Hansen, and D.J. Hoffman, 1992: Climate forcing by anthropogenic aerosols. *Science*, **255**, 423-430.

**Christy**, J.R., R.W. Spencer, W.B. Norris, W.D. Braswell, and D.E. Parker, 2003: Error estimates of Version 5.0 of MSU/AMSU bulk atmospheric temperatures. *Journal of Atmospheric and Oceanic Technology*, 20, 613-629.

**Chung**, C., V. Ramanathan, and J. Kiehl, 2002: Effects of the South Asian absorbing haze on the northeast monsoon and surface-air heat exchange. *Journal of Climate*, **15**, 2462-2476.

**Cubasch**, U., 2001: Projections of future climate change. In: *Climate Change 2001 The Scientific Basis. Contribution of Working Group I to the Third Assessment Report of the Intergovernmental Panel on Climate Change* [Houghton, J.T., Y. Ding, M. Noguer, P.J. van der Linden, X. Dai, K. Maskell, and C.A. Johnson, (eds.)]. Cambridge University Press, Cambridge, UK, and New York, NY, 881 pp.

**Dai**, A., K.E. Trenberth, and T.R. Karl, 1999: Effects of clouds, soil moisture, precipitation, and water vapor on diurnal temperature range. *Journal of Climate*, **12**, 2451-2473.

**Donner**, L., C.J. Seman, R.S. Hemler, and S. Fan, 2001: A cumulus parameterization including mass fluxes, convective vertical velocities, and mesoscale effects: thermodynamic and hydrological aspects in a general circulation model. *Journal of Climate*, **14**, 3444-3463.

**Easterling**, D.R., B. Horton, P.D. Jones, T.C. Peterson, T.R. Karl, D.E. Parker, M.J. Salinger, V. Razuvayev, N. Plummer, P. Jamason, and C.K. Folland, 1997: Maximum and minimum temperature trends for the globe. *Science*, **277**, 364–367.

**Erlick**, C.E. and V. Ramaswamy, 2003: Sensitivity of the atmospheric lapse rate to solar cloud absorption in a radiative-convective model. *Journal of Geophysical Research*, **108(D16)**, 4522, doi:10.1029/2002JD002966.

**Frohlich**, C. and J. Lean, 2004: Solar radiative output and its variability: evidence and mechanisms. *Astronomy and Astrophysics Review*, **12**, 273-320.

**Gillett**, N., and D.W.J. Thompson, 2003: Simulation of recent Southern Hemisphere climate change. *Science*, **302**, 273-275.

**Goody**, R.M. and Y.L. Yung, 1989: *Atmospheric Radiation A Theoretical Basis*. Oxford University Press, New York, 519 pp.

**Hansen**, J., H. Wilson, M. Sato, R. Ruedy, K. Shah, and E. Hansen, 1995: Satellite and surface temperature data at odds? *Climatic Change*, **30**, 103-117.

**Hansen**, J., M. Sato, L. Nazarenko, R. Ruedy, A. Lacis, D. Koch, I. Tegen, T. Hall, D. Shindell, B. Santer, P. Stone, T. Novakov, L. Thomason, R. Wang, Y. Wang, D. Jacob, S. Hollandsworth, L. Bishop, J. Logan, A. Thompson, R. Stolarski, J. Lean, R. Willson, S. Levitus, J. Antonov, N. Rayner, D. Parker, and J. Christy, 2002: Climate forcings in Goddard Institute for Space Studies SI2000 simulations. *Journal of Geophysical Research*, **107(D18)**, 4347, doi:10.1029/2001JD001143.

**Hansen**, J.E., M. Sato, and R. Ruedy, 1997: Radiative forcing and climate response, *Journal of Geophysical Research*, **102**, 6831-6864.

**Hartmann**, D.L., 1994: *Global Physical Climatology*. Academic Press, San Diego, CA, 411 pp.

**Hegerl**, G.C., and J. M. Wallace, 2002: Influence of patterns of climate variability on the difference between satellite and surface temperature trends. *Journal of Climate*, **15**, 2412-2428.

**Held**, I.M., 1982: On the height of the tropopause and the static stability of the troposphere. *Journal of the Atmospheric Sciences*, **39**, 412-417.

**Holton**, J.R., 1979: *An Introduction to Dynamic Meteorology*. Academic Press, New York, 391 pp.

**Holton**, J.R., P.H. Haynes, M.E. McIntyre, A.R. Douglass, R.B. Rood, and L. Pfister, 1995: Stratosphere-troposphere exchange. *Reviews of Geophysics*, **33**(4), 403-439.

**Houghton**, J.T., 1977: *The Physics of Atmospheres*. Cambridge University Press, UK, 203 pp.

**Hurrell**, J.W., 1996: Influence of variations in extratropical wintertime teleconnections on Northern Hemisphere temperatures. *Geophysical Research Letters*, **23**, 665-668.

**Hurrell**, J.W. and K.E. Trenberth, 1996: Satellite versus surface estimates of air temperature since 1979. *Journal of Climate*, **9**, 2222-2232.

**IPCC**, 1990: *Scientific Assessment of Climate Change – Report of Working Group I* [Houghton, J.T., G.J. Jenkins and J.J. Ephraums (eds.)]. Cambridge University Press, Cambridge, UK, 365 pp.

**Jones**, P. D., 1994: Hemispheric surface air temperature variations: a reanalysis and an update to 1993. *Journal of Climate*, **7**, 1794-1802.

**Jones**, P.D., A. Moberg, T.J. Osborn, and K.R. Briffa, 2003: Surface climate responses to explosive volcanic eruptions seen in long European temperature records and mid-to-high latitude tree-ring density around the Northern Hemisphere, In: *Volcanism and the Earth's Atmosphere* [Robock, A. and C. Oppenheimer, (eds.)]. American Geophysical Union, Washington D.C , pp. 239-254.

**Kalnay**, E., M Kanamitsu, R. Kistler, W. Collins, D. Deaven, L. Gandin, M. Iredell, S. Saha, G. White, J. Woollen, Y. Zhu, M. Chelliah, W. Ebisuzaki, W. Higgins, J. Janowiak, K.C. Mo, C. Ropelewski, J. Wang, A. Leetmaa, R. Reynolds, R. Jenne, and D. Joseph, 1996: The NCEP/NCAR 40-year reanalysis project. *Bulletin of the American Meteorological Society,* **77**, 437-471.

**Kiehl**, J.T., 1992: Atmospheric general circulation modeling. *In Climate System Monitoring* [K. Trenberth, (ed.)]. Cambridge University Press, UK, pp. 319-369.

**Langematz**, U., M. Kunze, K. Kruger, K. Labitzke, and G. Roff, 2003: Thermal and dynamical changes of the stratosphere since 1979 and their link to ozone and $CO_2$ changes. *Journal of Geophysical Research*, **108**, 4027, doi: 1029/2002JD002069.

**Lean**, J., G. Rottman, J. Harder and G. Kopp, 2005: SORCE contributions to new understanding of global change and solar variability. *Solar Physics*, **230**, 27-53.

**Lindzen**, R.S. and K. Emanuel, 2002: The greenhouse effect. In: *Encyclopedia of Global Change, Environmental Change and Human Society*, Vol. 1 [A.S. Goudie (ed.)]. Oxford University Press, New York, NY, pp. 562-566.

**Lindzen**, R.S. and C. Giannitsis, 2002: Reconciling observations of global temperature change. *Geophysical Research Letters*, **29(12)**, 1583, doi:10.1029/2001GL014074.

**Lohmann**, U. and J. Feichter, 2005: Global indirect aerosol effects: a review. *Atmospheric Chemistry and Physics*, **5**, 715-737.

**Lohmann**, U., J. Feichter, J. Penner, and R. Leaitch, 2000: Indirect effect of sulfate and carbonaceous aerosols: a mechanistic treatment. *Journal of Geophysical Research*, **105**, 12193-12206.

**Mahlman**, J.D., J.P. Pinto, and L.J. Umscheid, 1994: Transport, radiative, and dynamical effects of the Antarctic ozone hole: a GFDL "SKYHI" experiment. *Journal of the Atmospheric Sciences*, **51**, 489-508.

**Manabe**, S. and R. Wetherald, 1967: Thermal equilibrium of the atmosphere with a given distribution of relative humidity. *Journal of the Atmospheric Sciences*, **24**, 241-259.

**McClatchey**, R.A., W. Fenn, J.E.A. Selby, F.E. Volz, J.S. Garing, 1972: *Optical Properties of the Atmosphere*, 3rd ed., AFCRL-72-0497, Air Force Cambridge Research Labs, Hanscom, MA, 110 pp.

**Mears**, C.A., M.C. Schabel, and F.W. Wentz, 2003: A reanalysis of the MSU channel tropospheric temperature record. *Journal of Climate*, **16**, 3650-3664.

**Meehl**, G.A., W.M. Washington, T.M.L. Wigley, J.M. Arblaster, and A. Dai, 2003: Solar and greenhouse gas forcing and climate response in the 20th century. *Journal of Climate*, **16**, 426-444.

**Meehl**, G.A., W.M. Washington, C. Ammann, J.M. Arblaster, T.M.L. Wigley, and C. Tebaldi, 2004: Combinations of natural and anthropogenic forcings and 20th century climate. *Journal of Climate*, **17**, 3721-3727.

**Menon**, S., J. Hansen, L. Nazarenko, and Y. Luo, 2002: Climate effects of back carbon aerosols in China and India. *Science*, **297**, 2250-2253.

**Mitchell**, J.F.B, D.J. Karoly, G.C. Hegerl, F.W. Zwiers, M.R. Allen, and J. Marengo, 2001: Detection of climate change and attribution of causes. In *Climate Change 2001 The Scientific Basis. Contribution of Working Group I to the Third Assessment Report of the Intergovernmental Panel on Climate Change* [Houghton, J.T., Y. Ding, M. Noguer, P.J. van der Linden, X. Dai, K. Maskell, and C.A. Johnson, (eds.)]. Cambridge University Press, Cambridge, UK, and New York, NY, 881 pp.

**NRC** (National Research Council), 2003: *Estimating Climate Sensitivity. Report of a Workshop*, [J. Mahlman, (Chair)]. National Academies Press, Washington, DC, 62 pp.

**NRC** (National Research Council), 2003: *Understanding Climate Change Feedbacks*. The National Academies Press, Washington, DC, 152 pp.

**NRC** (National Research Council), 2005: *Radiative Forcing of Climate Change Expanding the Concept and Addressing Uncertainties.* [D. Jacob, (Chair)]. National Academies Press, Washington, DC, 207 pp.

**Oort**, A.H. and J. Peixoto, 1992: *Physics of Climate*. American Institute of Physics, New York, 520 pp.

**Pielke**, R.A., Sr., 2001: Influence of the spatial distribution of vegetation and soils on the prediction of cumulus convective rainfall. *Review of Geophysics*, **39**, 151-177.

**Pielke**, R.A., Sr., C. Davey and J. Morgan, 2004: Assessing "global warming" with surface heat content. *EOS*, **85(21)**, 210-211.

**Pyle**, J., T. Shepherd, G.E. Bodeker, P. Caziani, M. Dameris, P.M. Forster, A. Gruzdev, R. Müller, N. Muthama, G. Pitari, and W.J. Randel, 2005: Ozone and climate: a review of interconnections (Chapter 1). In: *Safeguarding the Ozone Layer and the Global Climate System*, IPCC/TEAP Special Report [B. Metz, L. Kuijpers, S. Solomon, S.O. Andersen, O. Davidson, J. Pons, D. de Jager, T. Kestin, M. Manning, and L. Meyer (eds.)]. Cambridge University Press, Cambridge UK, pp 83-132.

**Ramanathan**, V., 1981: The role of ocean-atmosphere interactions in the $CO_2$ climate problem. *Journal of the Atmospheric Sciences*, **38**, 918-930.

**Ramanathan**, V. and J.A. Coakley, 1978: Climate modeling through radiative-convective models. *Reviews of Geophysics and Space Physics*, **16**, 465-489.

**Ramanathan**, V. and R. Dickinson, 1979: The role of stratospheric ozone in the zonal and seasonal radiative energy balance of the Earth-troposphere system. *Journal of the Atmospheric Sciences*, **36**, 1084-1104.

**Ramanathan**, V., R.D. Cess, E.F. Harrison, P. Minnis, B.R. Barkstrom, E. Ahmad, and D. Hartmann, 1989: Cloud radiative forcing and climate: Results from the Earth budget radiation experiment. *Science*, **243**, 57-63.

**Ramanathan**, V., C. Chung, D. Kim, T. Bettge, L. Buja, J.T. Kiehl, W.M. Washington, Q. Fu, D.R. Sikka, and M. Wild, 2005: Atmospheric brown clouds: impacts on South Asian climate and hydrological cycle. *Proceedings of the National Academy of Sciences*, **102**, 5326-5333.

**Ramaswamy**, V. and C-T. Chen, 1997: Climate forcing-response relationships for greenhouse and shortwave radiative perturbations. *Geophysical Research Letters*, **24**, 667-670.

**Ramaswamy**, V. and M.D. Schwarzkopf, 2002: Effects of ozone and well-mixed gases on annual-mean stratospheric temperature trends. *Geophysical Research Letters*, **29**, 2064, doi: 10.1029/2002GL015141.

**Ramaswamy**, V. and V. Ramanathan, 1989: Solar absorption by cirrus clouds and the maintenance of the tropical upper troposphere thermal structure. *Journal of the Atmospheric Sciences*, **46**, 2293-2310.

**Ramaswamy**, V.M., D. Schwarzkopf and W. Randel, 1996: Fingerprint of ozone depletion in the spatial and temporal pattern of recent lower-stratospheric cooling. *Nature*, **382**, 616-618.

**Ramaswamy**, V., O. Boucher, J. Haigh, D. Hauglustaine, J. Haywood, G. Myhre, T. Nakajima, G.Y. Shi, and S. Solomon, 2001: Radiative forcing of climate change. In: *Climate Change 2001 The Scientific Basis. Contribution of Working Group I to the Third Assessment Report of the Intergovernmental Panel on Climate Change* [Houghton, J.T., Y. Ding, D.J. Griggs, M. Noguer, P.J. van der Linden, X Dai, K. Maskell, and C.A. Johnson (eds.)]. Cambridge University Press, Cambridge, UK, and New York, NY, 881 pp.

**Ramaswamy**, V., M.L. Chanin, J. Angell, J. Barnett, D. Gaffen, M. Gelman, P. Keckhut, Y. Koshelkov, K. Labitzke, J.J.R. Lin, A. O'Neill, J. Nash, W. Randel, R. Rood, K. Shine, M. Shiotani, R. Swinbank, 2001a: Stratospheric temperature trends: observations and model simulations. *Reviews of Geophysics*, **39**, 71-122.

**Robock**, A., 2000: Volcanic eruptions and climate. *Reviews of Geophysics*, **38**, 191-219.

**Robock**, A. and C. Oppenheimer, 2003: *Volcanism and the Earth's Atmosphere*. AGU Geophysical Monograph Series 139, American Geophysical Union, Washington DC, 360 pp.

**Salby**, M.L., 1996: *Fundamentals of Atmospheric Physics*. Academic Press, San Diego, CA, 627 pp.

**Santer**, B.D., K.E. Taylor, T.M.L. Wigley, T.C. Johns, P.D. Jones, D.J. Karoly, J.F.B. Mitchell, A.H. Oort, J.E. Penner, V. Ramaswamy, M.D. Schwarzkopf, R.J. Stouffer, and S. Tett, 1996: A search for human influences on the thermal structure of the atmosphere. *Nature*, **382(6586)**, 39-46.

**Santer**, B.D., T.M.L. Wigley, D.J. Gaffen, L. Bengtsson, C. Doutriaux, J.S. Boyle, M. Esch, J.J. Hnilo, P.D. Jones, G.A. Meehl, E. Roeckner, K.E. Taylor, and M.F. Wehner, 2000: Interpreting differential temperature trends at the surface and in the lower troposphere. *Science*, **287**, 1227-1232.

**Santer**, B.D., M.F. Wehner, T.M.L. Wigley, R. Sausen, G.A. Meehl, K.E. Taylor, C. Ammann, J. Arblaster, W.M. Washington, J.S. Boyle, W. Bruggemann, 2003: Contributions of anthropogenic and natural forcing to recent tropopause height changes. *Science*, **301**, 479-483.

**Sarachik**, E., 1985: A simple theory for the vertical structure of the tropical atmosphere. *Pure and Applied Geophysics*, **123**, 261-271.

**Schwarzkopf**, M.D. and V. Ramaswamy, 2002: Effects of changes in well-mixed gases and ozone on stratospheric seasonal temperatures. *Geophysical Research Letters*, **29**, 2184, doi:10.1029/2002GL015759.

**Sherwood**, S.C., 2002: A microphysical connection among biomass burning, cumulus clouds, and stratospheric moisture. *Science*, **295**, 1271-1275.

**Sherwood**, S.C. and A.E. Dessler, 2003: Convective mixing near the tropical tropopause: insights from seasonal variations. *Journal of the Atmospheric Sciences*, **60**, 2674-2685.

**Shindell**, D.T., G A. Schmidt, R.L. Miller and D. Rind, 2001: Northern Hemisphere winter climate response to greenhouse gas, ozone, solar and volcanic forcing. *Journal of Geophysical Research*, **106**, 7193-7210.

**Shine**, K.P., M.S. Bourqui, P.M.D. Forster, S.H.E. Hare, U. Langematz, P. Braesicke, V. Grewe, M. Ponater, C. Schnadt, C.A. Smiths, J.D. Haighs, J. Austin, N. Butchart, D.T. Shindell, W.J. Randels, T. Nagashima, R.W. Portmann, S. Solomon, D.J. Seidel, J. Lanzante, S. Klein, V. Ramaswamy, and M.D..Schwarzkopf, 2003: A comparison of model-simulated trends in stratospheric temperatures. *Quarterly Journal of the Royal Meteorological Society*, **129**, 1565-1588.

**Shine**, K.P., J. Cook, E.J. Highwood, and M.M. Joshi, 2003a: An alternative to radiative forcing for estimating the relative importance of climate change mechanisms. *Geophysical Research Letters*, **30**, 2047, doi:10.1029/2003GL018141.

**Stenchikov**, G , A. Robock, V. Ramaswamy, M.D. Schwarzkopf, K. Hamilton, and S. Ramachandran, 2002: Arctic oscillation response to the 1991 Mount Pinatubo eruption: effects of volcanic aerosols and ozone depletion. *Journal of Geophysical Research*, **107**, 4803, doi:10.1029/2002JD002090.

**Stephens**, G.L. and P.J. Webster, 1981: Clouds and climate: sensitivity of simple systems. *Journal of the Atmospheric Sciences*, **38**, 235-247.

**Stocker,** T.F., G.K.C. Clarke, H. LeTreut, R.S. Lindzen, V.P. Melesh-ko, R.K. Mugara, T.N. Palmer, R.T. Pierrehumbert, P.J. Sellers, K E.Trenberth, and J. Willebrand, 2001: Physical climate processes and feedbacks. In *Climate Change 2001 The Scientific Basis. Contribution of Working Group I to the Third Assessment Report of the Intergovernmental Panel on Climate Change* [Houghton, J.T., Y. Ding, D.J. Griggs, M. Noguer, P.J. van der Linden, X Dai, K. Maskell, and C.A. Johnson (eds.)]. Cambridge University Press, Cambridge, UK, and New York, NY, USA, 881 pp.

**Stott,** P.A, S.F.B. Tett, G.S. Jones, M.R. Allen, J.F.B. Mitchell, G.J. Jenkins, 2000: External control of 20th century temperature by natural and anthropogenic forcings. *Science,* **290,** 2133-2137.

**Stouffer,** R.J., G. Hegerl and S. Tett, 2000: A comparison of surface air temperature variability in three 1000-year coupled ocean-atmosphere model integrations. *Journal of Climate,* **13,** 513-537.

**Thompson,** D.W.J. and J.M. Wallace, 2000: Annular modes in the extratropical circulation: 1; month-to-month variability. *Journal of Climate,* **13,** 1000-1016.

**Thompson,** D.W. and S. Solomon, 2002: Interpretation of recent Southern Hemisphere climate change, *Science,* **296,** 895-899.

**Trenberth,** K.E. and D.E. Stepaniak, 2003: Covariability of components of poleward atmospheric energy transports on seasonal and interannual timescales. *Journal of Climate,* **16,** 3706-3722.

**Trenberth,** K.E. and J.W. Hurrell, 1994: Decadal atmosphere-ocean variations in the Pacific. *Climate Dynamics,* **9,** 303-319.

**Trenberth,** K.E. and L. Smith, 2006: The vertical structure of temperature in the tropics: different flavors of El Nino. *Journal of Climate,* (in press).

**Trenberth,** K.E., J.R. Christy and J.W. Hurrell, 1992: Monitoring global monthly mean surface temperatures. *Journal of Climate,* **5,** 1405-1423.

**Vose,** R.S., D.R. Easterling, and B. Gleason, 2005: Maximum and minimum temperature trends for the globe; an update through 2004. *Geophysical Research Letters,* **33,** doi:10.1029/2005GL024379.

**Wielicki,** B.A., T.M. Wong, R.P. Allan, A. Slingo, J.T. Kiehl, B.J. Soden, C.T. Gordon, A.J. Miller, S.K. Yang, D.A. Randall, F. Robertson, J. Susskind, and H. Jacobowitz, 2002: Evidence for large decadal variability in the tropical mean radiative energy budget. *Science,* **295(5556),** 841-844.

**Wigley,** T.M.L., C.M. Ammann, B.D. Santer, and S.C.B. Raper, 2005: The effect of climate sensitivity on the response to volcanic forcing. *Journal of Geophysical Research,* **110,** D09107, doi:10.1029/2004JD005557.

**WMO** (World Meteorological Organization), 1986: *Atmospheric ozone.* Global Ozone Research and Monitoring Project, Report no. 16, WMO, Geneva, 1095 pp.

**WMO** (World Meteorological Organization), 1999: *Scientific assessment of ozone depletion 1998.* Global Ozone Research and Monitoring Project, Report no. 44, WMO, Geneva, chapter 5.

**WMO** (World Meteorological Organization), 2003: *Scientific assessment of ozone depletion 2002.* Global Ozone Research and Monitoring Project, Report no. 47, WMO, Geneva, 498 pp.

## CHAPTER 2 REFERENCES

**Agudelo,** P.A. and J.A. Curry, 2004: Analysis of spatial distribution in tropospheric temperature trends. *Geophysical Research Letters,* **31,** L222207.

**Aires,** F., C. Prigent, and W.B. Rossow, 2004: Temporal interpolation of global surface skin temperature diurnal cycle over land under clear and cloudy conditions. *Journal of Geophysical Research,* **109,** doi:10.1029/2003JD003527.

**Andrae,** U., N. Sokka and K. Onogi, 2004: The radiosonde temperature bias corrections used in ERA-40. ERA-40 Project Series no. 15, European Centre for Medium Range Weather Forecasts, Reading, UK, 37 pp. (Available at: http://www.ecmwf.int/publications/)

**Angell,** J.K., 2000: Tropospheric temperature variations adjusted for El Niño, 1958-1998. *Journal of Geophysical Research,* **105,** 11841-11849.

**Andronova,** N.G. and M.E. Schlesinger, 2000: Causes of global temperature changes during the 19th and 20th centuries. *Geophysical Research Letters,* **27,** 2137-2140.

**Baldwin,** M.P., L.J. Gray, T.J. Dunkerton, K. Hamilton, P.H. Haynes, W.J. Randel, J.R. Holton, M.J. Alexander, I. Hirota, T. Horinouchi, D.B.A. Jones, J.S. Kinnersley, C. Marquardt, K. Sato, and M. Takahashi, 2001: The quasi-biennial oscillation. *Reviews of Geophysics,* **39,** 179-229.

**Bengtsson,** L., S. Hagemann and K.I. Hodges, 2004: Can climate trends be calculated from reanalysis data? *Journal of Geophysical Research,* **109,** D11111, doi:10.1029/2004/JD004536.

**CCSP,** 2003: *Strategic Plan for the U.S. Climate Change Science Program, a Report by the Climate Change Science Program and the Subcommittee on Global Change Research.* U.S. Climate Change Science Program and the Subcommittee on Global Change Research, Washington, DC.

**Chelton,** D.B., 2005: The impact of SST specification on ECMWF surface wind stress fields in the eastern tropical Pacific. *Journal of Climate,* **18,** 530-550.

**Christy,** J.R., R.W. Spencer, and E.S. Lobl, 1998: Analysis of the merging procedure for the MSU daily temperature time series. *Journal of Climate,* **11,** 2016-2041.

**Christy,** J.R., R.W. Spencer, and R.T. McNider, 1995: Reducing noise in the MSU daily lower-tropospheric global temperature data set. *Journal of Climate,* **8,** 888-896.

**Christy,** J.R., R.W. Spencer, and W.D. Braswell, 2000: MSU Tropospheric temperatures: data set construction and radiosonde comparisons. *Journal of Atmospheric and Oceanic Technology* **17,** 1153-1170.

**Christy**, J.R., R.W. Spencer, W.B. Norris, W.D. Braswell and D.E. Parker, 2003: Error estimates of Version 5.0 of MSU/AMSU bulk atmospheric temperatures. *Journal of Atmospheric and Oceanic Technology* **20**, 613-629.

**Christy**, J.R. and R.T. McNider, 1994: Satellite greenhouse signal. *Nature*, **367**, 325.

**Christy**, J.R. and S.J. Drouilhet, 1994: Variability in daily, zonal mean lower-stratospheric temperatures. *Journal of Climate*, **7**, 106-120.

**Christy**, J.R. and W.B. Norris, 2004: What may we conclude about tropospheric temperature trends? *Geophysical Research Letters* **31**, L06211.

**Dai**, A. and K. E. Trenberth, 2004: The diurnal cycle and its depiction in the Community Climate System Model, *Journal of Climate*, *17*, 930-951.

**Douglass**, D.H. and B.D. Clader, 2002: Determination of the climate sensitivity of the earth to solar irradiance. *Geophysical Research Letters*, *29*, 331-334.

**Duvel**, J.P., R. Roca and J. Vialard, 2004: Ocean mixed layer temperature variations induced by intraseasonal convective perturbations of the Indian Ocean. *Journal of the Atmospheric Sciences*, **9**, 1004-1023.

**Enfield**, D.B., Mestas-Nuñez, A.M. and P.J Trimble, 2001: The Atlantic multidecadal oscillation and its relation to rainfall and river flows in the continental U.S. *Geophysical Research Letters,* **28**, 2077-2080.

**Folland**, C.K. 2005: Assessing bias corrections in historical sea surface temperature using a climate model. *International Journal of Climatology,* **25** (Special issue, CLIMAR II Conference), 895-911.

**Folland**, C.K. and D.E. Parker, 1995: Correction of instrumental biases in historical sea surface temperature data. *Quarterly Journal of the Royal Meteorological Society* **121,** 319-367.

**Folland**, C.K., D.E. Parker, A. Colman, and R. Washington, 1999: Large scale modes of ocean surface temperature since the late nineteenth century (Chapter 4). In: *Beyond El Nino Decadal and Interdecadal Climate Variability* [Navarra, A. (ed.)]. Springer-Verlag, Berlin, pp. 73-102.

**Folland**, C.K., J.A. Renwick, M J. Salinger and A.B. Mullan, 2002: Relative influences of the interdecadal Pacific oscillation and ENSO on the South Pacific convergence zone. *Geophysical Research Letters*, **29(13)**, doi:10.1029/2001GL014201.

**Folland**, C.K., N.A. Rayner, S.J. Brown, T.M. Smith, S.S.P. Shen, D.E. Parker, I. Macadam, P.D. Jones, R.N. Jones, N. Nicholls and D.M.H. Sexton, 2001b: Global temperature change and its uncertainties since 1861. *Geophysical Research Letters*, **28**, 2621-2624.

**Folland**, C.K., T.R. Karl, J.R. Christy, R.A. Clarke, G.V. Gruza, J. Jouzel, M.E. Mann, J. Oerlemans, M.J. Salinger, and S.-W. Wang, 2001a: Observed climate variability and change. In: *Climate Change 2001 The Scientific Basis. Contribution of Working Group I to the Third Assessment Report of the Intergovernmental Panel on Climate Change* [Houghton, J.T., Y. Ding, D.J. Griggs, M. Noguer, P.J. van der Linden, X Dai, K. Maskell, and C.A. Johnson (eds.)]. Cambridge University Press, Cambridge, United Kingdom and New York, NY, 881 pp.

**Free**, M. and J.K. Angell, 2002: Effect of volcanoes on the vertical temperature profile in radiosonde data. *Journal of Geophysical Research*, **107(D10)**, doi:10.1029/2001JD001128.

**Free**, M., I. Durre, E.Aguilar, D. Seidel, T.C. Peterson, R.E. Eskridge, J.K. Luers, D. Parker, M. Gordon, J. Lanzante, S. Klein, J. Christy, S. Schroeder, B. Soden, and L.M. McMillin, 2002: CARDS workshop on adjusting radiosonde temperature data for climate monitoring: meeting summary. *Bulletin of the American Meteorological Society,* **83**, 891-899.

**Fu**, Q. and C.M. Johanson, 2005: Satellite-derived vertical dependence of tropical tropospheric temperature trends. *Geophysical Research Letters*, **32**, L10703, doi: 10.1029/2004GL022266.

**Fu**, Q., C.M. Johanson, S.G. Warren, and D.J. Seidel, 2004: Contribution of sratospheric cooling to satellite-inferred troposheric temperature trends. *Nature*, **429**, 55-58.

**Gillett**, N.P., B.D. Santer, and A.J. Weaver, 2004, Stratospheric cooling and the troposphere, *Nature*, **432**, doi:10.1038/nature03209.

**Goldenberg**, S.B, Landsea, C.W., Mestas Nunez, A.M. and W.M. Gray, 2001: The recent increase in Atlantic hurricane activity: causes and implications. *Science*, **293**, 474-479.

**Grody**, N.C., K.Y. Vinnikov, M.D. Goldberg, J.T. Sullivan, and J.D. Tarpley, 2004. Calibration of multi-satellite observations for climatic studies: Microwave Sounding Unit (MSU), *Journal of Geophysical Research, Atmospheres.*, **109**, D24104, doi:10.1029/2004JD005079.

**Hasselmann**, K., 1999: Linear and nonlinear signatures. *Nature*, **398**, 755-756.

**Haimberger**, L., 2004: *Homogenization of Radiosonde Temperature Time Series Using ERA-40 Analysis Feedback Information.* ERA-40 Project Report Series no. 22, European Centre for Medium Range Weather Forecasts, Reading, UK, 67 pp.

**Hurrell**, J., S.J. Brown, K.E. Trenberth, and J.R. Christy, 2000: Comparison of tropospheric temperatures from radiosondes and satellites: 1979-1998. *Bulletin of the American Meteorological Society,* **81**, 2165-2177.

**Hurrell**, J.W., Y. Kushnir, G. Ottersen, and M. Visbeck, 2003: An overview of the North Atlantic Oscillation. In *The North Atlantic Oscillation Climatic Significance and Environmental Impacts.* Hurrell, J.W., Y. Kushnir, G. Ottersen, and M. Visbeck, eds., Geophysical Monograph Series **134**, American Geophysical Union, Washington, DC, pp. 1-35.

**Jin**, M., 2004: Analysis of land skin temperature using AVHRR observations, *Bulletin of the American Meteorological Society,* **85**, 587-600, doi:10.1175/BAMS-85-4-587).

**Johanson**, C.M. and Q. Fu, 2006: Robustness of tropospheric temperature trends from MSU channels 2 and 4. *Journal of Climate (in press).*

**Jones**, P.D and A. Moberg, 2003: Hemispheric and large scale surface air temperature variations: an extensive revision and an update to 2001. *Journal of Climate*, **16**, 206-223.

**Jones**, P.D., T.J. Osborn, K.R. Briffa, C.K. Folland, E.B. Horton, L.V. Alexander, D.E. Parker and N.A. Rayner, 2001: Adjusting for sampling density in grid box land and ocean surface temperature time series. *Journal of Geophysical Research*, **106**, 3371-3380.

**Kalney**, E., M. Kanamitsu, R. Kistler, W. Collins, D. Deaven, L. Gandin, M. Iredell, S. Saha, G. White, J. Woollen, Y. Zhu, M. Chelliah, W. Ebisuzaki, W. Higgins, J. Janowiak, K.C. Mo, C. Ropelewski, J. Wang, A. Leetmaa, R. Reynolds, R. Jenne, and D. Joseph, 1996: The NCEP/NCAR 40-year reanalysis project. *Bulletin of the American Meteorological Society*, **77**, 437-471.

**Kalnay**, E. and M. Cai, 2003: Impact of urbanization and land-use change on climate. *Nature*, **423**, 528-531.

**Kent**, E.C. and A. Kaplan, 2006: Toward estimating climatic trends in SST data, part 3: systematic biases. *Journal of Atmospheric and Oceanic Technology*, (in press).

**Kent**, E.C. and P.G. Challenor, 2006: Toward estimating climatic trends in SST data, part 2: random errors. *Journal of Atmospheric and Oceanic Technology*, (in press).

**Kent**, E.C. and P.K. Taylor, 2006: Toward estimating climatic trends in SST data, part 1: methods of measurement. *Journal of Atmospheric and Oceanic Technology*, (in press).

**Kiehl**, J.T, J.M. Caron and J.J. Hack, 2005: On using global climate model simulations to assess the accuracy of MSU retrieval methods for tropospheric warming trends. *Journal of Climate*, **18**, 2533-2539.

**Kilpatrick**, K.A., G.P. Podesta, and R. Evans, 2001: Overview of the NOAA/NASA advanced very high resolution radiometer pathfinder algorithm for sea surface temperature and associated matchup database. *Journal of Geophysical Research*, **106(C5),** 9179-9198.

**Knight**, J., Allan, R.J., Folland, C.K., Vellinga, M. and Mann, M.E., 2005: A signature of persistent natural thermohaline circulation cycles in observed climate. *Geophysical Research Letters*, **32**, L20708, doi:10 1029/1005GL024233

**Kursinski** E.R., G.A. Hajj, J.T. Schofiled, R.P. Linfield, and K.R. Hardy, 1997: Observing the Earth's atmosphere with radio occultation measurements using the global positioning system. *Journal of Geophysical Research* **102**, 23429-23465.

**Labitzke**, K., 1987: Sunspots, the QBO, and the stratospheric temperature in the north polar region. *Geophysical Research Letters*, **14**, 535-537.

**Lean**, J., J. Beer, and R. Bradley, 1995: Reconstruction of solar irradiance since 1610: implications for climate change. *Geophysical Research Letters*, **22**, 3195-3198.

**Mann**, M.E , R.S. Bradley and M.K. Hughes, 1998: Global-scale temperature patterns and climate forcing over the past six centuries. *Nature*, **392**, 779-787.

**McPhaden**, M.J., 1995: The Tropical Atmosphere Ocean array is completed. *Bulletin of the American Meteorological Society,* **76**: 739-741.

**Mears**, C.A., M.C. Schabel, and F.J. Wentz, 2003: A reanalysis of the MSU channel 2 tropospheric temperature record. *Journal of Climate*, **16**, 3650-3664.

**Michaels**, P.J., and P.C. Knappenberger, 2000: Natural signals in the MSU lower tropospheric temperature record. *Geophysical Research Letters* **27**, 2905-2908.

**Morgan**, M.G., 1990: *Uncertainty  A Guide to Dealing with Uncertainty in Quantitative Risk and Policy Analysis.* Cambridge University Press, UK, 332 pp.

**NRC** (National Research Council), 2000a: *Ensuring the Climate Record from the NPP and NPOESS Meteorological Satellites.* National Academy Press, Washington, DC, 51 pp.

**NRC** (National Research Council), 2000b: *Issues in the Integration of Research and Operational Satellite Systems for Climate Research II Implementation.* National Academy Press. Washington, DC, 82 pp.

**NRC** (National Research Council), 2004: *Climate data records from environmental satellites.* National Academy Press, Washington, DC, 150 pp.

**Palmen**, E. and C. Newton, 1969: *Atmospheric Circulation Systems Their Structure and Interpretation.* Academic Press. New York, NY, 606 pp.

**Pan**, Y.-H. and A.H. Oort, 1983: Global climate variations connected with sea surface temperature anomalies in the eastern equatorial Pacific Ocean for the 1958-1973 period. *Monthly Weather Review*, **111**, 1244-1258.

**Parker**, D.E. and J.L. Brownscombe, 1983: Stratospheric warming following the El Chichón volcanic eruption. *Nature*, **301**, 406-408.

**Parker**, D.E., M. Gordon, D.P.N. Cullum, D.M.H. Sexton, C.K. Folland and N. Rayner, 1997: A new global gridded radiosonde temperature data base and recent temperature trends. *Geophysical Research Letters*, **24**, 1499-1502.

**Parker**, D.E., H. Wilson, P.D. Jones, J.R. Christy and C.K. Folland, 1996: The impact of Mount Pinatubo on world-wide temperatures. *International Journal of Climatology*, **16**, 487-497.

**Pawson**, S. and M. Fiorino, 1999: A comparison of reanalyses in the tropical stratosphere, part 3: inclusion of the pre-satellite data era, *Climate Dynamics*, **15**, 241-250.

**Power**, S., Casey, T., Folland, C.K., Colman, A and V. Mehta, 1999: Inter-decadal modulation of the impact of ENSO on Australia. *Climate Dynamics,* **15**, 319-323.

**Ramaswamy**, V., M.-L. Chanin, J. Angell, J. Barnett, D. Gaffen, M. Gelman, P. Kekhut, Y. Koshelkov, K. Labitzke, J.-J. R. Lin, A. O'Neill, J. Nash, W. Randel, R.Rood, K. Shine, M. Shiotani, and R. Swinbank, 2001: Stratospheric temperature trends: observations and model simulations. *Reviews of Geophysics,* **39**, 71-122.

**Randel**, W.J., F. Wu, R. Swinbank, J. Nash, and A. O'Neill, 1999: Global QBO circulation derived from UKMO stratospheric analyses. *Journal of the Atmospheric Sciences,* **56**, 457-474.

**Randel**, W.J. and F. Wu, 2006: Biases in stratospheric and tropospheric temperature trends derived from historical radiosonde data. *Journal of Climate,* (in press).

**Rayner,** N.A., D E. Parker, E B. Horton, C.K. Folland, L.V. Alexander, D.P. Rowell, E.C. Kent, A. Kaplan, 2003: Global analyses of sea surface temperature, sea ice, and night marine air temperature since the late nineteenth century, *Journal of Geophysical Research,* **108(D14)**, 4407, doi:10.1029/2002JD002670.

**Reynolds**, R.W., 1993: Impact of Mt Pinatubo aerosols on satellite-derived sea surface temperatures. *Journal of Climate,* **6**, 768-774.

**Reynolds**, R.W., Rayner, N.A. Smith, T.H. Stokes, D.C. and Wang W., 2002: An improved in situ and satellite SST analysis for climate. *Journal of Climate,* **15**, 1609-1625.

**Robock**, Alan, 2000: Volcanic eruptions and climate. *Reviews of Geophysics,* **38**, 191-219.

**Santer**, B.D., J.J. Hnilo, T.M.L. Wigley, J.S. Boyle, C. Doutriaux, M. Fiorino, D.E. Parker, and K.E. Taylor, 1999: Uncertainties in observationally based estimates of temperature change in the free atmosphere. *Journal of Geophysical Research,* **104**, 6305-6333.

**Santer**, B.D., T.M.L. Wigley, C. Doutriaux, J.S. Boyle, J.E. Hansen, P.D. Jones, G.A. Meehl, E. Roeckner, S. Sengupta, and K.E. Taylor, 2001: Accounting for the effects of volcanoes and ENSO in comparisons of modeled and observed temperature trends. *Journal of Geophysical Research,* **106**, 28033-28059.

**Scaife**, A.A., Knight, J.R., Vallis, G.K. and Folland, C.K., 2005: A stratospheric influence on the winter NAO and North Atlantic surface climate. *Geophysical Research Letters,* **32**, L18715, doi:10.1029/2005GL023226.

**Schlesinger**, M.E. and N. Ramankutty, 1994: An oscillation in the global climate system of period 65-70 years. *Nature,* **367**, 723-726.

**Seidel**, D.J., J.K. Angell, J.R. Christy, M. Free, S.A. Klein, J.R. Lanzante, C. Mears, D. Parker, M. Schabel, R. Spencer, A. Sterin, P. Thorne and F. Wentz, 2004: Uncertainty in signals of large-scale climate variations in radiosonde and satellite upper-air temperature data sets. *Journal of Climate,* **17**, 2225-2240.

**Seidel**, D.J., M. Free, and J. Wang, 2005: The diurnal cycle of temperature in the free atmosphere estimated from radiosondes, *Journal of Geophysical Research* **110**, D09102, doi:10.1029/2004JD005526.

**Seidel**, D.J. and J.R. Lanzante, 2004: An assessment of three alternatives to linear trends for characterizing global atmospheric temperature changes. *Journal of Geophysical Research,* **109**, doi:10.1029/2003JD004414.

**Sherwood**, S.C , 2000: Climate signal mapping and an application to atmospheric tides, *Geophysical Research Letters,* **27(21)**, 3525-3528.

**Sherwood**, S.C., J. Lanzante, and C. Meyer, 2005: Radiosonde daytime biases and late 20th century warming. *Science,* **309**, 1556-1559.

**Simmons**, A.J., Jones, P.D., da Costa Bechtold, V., Beljaars, A.C.M., Kållberg, P., Saarinen, S., Uppala, S.M., Viterbo, P. and N. Wedi, N. 2004: Comparison of trends and variability in CRU, ERA-40 and NCEP/NCAR analyses of monthly-mean surface air temperature. *Journal of Geophysical Research-Atmospheres,* **109(D24)** D24115 doi.org/10.1029/ 2004JD005306.

**Smith**, T.M. and R.W. Reynolds, 2004: Improved extended reconstruction of SST (1854-1997). *Journal of Climate,* **17**, 2466-2477.

**Smith**, T.M. and R.W. Reynolds, 2005: A global merged land and sea surface temperature reconstruction based on historical observations (1880-1997). *Journal of Climate,* **18**, 2021-2036.

**Spencer**, R.W., J R. Christy and N.C. Grody, 1990: Global atmospheric temperature monitoring with satellite microwave measurements: method and results 1979-1985. *Journal of Climate,* **3**, 1111-1128.

**Spencer**, R.W. and J.R. Christy, 1992: Precision and radiosonde validation of satellite gridpoint temperature anomalies, part II: a tropospheric retrieval and trends during 1979-90. *Journal of Climate,* **5**, 858-866.

**Spencer**, R.W., J.R. Christy and W.D. Braswell, 2006: On the estimation of tropospheric temperature trends from MSU channels 2 and 4. *Journal of Atmospheric and Oceanic Technology* (in press).

**Tett**, S.F.B. and P.W. Thorne, 2004: Comment on tropospheric temperature series from satellites. *Nature,* **432**, doi:10.1038/nature03208.

**Thompson**, D.W.J., J.M. Wallace and G.C. Hegerl, 2000: Annual modes in the extratropical circulation, part II: trends. *Journal of Climate,* **13**, 1018-1036.

**Thorne**, P.W., D.E. Parker, J.R. Christy and C.A. Mears, 2005a: Causes of differences in observed climate trends. *Bulletin of the American Meteorological Society,* **86**, 1437-1442.

**Thorne**, P.W., D.E. Parker, S.F.B. Tett, P.D. Jones, M. McCarthy, H. Coleman, and P. Brohan, 2005b: Revisiting radiosonde upper-air temperatures from 1958 to 2002. *Journal of Geophysical Research,* **110**, D18105, doi:10.1029/2004JD005753.

**Trenberth**, K.E. and J.W. Hurrell, 1994: Decadal atmosphere-ocean variations in the Pacific. *Climate Dynamics,* **9**, 303-319.

**Trenberth**, K.E., Carron, J M , Stepaniak, D.P. and S. Worley, 2002: Evolution of the El Nino-southern oscillation ad global atmospheric surface temperatures. *Journal of Geophysical Research*, **107(D8)**, 10.1029/2000JD000298.

**Uppala**, S.M., P.W. Kållberg, A.J. Simmons, U. Andrae, V. da Costa Bechtold, M. Fiorino, J.K. Gibson, J. Haseler, A. Hernandez, G.A. Kelly, X. Li, K. Onogi, S. Saarinen, N. Sokka, R.P. Allan, E. Andersson, K. Arpe, M.A. Balmaseda, A.C.M. Beljaars, L. van de Berg, J. Bidlot, N. Bormann, S. Caires, F. Chevallier, A. Dethof, M. Dragosavac, M. Fisher, M. Fuentes, S. Hagemann, E. Hólm, B.J. Hoskins, L. Isaksen, P.A.E.M. Janssen, R. Jenne, A.P. McNally, J-F. Mahfouf, J.-J. Morcrette, N.A. Rayner, R.W. Saunders, P. Simon, A. Sterl, K.E. Trenberth, A. Untch, D. Vasiljevic, P. Viterbo, and J. Woollen, 2005: The ERA-40 reanalysis. *Quarterly Journal of the Royal Meteorological Society*, **131**, 2961-3012.

**Wallace**, J.M. and P.V. Hobbs, 1977: *Atmospheric Science An Introductory Survey*. Academic Press, New York, NY, 467 pp.

**Wallis**, T.W.R., 1998: A subset of core stations from the Comprehensive Aerological Reference Data Set (CARDS). *Journal of Climate*, **11**, 272-282.

**Vinnikov**, K.Y., and N.C. Grody, 2003. Global warming trend of mean tropospheric temperature observed by satellites, *Science*, **302**, 269-272.

**Vinnikov**, K.Y., N.C. Grody, A. Robock, R.J. Stouffer, P.D. Jones, and M.D. Goldberg, 2006: Temperature trends at the surface and in the troposphere. *Journal of Geophysical Research*, **111**, D03106, doi:10.1029/2005jd006392.

**Zhang**, Y., J.M. Wallace, and D.S. Battisti, 1997: ENSO-like interdecadal variability: 1900-93, *Journal of Climate*, **10**, 1004-1020.

## CHAPTER 3 REFERENCES

**Aguilar**, E , I. Auer, M. Brunet, T.C. Peterson and J. Wieringa, 2003: *Guidelines on Climate Metadata and Homogenization, WCDMP no. 53, WMO-TD no. 1186.* World Meteorological Organization, Geneva, 55 pp.

**Angell**, J.K., and J. Korshover, 1975: Estimate of the global change in tropospheric temperature between 1958 and 1973. *Monthly Weather Review*, **103**, 1007-1012.

**Angell**, J.K., 2003: Effect of exclusion of anomalous tropical stations on temperature trends from a 63-station radiosonde network, and comparison with other analyses. *Journal of Climate*, **16**, 2288-2295.

**Bottomley**, M., Folland, C.K., Hsiung, J., Newell, R.E. and D.E. Parker, 1990: *Global Ocean Surface Temperature Atlas (GOSTA)*. HMSO, London, 20 pp., 313 plates.

**Brown**, S.J., D.E. Parker, C.K. Folland, and I. Macadam, 2000: Decadal variability in the lower-tropospheric lapse rate. *Geophysical Research Letters*, **27**, 997-1000.

**Christy**, J.R., D.E. Parker, S.J. Brown, I. Macadam, M. Stendel, and W.B. Norris, 2001: Differential trends in tropical sea surface and atmospheric temperature since 1979. *Geophysical Research Letters*, **28**, 183-186.

**Christy**, J.R., R.W. Spencer, and E.S. Lobl, 1998: Analysis of the merging procedure for the MSU daily temperature time series. *Journal of Climate*, **11**, 2016-2041.

**Christy**, J.R., R.W. Spencer, and W.D. Braswell, 2000: MSU tropospheric temperatures: data set construction and radiosonde comparisons. *Journal of Atmospheric and Oceanic Technology*, **17**, 1153-1170.

**Christy**, J.R., R.W. Spencer, W.B. Norris, W D. Braswell, and D.E. Parker, 2003: Error estimates of version 5.0 of MSU-AMSU bulk atmospheric temperatures. *Journal of Atmospheric and Oceanic Technology*, **20**, 613-629.

**Dai**, A., Trenberth, K.E. and Karl, T.R., 1999: Effects of clouds, soil moisture, precipitation, and water vapor on diurnal temperature range. *Journal of Climate*, **12**, 2451-2473.

**Deser**, C., A.S. Phillips, and J.W. Hurrell, 2004: Pacific interdecadal climate variability: Linkages between the tropics and the north Pacific during boreal winter since 1900. *Journal of Climate*, **17**, 3109-3124.

**Diaz**, H.F., C.K. Folland, T. Manabe, D.E. Parker, R.W. Reynolds, and S.D. Woodruff, 2002: Workshop on Advances in the Use of Historical Marine Climate Data (Boulder, CO., USA, 29th Jan - 1st Feb 2002). *WMO Bulletin*, **51**, 377-380

**Durre**, I., R.S. Vose, and D.B. Wuertz, 2006: Overview of the integrated global radiosonde archive. *Journal of Climate*, **19**, 53-68.

**Easterling**, D.R., T.R. Karl, E.H. Mason, P.Y. Hughes, and D.P. Bowman, 1996: *United States Historical Climatology Network (U.S.HCN), monthly temperature and precipitation data*, Environmental Sciences Division publication no. 4500, ORNL/CDIAC 87, NDP-0l9/R3, Carbon Dioxide Information Analysis Center, Oak Ridge, 263 pp.

**Easterling**, D. R., B. Horton, P.D. Jones, T.C. Peterson, T.R. Karl, D.E. Parker, M.J. Salinger, V. Razuvayev, N. Plummer, P. Jamason, and C.K. Folland, 1997: Maximum and minimum temperature trends for the globe. *Science*, **277**, 364-367.

**Folland**, C.K., N.A. Rayner, S.J. Brown, T.M. Smith, S.S.P. Shen, D.E. Parker, I. Macadam, P.D. Jones, R.N. Jones, N. Nicholls, and D.M.H. Sexton, 2001a: Global temperature change and its uncertainties since 1861. *Geophysical Research Letters*, **28**, 2621-2624.

**Folland**, C.K., T.R. Karl, J.R. Christy, R.A. Clarke, G.V. Gruza, J. Jouzel, M.E. Mann, J. Oerlemans, M.J. Salinger and S-W. Wang, and 142 other authors, 2001b: Observed climate variability and change. In: *Climate Change 2001 The Scientific Basis. Contribution of Working Group I to the Third Assessment Report of the Intergovernmental Panel on Climate Change* [Houghton, J. T., Y Ding, D. J. Griggs, M. Noguer, P. van der Linden, X. Dai, K. Maskell, and C. I. Johnson (eds.)]. Cambridge University Press, Cambridge, UK, and New York, NY, pp. 99-181.

**Folland**, C.K., M.J. Salinger, N. Jiang, and N.A. Rayner, 2003: Trends and variations in South Pacific Island and ocean surface temperatures. *Journal of Climate*, **16**, 2859-2874.

**Free**, M., and J.K. Angell, 2002: Effect of volcanoes on the vertical temperature profile in radiosonde data. *Journal of Geophysical Research*, **107**, 4101, doi:10.1029/2001JD001128.

**Free**, M., J K. Angell, I. Durre, J.R. Lanzante, T.C. Peterson, and D.J. Seidel, 2003: Using first differences to reduce inhomogeneity in radiosonde temperature data sets. *Journal of Climate*, **21**, 4171-4179.

**Free**, M., D.J. Seidel, J.K. Angell, J.R. Lanzante, I. Durre, and T.C. Peterson, 2005: Radiosonde Atmospheric Temperature Products for Assessing Climate (RATPAC): a new data set of large-area anomaly time series. *Journal of Geophysical Research*, **110**, D22101, doi:10.1029/2005JD006169.

**Fu**, Q., and C.M. Johanson, 2005: Satellite-derived vertical dependence of tropical tropospheric temperature trends. *Geophysical Research Letters*, **32**, L10703, doi:10.1029/2004GL022266.

**Fu**, Q., C.M. Johanson, S. Warren, and D. Seidel, 2004: Contribution of stratospheric cooling to satellite-inferred tropospheric temperature trends. *Nature*, **429**, 55-58.

**Gaffen**, D, B. Santer, J. Boyle, J. Christy, N. Graham, and R. Ross, 2000: Multi-decadal changes in the vertical temperature structure of the tropical troposphere. *Science*, **287**, 1239-1241.

**Gettelman**, A., D.J. Seidel, M.C. Wheeler, and R.J. Ross, 2002: Multidecadal trends in tropical convective available potential energy. *Journal of Geophysical Research*, **107**, 4606, doi:10.1029/2001JD001082.

**Grody**, N.C., K.Y. Vinnikov, M.D. Goldberg, J.T. Sullivan, and J.D. Tarpley, 2004: Calibration of multi-satellite observations for climatic studies: Microwave Sounding Unit (MSU). *Journal of Geophysical Research*, **109**, D24104, doi:10.1029/2004JD005079.

**Groisman**, P.Y., R.W. Knight, T.R. Karl, D.R. Easterling, B. Sun, and J.H. Lawrimore, 2004: Contemporary changes of the hydrological cycle over the contiguous United States, trends derived from in situ observations. *Journal of Hydrometeorology*, **5,** 64-85.

**Hansen**, J., R. Ruedy, M. Sato, and R. Reynolds 1996. Global surface air temperature in 1995: return to pre-Pinatubo level. *Geophysical Research Letters*, **23**, 1665-1668.

**Hansen**, J., R. Ruedy, M. Sato, M. Imhoff, W. Lawrence, D. Easterling, T. Peterson, and T. Karl, 2001: A closer look at United States and global surface temperature change. *Journal of Geophysical Research*, **106**, 23947-23963.

**Hegerl**, G. and J. Wallace, 2002: Influence of patterns of climate variability on the difference between satellite and surface temperature trends. *Journal of Climate*, **15**, 2412-2428.

**Henderson-Sellers**, A., 1992: Continental cloudiness changes this century. *GeoJournal*, **27**, 255–262.

**Houghton**, J.T., Y. Ding, D. J. Griggs, M. Noguer, P.J. van der Linden, X. Dai, K. Maskell, and C.A. Johnson, (eds.), 2001: *Climate Change 2001 The Scientific Basis. Contribution of Working Group I to the Third Assessment Report of the Intergovernmental Panel on Climate Change.* Cambridge University Press, Cambridge, UK, and New York, NY, 881 pp.

**Hurrell**, J.W., 1996: Influence of variations in extratropical wintertime teleconnections on Northern Hemisphere temperature. *Geophysical Research Letters*, **23**, 665-668.

**Johanson**, C.M., and Q. Fu, 2006: Robustness of tropospheric temperature trends from MSU channels 2 and 4. *Journal of Climate* (accepted).

**Jones**, P.D., 1995: Land surface temperatures - is the network good enough? *Climatic Change* **31**, 545-558.

**Jones**, P.D., P.Y. Groisman, M. Coughlan, N. Plummer, W-C. Wang, and T.R. Karl, 1990: Assessment of urbanization effects in time series of surface air temperature over land. *Nature*, **347**, 169-172.

**Jones**, P.D., T.J. Osborn, and K.R. Briffa, 1997: Estimating sampling errors in large-scale temperature averages. *Journal of Climate*, **10**, 2548-2568.

**Jones**, P.D., T.J. Osborn, K.R. Briffa, C.K. Folland, B. Horton, L.V. Alexander, D.E. Parker, and N.A. Rayner, 2001: Adjusting for sampling density in grid-box land and ocean surface temperature time series. *Journal of Geophysical Research*, **106**, 3371-3380.

**Jones**, P.D. and A. Moberg, 2003: Hemispheric and large-scale surface air temperature variations: An extensive revision and an update to 2001. *Journal of Climate*, **16**, 206-223.

**Karl**, T.R, P.D. Jones, R.W. Knight, G. Kukla, N. Plummer, V. Razuvayev, K.P. Gallo, J. Lindseay, R.J. Charlson, and T.C. Peterson, 1993: A new perspective on recent global warming: asymmetric trends of daily maximum and minimum temperature. *Bulletin of the American Meteorological Society,* **74**, 1007-1024.

**Karl**, T.R., R.W. Knight, and B. Baker, 2000: The record breaking global temperatures of 1997 and 1998: evidence for an increase in the rate of global warming? *Geophysical Research Letters*, **27**, 719-722.

**Lanzante**, J.R., S.A. Klein, and D.J. Seidel, 2003a: Temporal homogenization of monthly radiosonde temperature data, part I: methodology. *Journal of Climate*, **16**, 224-240.

**Lanzante**, J.R., S.A. Klein, and D.J. Seidel, 2003b: Temporal homogenization of monthly radiosonde temperature data, part II: trends, sensitivities, and MSU comparison. *Journal of Climate*, **16**, 241-262.

**Mears**, C.A. and F.J. Wentz, 2005: The effect of diurnal correction on satellite-derived lower tropospheric temperature. *Science*, **309**, 1548-1551.

**Mears**, C.A., M.C. Schabel, and F.J. Wentz, 2003: A Reanalysis of the MSU channel 2 tropospheric temperature record. *Journal of Climate*, **16**, 3650-3664.

**Oleson**, K.W., G.B. Bonan, S. Levis and M. Vertenstein, 2004: Effects of land use change on North American climate: impact of surface data sets and model biogeophysics. *Climate Dynamics*, **23**, 117-132.

**Oort**, A.H., 1983: *Global Atmospheric Circulation Statistics, 1958-1973*, NOAA Professional Paper no. 4, U.S. Government Printing Office, Washington, DC, 180 pp.

**Parker**, D.E., 2004: Large-scale warming is not urban. *Nature*, **432**, 290-291, (Available also online: doi:10.1038/432290b).

**Parker** D.E., C.K. Folland, and M. Jackson, 1995: Marine surface temperature: observed variations and data requirements. *Climatic Change*, **31(2-4)**, 559-600 (Also in: *Long-Term Climate Monitoring by the Global Climate Observing System* [T.R. Karl (ed.)]. Kluwer Academic, Dordrecht, NL, and Boston, MA, pp. 429-470, a 1996 reprint of *Climatic Change* **31(2-4)**.

**Parker**, D.E., L.V. Alexander, and J. Kennedy, 2004: Global and regional climate in 2003. *Weather*, **59**, 145-152.

**Parker**, D.E , M. Gordon, D. P.N. Cullum, D.M.H. Sexton, C.K. Folland, and N. Rayner, 1997: A new global gridded radiosonde temperature data base and recent temperature trends. *Geophysical Research Letters*, **24**, 1499-1502.

**Parker**, D.E., P.D. Jones, C.K. Folland, and A. Bevan, 1994: Interdecadal changes of surface temperature since the late nineteenth century. *Journal of Geophysical Research*, **99**, 14373-14399.

**Pawson**, S., K. Labitzke, and S. Leder, 1998: Stepwise changes in stratospheric temperature. *Geophysical Research Letters*, **25**, 2157-2160.

**Peterson**, T.C., 2003: Assessment of urban versus rural in situ surface temperatures in the contiguous U.S.: no difference found. *Journal of Climate*, **18**, 2941-2959.

**Peterson**, T.C , 2006: Examination of potential biases in air temperature caused by poor station locations. *Bulletin of the American Meteorological Society*, (submitted).

**Peterson**, T.C., D.R. Easterling, T.R. Karl, P.Y. Groisman, N. Nicholls, N. Plummer, S. Torok, I. Auer, R. Boehm, D. Gullett, L. Vincent, R. Heino, H. Tuomenvirta, O. Mestre, T. Szentimre, J. Salinger, E. Førland, I. Hanssen-Bauer, H. Alexandersson, P. Jones, and D. Parker, 1998: Homogeneity adjustments of in situ atmospheric climate data: a review. *International Journal of Climatology*, **18**, 1493-1517.

**Peterson**, T.C., T.R. Karl, P.F. Jamason, R. Knight, and D.R. Easterling, 1998b: The first difference method: maximizing station density for the calculation of long-term global temperature change. *Journal of Geophysical Research*, **103**, 25,967-25,974.

**Peterson**, T.C., K.P. Gallo, J. Lawrimore, T.W. Owen, A. Huang, and D.A. McKittrick, 1999: Global rural temperature trends. *Geophysical Research Letters*, **26**, 329-332.

**Peterson**, T.C., T.W. Owen, 2005. Urban heat island assessment: metadata are important. *Journal of Climate*, **18**, 2637-2646.

**Peterson**, T.C. and R.S. Vose, 1997: An overview of the Global Historical Climatology Network temperature database. *Bulletin of the American Meteorological Society*, **78**, 2837-2849.

**Prabhakara**, C., R. Iacovazzi, Jr, J.-M. Yoo, and G. Dalu., 2000: Global warming: estimation from satellite observations. *Geophysical Research Letters*, **27**, 3517-3520.

**Randel**, W.J. and F. Wu, 2006: Biases in stratospheric and tropospheric temperature trends derived from historical radiosonde data. *Journal of Climate*, (accepted).

**Rayner**, N.A., D.E. Parker, E.B. Horton, C.K. Folland, L.V. Alexander, D.P. Rowell, E.C. Kent and A. Kaplan, 2003: Global analyses of sea surface temperature, sea ice, and night marine air temperature since the late nineteenth century. *Journal of Geophysical Research*, **108**, 4407, doi:10.1029/2002JD002670.

**Reynolds**, R.W. and T.M. Smith, 1994: Improved global sea surface temperature analyses using optimum interpolation. *Journal of Climate*, **7**, 929-948.

**Seidel**, D.J., J.K, Angell, M. Free, J. Christy, R. Spencer, S.A. Klein, J.R. Lanzante, C. Mears, M. Schabel, F. Wentz, D. Parker, P. Thorne, and A. Sterin, 2004: Uncertainty in signals of large-scale climate variations in radiosonde and satellite upper-air temperature data sets. *Journal of Climate*, **17**, 2225-2240.

**Seidel**, D.J., and J.R. Lanzante, 2004: An assessment of three alternatives to linear trends for characterizing global atmospheric temperature changes. *Journal of Geophysical Research*, **109**, doi:10 1029/2003JD004414.

**Sherwood**, S., J.R. Lanzante, and C. Meyer, 2005: Radiosonde daytime biases and late-20th century warming. *Science*, **309**, 1556-1559.

**Slutz**, R.J., S.J. Lubker, J.D. Hiscox, S.D. Woodruff, R.L. Jenne, D.H. Joseph, P.M. Steurer, and J.D. Elms, 1985: *COADS Comprehensive Ocean-Atmosphere Data Set. Release 1.* Climate Research Program, Environmental Research Laboratories, Boulder, CO, 262 pp.

**Smith**, T.M., R.W. Reynolds, R.E. Livezey, and D.C. Stokes, 1996: Reconstruction of historical sea surface temperatures using empirical orthogonal functions. *Journal of Climate*, **9**, 1403-1420.

**Smith**, T.M. and R.W. Reynolds, 2003: Extended reconstruction of global sea surface temperatures based on COADS Data (1854-1997). *Journal of Climate*, **16**, 1495-1510.

**Smith**, T.M. and R.W. Reynolds, 2005: A global merged land and sea surface temperature reconstruction based on historical observations (1880-1997). *Journal of Climate*, **18**, 2021-2036.

**Smith**, T.M., T.C. Peterson, J. Lawrimore, and R.W. Reynolds, 2005: New surface temperature analyses for climate monitoring. *Geophysical Research Letters*, (submitted).

**Spencer**, R.W. and J.R. Christy, 1992: Precision and radiosonde validation of satellite gridpoint temperature anomalies. Part I: MSU channel 2. *Journal of Climate*, **5**, 847-857.

Stendel, M., J. Christy, and L. Bengtsson, 2000: Assessing levels of uncertainty in recent temperature time series. *Climate Dynamics*, **16**, 587-601.

Stott, P.A. and S.F.B. Tett, 1998: Scale-dependent detection of climate change. *Journal of Climate*, **11**, 3282-3294.

Sun, B. and P.Y. Groisman, 2000: Cloudiness variations over the former Soviet Union. *International Journal of Climatology*, **20**, 1097–1111.

Thorne, P.W., D.E. Parker, S.F.B. Tett, P.D. Jones, M. McCarthy, H. Coleman, and P. Brohan, 2005: Revisiting radiosonde upper-air temperatures from 1958 to 2002. *Journal of Geophysical Research*, **110**, D18105, doi:10.1029/2004JD00575.

Trenberth, K.E., 1990: Recent observed interdecadal climate changes in the Northern Hemisphere. *Bulletin of the American Meteorological Society*, **71**, 988-993.

Trenberth, K.[E.] and J. Hurrell, 1994: Decadal atmosphere-ocean variations in the Pacific. *Climate Dynamics*, **9**, 303-319.

Trenberth, K.E., J.M. Carron, D.P. Stepaniak, and S. Worley, 2002: Evolution of the El Nino-southern oscillation and global atmospheric surface temperatures. *Journal of Geophysical Research-Atmospheres*, **107(D8)**, doi:10.1029/2000JD000298.

Van den Dool, H.M., S. Saha, and A. Johansson, 2000: Empirical orthogonal teleconnections. *Journal of Climate*, **13**, 1421-1435.

Vinnikov, K.Y. and N.C. Grody, 2003: Global warming trend of mean tropospheric temperature observed by satellites. *Science*, **302**, 269-272.

Vinnikov, K.Y., A. Robock, N.C. Grody, and A. Basist, 2004: Analysis of diurnal and seasonal cycles and trends in climatic records with arbitrary observation times. *Geophysical Research Letters*, **31**, L06205, doi:10.1029/2003GL019196, 2004.

Vinnikov, K.Y., N.C. Grody, A. Robock, R.J. Stouffer, P.D. Jones, and M.D. Goldberg, 2006: Temperature trends at the surface and in the troposphere. *Journal of Geophysical Research*, **111**, doi:10.1029/ 2005jd006392, 2006.

Vose, R.S., C.N. Williams, T.C. Peterson, T.R. Karl, and D.R. Easterling, 2003: An evaluation of the time of observation bias adjustment in the U.S. historical climate network. *Geophysical Research Letters*, **30**, doi:10.1029/2003GL018111, 2003.

Vose, R.S., D.R. Easterling, and B. Gleason, 2005a: Maximum and minimum temperature trends for the globe: an update through 2004. *Geophysical Research Letters*, **32**, doi:10.1029/2005GL024379, 2005.

Vose, R.S., D. Wuertz, T.C. Peterson, and P.D. Jones, 2005b: An intercomparison of trends in surface air temperature analyses at the global, hemispheric, and grid-box scale. *Geophysical Research Letters*, **32**, doi:10.1029/2005GL023502, 2005.

Wallis, T.W.R., 1998: A subset of core stations from the Comprehensive Aerological Reference Data Set (CARDS). *Journal of Climate*, **11**, 272-282.

Waple, A.M. and J.H. Lawrimore, (eds.), 2003: State of the climate in 2002. *Bulletin of the American Meteorological Society*, **84(6)**, S1-S68.

Woodruff, S.D., H.F. Diaz, J.D. Elms, and S.J. Worley, 1998: CO-ADS release 2 data and metadata enhancements for improvements of marine surface flux fields. *Physics and Chemistry of the Earth*, **23**, 517-527.

## CHAPTER 4 REFERENCES

Agudelo, P.A. and J.A. Curry, 2004: Analysis of spatial distribution in tropospheric temperature trends. *Geophysical Research Letters*, **31**, L22207.

Christy, J.R. and W.B. Norris, 2004: What may we conclude about global tropospheric temperature trends? *Geophysical Research Letters*, **31**, L06211.

Christy, J.R., R. Spencer, and W. Braswell., 2000: MSU tropospheric temperatures: data set construction and radiosonde comparisons. *Journal of Atmospheric and Oceanic Technology*, **17(9)**, 1153-1170.

Christy, J.R., R.W. Spencer, W. Norris, W. Braswell, D. Parker, 2003: Error estimates of version 5.0 of MSU/AMSU bulk atmospheric temperatures. *Journal of Atmospheric and Oceanic Technology*, **20**, 613-629.

Dai, A. and K.E. Trenberth, 2004: The Diurnal Cycle and Its Depiction in the Community Climate System Model. *Journal of Climate*, **17**, 930-951.

Durre, I., T.C. Peterson, R. Vose, 2002: Evaluation of the effect of the Luers-Eskridge radiation adjustments on radiosonde temperature homogeneity. *Journal of Climate*, **15(6)**, 1335-1347.

Easterling, D.R., B. Horton, P.D. Jones, T.C. Peterson, T.R. Karl, D.E. Parker, M.J. Salinger, V. Razuvayev, N. Plummer, P. Jamason, C.K. Folland, 1997: Maximum and minimum temperature trends for the globe, *Science*, **277**, 364-367.

Folland, C.K. and D.E. Parker, 1995: Correction of instrumental biases in historical sea surface temperature data. *Quarterly Journal of the Royal Meteorological Society* **121**, 319-367.

Folland, C.K., N.A. Rayner, S.J. Brown, T.M. Smith, S.S.P. Shen, D.E. Parker, I. Macadam, P.D. Jones, R.N. Jones, N. Nicholls, D.M.H. Sexton, 2001: Global temperature change and its uncertainties since 1861. *Geophysical Research Letters* **28(13)**. 2621-2624.

Free, M., I. Durre, E. Aguilar, D. Seidel, T.C. Peterson, R.E. Eskridge, J.K. Luers, D. Parker, M. Gordon, J. Lanzante, S. Klein, J. Christy, S. Schroeder, B. Soden, L.M. McMillin, 2002: Creating climate reference data sets: CARDS workshop on adjusting radiosonde temperature data for climate monitoring. *Bulletin of the American Meteorological Society*, **83**, 891-899.

Free, M. and D. Seidel, 2005: Causes of differing temperature trends in radiosonde upper air data sets. *Journal of Geophysical Research*, **110**, D07101, doi:10.1029/2004JD005481.

**Fu**, Q. and C.M. Johanson, 2005: Satellite-derived vertical dependence of tropospheric temperature trends. *Geophysical Research Letters* **32**, L10703.

**Fu**, Q., C.M. Johanson, F. Warren, D. Seidel, 2004: Contribution of stratospheric cooling to satellite-inferred tropospheric temperature trends. *Nature*, **429**, 55-58.

**Gentemann**, C.L., F.J. Wentz, C. Mears, D. Smith, 2004: In situ validation of tropical rainfall measuring mission microwave sea surface temperatures. *Journal of Geophysical Research,* **109**, C04021.

**Grody**, N.C., K.Y. Vinnikov, M. Goldberg, J. Sullivan, J. Tarpley, 2004: Calibration of multi-satellite observations for climatic studies: Microwave Sounding Unit (MSU). *Journal of Geophysical Research*, **109**, D24104, doi:10.1029/2004JD005079.

**Hansen**, J.E., R. Ruedy, M. Sato, M. Imhoff, W. Lawrence, D. Easterling, T. Peterson, T. Karl, 2001: A closer look at United States and global surface temperature change. *Journal of Geophysical Research,* **106**, 23947-23963.

**Hurrell**, J.W. and Trenberth, K E. 1998: Difficulties in obtaining reliable temperature trends: reconciling the surface and satellite Microwave Sounding Unit records, *Journal of Climate*, **11**, 945-967.

**Hurrell**, J.W., S.J. Brown, K.E. Trenberth, J.R. Christy, 2000: Comparison of tropospheric temperatures from radiosondes and satellites: 1979-1998, *Bulletin of the American Meteorological Society,* **81**, 2165-2177

**Jones**, P.D. and A. Moberg, 2003: Hemispheric and large-scale surface air temperature variations: an extensive revision and an update to 2001. *Journal of Climate*, **16,** 206–223.

**Jones**, P.D., T.J. Osborn, K. Briffa, 1997: Estimating sampling errors in large-scale temperature averages. *Journal of Climate* **10(10)**, 2548–2568.

**Jones**, P.D., P.Y. Groisman, M. Coughlan, N. Plummer, W-C. Wang, and T.R. Karl, 1990: Assessment of urbanization effects in time series of surface air temperature over land, *Nature,* **347**, 169-172.

**Kilpatrick**, K.A., G.P. Podesta, R. Evans, 2001: Overview of the NOAA/NASA advanced very high resolution radiometer pathfinder algorithm for sea surface temperature and associated matchup database. *Journal of Geophysical Research,* **106(C5)**, 9179-9198.

**Lanzante**, J., S. Klein, D. Seidel, 2003: Temporal homogenization of monthly radiosonde temperature data, part II, trends, sensitivities, and MSU comparison. *Journal of Climate*, **16,** 241-262.

**Li**, Q.X., X.N. Liu, H.Z. Zhang, T.C. Peterson, and D.R. Easterling, Detecting and Adjusting Temporal Inhomogeneity in Chinese Mean Surface Air Temperature Data, 2004: *Advances in Atmospheric Sciences* **21(2)**, 260-268.

**Luers**, J.K. and R.E. Eskridge, 1998: Use of radiosonde temperature data in climate studies. *Journal of Climate*, **11**, 1002-1019.

**Mears**, C. and F. Wentz, 2005: The effect of diurnal correction on satellite-derived lower tropospheric temperature. *Science* **309**, 1548-1551.

**Mears**, C.A., M.C. Schabel, F.W. Wentz, 2003: A reanalysis of the MSU channel 2 tropospheric temperature record. *Journal of Climate,* **16**, 3650-3664.

**Nash,** J., R. Smout, T. Oakley, B. Pathack, and S. Kurnosenko, 2005: WMO Intercomparison of High Quality Radiosonde Systems, Vacoas, Mauritius, 2-25 February 2005, Final Report, WMO Commission on Instruments and Methods of Observation, 118 pp., www.wmo.ch/web/www/IMOP/reports/2003-2007/RSO-IC-2005_Final_Report.pdf **Parker**, D.E., 2004: Large scale warming is not urban. *Nature* **432**, 290, doi:10.1038/432290b.

**Parker**, D.E. and Cox, D.I., 1995: Towards a consistent global climatological rawinsonde data base. *International Journal of Climatology*, **15**, 473-496.

**Parker**, D.E., M. Gordon, D.P.N. Cullum, D.M.H. Sexto, C.K. Folland, N. Rayner, 1997: A new global gridded radiosonde temperature data base and recent temperature trends. *Geophysical Research Letters* **24**, 1499-1502.

**Peterson**, T.C., 2003: Assessment of urban versus rural in situ surface temperatures in the contiguous United States: no difference found. *Journal of climate*, 16(18), 2941-2959.

**Peterson**, T.C., K.P. Gallo, J. Lawrimore, T.W. Owen, A. Huang, D.A. McKittrick, 1999: Global rural temperature trends. *Geophysical Research Letters* **26(3)**, 329-332.

**Randel**, W.J. and F. Wu, 2006: Biases in stratospheric temperature trends derived from historical radiosonde data. *Journal of Climate,* (in press).

**Rayner**, N.A., D.E. Parker, E.B. Horton, C.K. Folland, L.V. Alexander, D.P. Rowell, 2003: Global analyses of sea surface temperature, sea ice, and night marine air temperature since the late nineteenth century. *Journal of Geophysical Research, Atmospheres,* **108(D14)**, 4407, doi:10.1029/2002JD002670.

**Reynolds**, R.W., 1993: Impact of Mt Pinatubo aerosols on satellite-derived sea surface temperatures. *Journal of Climate*, **6**, 768-774.

**Reynolds**, R.W. and T.M. Smith, 1994: Improved global sea surface temperature analyses using optimum interpolation. *Journal of Climate,* **7**, 929-948.

**Reynolds**, R.W., C.L. Gentemann, and F.W. Wentz, 2004: Impact of TRMM SSTs on a climate-scale SST analysis. *Journal of Climate,* **17,** 2938-2952.

**Reynolds**, R.W., N.A. Rayner, T.M. Smith, D.C. Stokes, and W. Wang, 2002: An improved in situ and satellite SST analysis for climate. *Journal of Climate,* **15**, 1609-1625.

**Sherwood**, S., J. Lanzante, and C. Meyer, 2005: Radiosonde daytime biases and late 20th century warming. *Science*, **309**, 1556-1559.

**da Silveira**, R B., G. Fisch, L. Machato, A. Dall'Antonia, L. Sapuchi, D. Fernandes, J. Nash, 2003: Executive Summary. In: *WMO Intercomparison of GPS Radiosondes*. World Meteorological Organization, Geneva, p.17.

**Smith**, T.M. and R.W. Reynolds, 2004: Improved extended reconstruction of SST (1854-1997). *Journal of Climate,* **17**, 2466-2477.

Smith, T.M. and R.W. Reynolds, 2005: A global merged land and sea surface temperature reconstruction based on historical observations (1880-1997). *Journal of Climate*, **18(12)**, 2021-2036.

Spencer, R.W. and J.R. Christy, 1992: Precision and radiosonde validation of satellite gridpoint temperature anomalies, part II, a tropospheric retrieval and trends during 1979-1990. *Journal of Climate*, **5,** 858-866.

Spencer, R.W., J.R. Christy, and W.D. Braswell, 2006: New diurnal adjustments for MSU/AMSU lower tropospheric temperature monitoring. *Journal of Atmospheric and Oceanic Technology*, (in press).

Thorne, P.W., D.E. Parker, S.F.B. Tett, P.D. Jones, M. McCarthy, H. Coleman, and P. Brohan, 2005a: Revisiting radiosonde upper-air temperatures from 1958 to 2002. *Journal of Geophysical Research*, **110,** D18105, doi:10.1029/2004JD005753.

Thorne, P.W., D.E. Parker, J.R. Christy, and C.A. Mears, 2005b: Uncertainties in climate trends: lessons from upper-air temperature records. *Bulletin of the American Meteorological Society,* 86, 1437-1442.

Vinnikov, K.Y. and N.C. Grody, 2003: Global warming trend of mean tropospheric temperature observed by satellites. *Science* 302, 269-272.

Vinnikov, K.Y., N.C. Grody, A. Robock, R.J. Stouffer, P.D. Jones, M.D. Goldberg, 2006: Temperature trends at the surface and in the troposphere, *Journal of Geophysical Research*, **111,** D03106, doi:10.1029/2005JD006392.

Vose, R.S., D. Wuertz, T.C. Peterson, and P.D. Jones, 2004: An intercomparison of trends in surface air temperature analyses at the global, hemispheric, and grid-box scale. *Geophysical Research Letters* 32: L18718, doi:10.1029/2005GL023502.

Wallis, T.W.R., 1998: A subset of core station from the comprehensive aerological reference data set (CARDS). *Journal of Climate* 11, 272-282.

Wentz, F. and M. Schabel, 1998: Effects of satellite orbital decay on MSU lower tropospheric temperature trends. *Nature*, **394,** 661-664.

Wentz, F., C. Gentemann, D. Smith, and D. Chelton, 2000: Satellite measurements of sea surface temperature through clouds. *Science*, **288,** 847-850.

Wentz, F.J., P.D. Ashcroft, and C. Gentemann, 2001: Post-launch calibration of the TMI microwave radiometer. *IEEE Transactions on Geoscience and Remote Sensing*, **39,** 415-422.

## CHAPTER 5 REFERENCES

AchutaRao, K., and K.R. Sperber, 2006: ENSO simulation in coupled atmosphere-ocean models: are the current models better? *Climate Dynamics* (in press).

AchutaRao, K.M., B.D. Santer, P.J. Gleckler, K.E. Taylor, D.W. Pierce, T.P. Barnett, and T.M.L. Wigley, 2006: Variability of ocean heat uptake: reconciling observations and models. *Journal of Geophysical Research, Oceans* (in press).

Allen, M., 1999: Do-it-yourself climate prediction. *Nature*, **401,** 642.

Allen, M.R. and D.A. Stainforth, 2002: Towards objective probabilistic climate forecasting. *Nature*, **419,** 228.

Allen, M.R., and S.F.B. Tett, 1999: Checking for model consistency in optimal fingerprinting. *Climate Dynamics*, **15,** 419-434.

Allen, M.R., N.P. Gillett, J.A. Kettleborough, R. Schnur, G.S. Jones, T. Delworth, F. Zwiers, G. Hegerl, and T.P. Barnett, 2006: Quantifying anthropogenic influence on recent near-surface temperature. *Surveys in Geophysics* (in press).

Ammann, C.M., G.A. Meehl, W.M. Washington, and C.S. Zender, 2003: A monthly and latitudinally-varying forcing data set in simulations of 20th century climate. *Geophysical Research Letters*, **30,** 1657, doi:10.1029/2003GL016875.

Andronova, N.G., and M.E. Schlesinger, 2001: Objective estimation of the probability density function for climate sensitivity. *Journal of Geophysical Research, Atmospheres*, **106,** 22605-22611.

Andronova, N.G., E.V. Rozanov, F.L. Yang, M.E. Schlesinger, and G.L. Stenchikov, 1999: Radiative forcing by volcanic aerosols from 1850 to 1994. *Journal of Geophysical Research, Atmospheres*, **104,** 16807-16826.

Arendt, A.A., K.A. Echelmeyer, W.D. Harrison, C.S. Lingle, and V.B. Valentine, 2002: Rapid wastage of Alaska glaciers and their contribution to rising sea level. *Science*, **297,** 382-386.

Barnett, T.P., and M.E. Schlesinger, 1987: Detecting changes in global climate induced by greenhouse gases. *Journal of Geophysical Research, Atmospheres*, 92, 14772-14780.

Barnett, T.P., D.W. Pierce, and R. Schnur, 2001: Detection of anthropogenic climate change in the world's oceans. *Science*, **292,** 270-274.

Barnett, T.P., D.W. Pierce, K.M. AchutaRao, P.J. Gleckler, B.D. Santer, J.M. Gregory, and W.M. Washington, 2005: Penetration of human-induced warming into the world's oceans. *Science*, **309,** 284-287.

Bengtsson, L., E. Roeckner, and M. Stendel, 1999: Why is the global warming proceeding much slower than expected? *Journal of Geophysical Research, Atmospheres*, **104,** 3865-3876.

Brown, S.J., D.E. Parker, C.K. Folland, and I. Macadam, 2000: Decadal variability in the lower-tropospheric lapse rate. *Geophysical Research Letters*, **27,** 997-1000.

**Brovkin**, V., S. Sitch, W. von Bloh, M. Claussen, E. Bauer, and W. Cramer, 2004: Role of land cover changes for atmospheric $CO_2$ increase and climate change during the last 150 years. *Global Change Biology*, **10**, 1253-1266.

**Cabanes**, C., A. Cazenave, and C. Le Provost, 2001: Sea level rise during past 40 years determined from satellite and in situ observations. *Science*, **294**, 840-842.

**Chase**, T.N., R.A. Pielke, B. Herman, and X. Zeng, 2004: Likelihood of rapidly increasing surface temperatures unaccompanied by strong warming in the free troposphere. *Climate Research*, **25**, 185-190.

**Christy**, J.R., and R. Spencer, 2003: Reliability of satellite data sets. *Science*, **301**, 1046-1047.

**Christy**, J.R. and R.T. McNider, 1994: Satellite greenhouse signal. *Nature*, **367**, 325.

**Christy**, J.R., R.W. Spencer, W.B. Norris, W.D. Braswell, and D.E. Parker, 2003: Error estimates of version 5.0 of MSU-AMSU bulk atmospheric temperatures. *Journal of Atmospheric and Oceanic Technology*, **20**, 613-629.

**Church**, J.A., N.J. White, and J.M. Arblaster, 2005: Significant decadal-scale impact of volcanic eruptions on sea level and ocean heat content. *Nature*, **438**, doi:10.1038/nature04237.

**Collins**, W.D., C.M. Bitz, M.L. Blackmon, G.B. Bonan, C.S. Bretherton, J.A. Carton, P. Chang, S.C. Doney, J.J. Hack, T.B. Henderson, J.T. Kiehl, W.G. Large, D.S. McKenna, B.D. Santer, and R.D. Smith, 2006: The Community Climate System Model: CCSM3. *Journal of Climate* (accepted).

**Crowley**, T.J., 2000: Causes of climate change over the past 1,000 years. *Science*, **289**, 270-277.

**Dai**, A., B.A. Boville, J.T. Kiehl, and L.E. Buja, 2001: Climates of the twentieth and twenty-first centuries simulated by the NCAR Climate System Model. *Journal of Climate*, **14**, 485-519.

**Dameris**, M., V. Grewe, M. Ponater, R. Deckert, V. Eyring, F. Mager, S. Matthes, C. Schnadt, A. Stenke, B. Steil, C. Brühl, and M.A. Giorgetta, 2005: Long-term changes and variability in a transient simulation with a chemistry-climate model employing realistic forcing. *Atmospheric Chemistry and Physics*, **5**, 2121-2145.

**Delworth**, T.L., A.J. Broccoli, A. Rosati, R.J. Stouffer, V. Balaji, J.A. Beesley, W.F. Cooke, K.W. Dixon, J. Dunne, K.A. Dunne, J.W. Durachta, K.L. Findell, P. Ginoux, A. Gnanadesikan, C.T. Gordon, S.M. Griffies, R. Gudgel, M.J. Harrison, I.M. Held, R.S. Hemler, L.W. Horowitz, S.A. Klein, T.R. Knutson, P.J. Kushner, A.R. Langenhorst, H.-C. Lee, S.-J. Lin, J. Lu, S.L. Malyshev, P.C.D. Milly, V. Ramaswamy, J. Russell, M D. Schwarzkopf, E. Shevliakova, J.J. Sirutis, M.J. Spelman, W.F. Stem, M. Winton, A.T. Wittenberg, B. Wyman, F. Zeng, and R. Zhang, 2006: GFDL's CM2 global coupled climate models, part 1, formulation and simulation characteristics. *Journal of Climate* (in press).

**Domack**, E., D. Duran, A. Leventer, S. Ishman, S. Doane, S. McCallum, D. Amblas, J. Ring, R. Gilbert, and M. Prentice, 2005: Stability of the Larsen B ice shelf on the Antarctic Peninsula during the Holocene epoch. *Nature*, **436**, 681-685.

**Douglass**, D.H. and R.S. Knox, 2005: Climate forcing by the volcanic eruption of Mount Pinatubo. *Geophysical Research Letters*, **32**, L05710, doi:10.1029/2004GL022119.

**Douglass**, D.H., B.D. Pearson, and S.F. Singer, 2004: Altitude dependence of atmospheric temperature trends: climate models versus observation. *Geophysical Research Letters*, **31**, doi:10.1029/2004GL020103.

**Douglass**, D.H. and B.D. Clader, 2002: Climate sensitivity of the Earth to solar irradiance. *Geophysical Research Letters*, **29**, doi:10.1029/2002GL015345.

**Elliott**, W.P., 1995: On detecting long-term changes in atmospheric moisture. *Climatic Change*, **31**, 349-367.

**Feddema**, J., K. Oleson, G. Bonan, L.O. Mearns, W.M. Washington, G.A. Meehl, and D. Nychka, 2005: An comparison of a GCM response to historical anthropogenic land cover change and model sensitivity to uncertainty in present-day land cover representations. *Climate Dynamics*, **25**, 581-609, doi:10.1007/s00382-005-0038-z.

**Free**, M., and J.K. Angell, 2002: Effect of volcanoes on the vertical temperature profile in radiosonde data. *Journal of Geophysical Research, Atmospheres*, **107**, doi:10.1029/2001JD001128.

**Folland**, C.K., T.R. Karl, J.R. Christy, R.A. Clarke, G.V. Gruza, J. Jouzel, M.E. Mann, J. Oerlemans, M.J. Salinger, and S.-W. Wang, 2001: Observed climate variability and change. In: *Climate Change 2001 The Scientific Basis. Contribution of Working Group I to the Third Assessment Report of the Intergovernmental Panel on Climate Change* [Houghton, J.T., Y. Ding, D.J. Griggs, M. Noguer, P.J. van der Linden, X. Dai, K. Maskell, and C.A. Johnson, (Eds.)]. Cambridge University Press, Cambridge, United Kingdom and New York, NY, pp. 99-181.

**Folland**, C.K., D.M.H. Sexton, D.J. Karoly, C.E. Johnson, D.P. Rowell, and D.E. Parker, 1998: Influences of anthropogenic and oceanic forcing on recent climate change. *Geophysical Research Letters*, **25**, 353-356.

**Forest**, C.E., P.H. Stone, A.P. Sokolov, M.R. Allen, and M.D. Webster, 2002: Quantifying uncertainties in climate system properties with the use of recent climate observations. *Science*, **295**, 113-117.

**Forest**, C.E., M.R. Allen, A.P. Sokolov, and P.H. Stone, 2001: Constraining climate model properties using optimal fingerprint detection studies. *Climate Dynamics*, **18**, 277-295.

**Fu**, Q., and C.M. Johanson, 2005: Satellite-derived vertical dependence of tropical tropospheric temperature trends. *Geophysical Research Letters*, **32**, L10703, doi:10.1029/2004GL022266.

**Fu**, Q., C.M. Johanson, S.G. Warren, and D.J. Seidel, 2004a: Contribution of stratospheric cooling to satellite-inferred tropospheric temperature trends. *Nature*, **429**, 55-58.

**Fu**, Q., D.J. Seidel, C.M. Johanson, and S.G. Warren, 2004b: Reply to "Tropospheric temperature series from satellites" and "Stratospheric cooling and the troposphere". *Nature*, **432**, doi:10.1038/nature03208.

**Gaffen**, D.J., B.D. Santer, J.S. Boyle, J.R. Christy, N.E. Graham, and R.J. Ross, 2000: Multidecadal changes in the vertical structure of the tropical troposphere. *Science*, **287**, 1242-1245.

**Gates**, W.L., J.S. Boyle, C.C. Covey, C.G. Dease, C.M. Doutriaux, R.S. Drach, M. Fiorino, P.J. Gleckler, J.J. Hnilo, S.M. Marlais, T.J. Phillips, G.L. Potter, B.D. Santer, K.R. Sperber, K E. Taylor, and D.N. Williams, 1999: An overview of the results of the Atmospheric Model Intercomparison Project (AMIP I). *Bulletin of the American Meteorological Society*, **80**, 29-55.

**Gillett**, N.P., A.J. Weaver, F.W. Zwiers, M.F. Wehner, 2004c: Detection of volcanic influence on global precipitation. *Geophysical Research Letters*, **31**, doi:10.1029/2004GL020044.

**Gillett**, N.P., B.D. Santer, and A.J. Weaver, 2004b: Stratospheric cooling and the troposphere. *Nature*, **432**, doi:10.1038/nature03209.

**Gillett**, N.P., F.W. Zwiers, A.J. Weaver, and P.A. Stott, 2003: Detection of human influence on sea level pressure. *Nature*, **422**, 292-294.

**Gillett**, N.P., M.F. Wehner, S.F.B. Tett, A.J. Weaver, 2004a: Testing the linearity of the response to combined greenhouse gas and sulfate aerosol forcing. *Geophysical Research Letters*, **31**, L14201, doi:10.1029/2004GL020111.

**Gillett**, N.P., M.R. Allen, and S.F.B. Tett, 2000: Modelled and observed variability in atmospheric vertical temperature structure. *Climate Dynamics*, **16**, 49-61.

**Gregory**, J.M., H.T. Banks, P.A. Stott, J.A. Lowe, and M.D. Palmer, 2004: Simulated and observed decadal variability in ocean heat content. *Geophysical Research Letters*, **31**, L15312, doi:10.1029/2004GL020258.

**Gregory**, J.M., R.J. Stouffer, S.C.B. Raper, P.A. Stott, N.A. Rayner, 2002: An observationally based estimate of the climate sensitivity. *Journal of Climate*, **15**, 3117-3121.

**Grody**, N.C., K.Y. Vinnikov, M.D. Goldberg, J.T. Sullivan, and J.D. Tarpley, 2004: Calibration of multi-satellite observations for climatic studies: Microwave Sounding Unit (MSU). *Journal of Geophysical Research, Atmospheres*, **109**, D24104, doi:10.1029/2004JD005079.

**Haigh**, J.D., 1994: The role of stratospheric ozone in modulating the solar radiative forcing of climate. *Nature*, **370**, 544-546.

**Hansen**, J., 2002: A brighter future. *Climatic Change*, **52**, 435-440.

**Hansen**, J., M. Sato, R. Ruedy, L. Nazarenko, A. Lacis, G.A. Schmidt, G. Russell, I. Aleinov, M. Bauer, S. Bauer, N. Bell, B. Cairns, V. Canuto, M. Chandler, Y. Cheng, A. Del Genio, G. Faluvegi, E. Fleming, A. Friend, T. Hall, C. Jackman, M. Kelley, N. Kiang, D. Koch, J. Lean, J. Lerner, K. Lo, S. Menon, R. Miller, P. Minnis, T. Novakov, V. Oinas, J. Perlwitz, D. Rind, A. Romanou, D. Shindell, P. Stone, S. Sun, N. Tausnev, D. Thresher, B. Wielicki, T. Wong, M. Yao, and S. Zhang, 2005a: Efficacy of climate forcings. *Journal of Geophysical Research, Atmospheres*, **110**, D18104, doi:10.1029/2005JD005776.

**Hansen**, J., A. Lacis, R. Ruedy, M. Sato, and H. Wilson, 1993: How sensitive is the world's climate? *Research & exploration a scholarly publication of the National Geographic Society*, **9** (2, Special Issue, Global Warming Debate), 142-158.

**Hansen**, J. and L. Nazarenko, 2003: Soot climate forcing via snow and ice albedos. *Proceedings of the National Academy of Sciences*, **101**, 423-428, doi:10.1073/pnas.2237157100.

**Hansen**, J., B. Cairns, L. Druyan, X. Jiang, R. Miller, and I. Tegen, 1997: Forcings and chaos in interannual to decadal climate change. *Journal of Geophysical Research, Atmospheres*, **102**, 25679-25720.

**Hansen**, J., H. Wilson, M. Sato, R. Ruedy, K. Shah, E. Hansen, 1995: Satellite and surface temperature data at odds? *Climatic Change*, **30**, 103-117.

**Hansen**, J., L. Nazarenko, R. Ruedy, M. Sato, J. Willis, A. Del Genio, D. Koch, A. Lacis, K. Lo, S. Menon, T. Novakov, J. Perlwitz, G. Russell, G.A. Schmidt, and N. Tausnev, 2005b: Earth's energy imbalance: confirmation and implications. *Science*, **308**, 1431-1435.

**Hansen**, J., M. Sato, L. Nazarenko, R. Ruedy, A. Lacis, D. Koch, I. Tegen, T. Hall, D. Shindell, B. Santer, P. Stone, T. Novakov, L. Toamson, R. Wang, Y. Wang, D. Jacob, S. Hollingsworth, L. Bishop, J. Logan, A. Thompson, R. Stolarski, J. Lean, R., Willson, S. Levitus, J. Antoniv, N. Rayner, D. Parker, and J. Christy, 2002: Climate forcings in Goddard Institute for Space Studies SI2000 simulations. *Journal of Geophysical Research, Atmospheres*, **107**, doi:10.1029/2000JD001143.

**Hansen**, J., M. Sato, R. Ruedy, A. Lacis, and V. Oinas, 2000: Global warming in the twenty-first century: an alternative scenario. *Proceedings of the National Academy of Sciences*, **97**, 9875-9880.

**Harvey**, L.D.D. and R.K. Kaufmann, 2002: Simultaneously constraining climate sensitivity and aerosol radiative forcing. *Journal of Climate*, **15**, 2837-2861.

**Hasselmann**, K., 1979: On the signal-to-noise problem in atmospheric response studies. In: *Meteorology of Tropical Oceans* [Shaw, D.B. (ed.)]. Royal Meteorological Society of London, London, U.K., pp. 251-259.

**Hasselmann**, K., 1993: Optimal fingerprints for the detection of time dependent climate change. *Journal of Climate*, **6**, 1957-1971.

**Hasselmann**, K., 1997: Multi-pattern fingerprint method for detection and attribution of climate change. *Climate Dynamics*, **13**, 601-612.

**Hasselmann**, K., 1999: Climate change: linear and nonlinear signatures. *Nature*, **398**, 755-756, doi:10.1038/19635.

**Hegerl**, G.C. and J.M. Wallace, 2002: Influence of patterns of climate variability on the difference between satellite and surface temperature trends. *Journal of Climate*, **15**, 2412-2428.

**Hegerl**, G.C., K. Hasselmann, U. Cubasch, J.F.B. Mitchell, E. Roeckner, R. Voss, and J. Waszkewitz, 1997: Multi-fingerprint detection and attribution of greenhouse-gas- and aerosol-forced climate change. *Climate Dynamics*, **16**, 737-754.

**Hegerl**, G.C., H. vonStorch, K. Hasselmann, B.D. Santer, U. Cubasch, and P.D. Jones, 1996: Detecting greenhouse-gas-induced climate change with an optimal fingerprint method. *Journal of Climate*, **9**, 2281-2306.

**Hess**, S.L., 1959: *Introduction to Theoretical Meteorology*. Holt, Rinehart and Winston, New York, NY, 362 pp.

**Highwood**, E.J., B.J. Hoskins, and P. Berrisford, 2000: Properties of the Arctic tropopause. *Quarterly Journal of the Royal Meteorological Society*, **126**, 1515-1532.

**Hoffert**, M.I. and C. Covey, 1992: Deriving global climate sensitivity from paleoclimate reconstructions. *Nature*, **360**, 573-576.

**Horel**, J.D. and J.M. Wallace, 1981: Planetary-scale atmospheric phenomena associated with the southern oscillation. *Monthly Weather Review*, **109**, 813-829.

**Horowitz**, L.W., S. Walters, D.L. Mauzerall, L.K. Emmons, P.J. Rasch, C. Granier, X.X. Tie, J.F. Lamarque, M.G. Schultz, G.S. Tyndall, J.J. Orlando, and G.P. Brasseur, 2003: A global simulation of tropospheric ozone and related tracers: description and evaluation of MOZART, version 2, *Journal of Geophysical Research, Atmospheres*, **108**, 4784, doi:10.1029/2002JD002853.

**Houghton**, J.T., Y. Ding, D. J. Griggs, M. Noguer, P.J. van der Linden, X. Dai, K. Maskell, and C.A. Johnson, (eds.), 2001: *Climate Change 2001 The Scientific Basis. Contribution of Working Group I to the Third Assessment Report of the Intergovernmental Panel on Climate Change*. Cambridge University Press, Cambridge, UK, and New York, NY, 881 pp.

**Hoyt**, D.V., and K.H. Schatten, 1993: A discussion of plausible solar irradiance variations, 1700-1992. *Journal of Geophysical Research, Atmospheres*, **98**, 18895-18906.

**Hurrell**, J.W., Y. Kushnir, G. Ottersen, and M. Visbeck, 2003: *The North Atlantic Oscillation Climatic Significance and Environmental Impact*, Geophysical Monograph 134, American Geophysical Union, Washington, DC, 279 pp.

**Hurrell**, J.W., and K.E. Trenberth, 1998: Difficulties in obtaining reliable temperature records: Reconciling the surface and satellite Microwave Sounding Unit records. *Journal of Climate*, **11**, 945-967.

**IDAG** (International Detection and Attribution Group), 2005: Detecting and attributing external influences on the climate system: a review of recent advances. *Journal of Climate*, **18**, 1291-1314.

**Jacobson**, M.Z., 2004: Climate response of fossil fuel and biofuel soot, accounting for soot's feedback to snow and sea ice albedo and emissivity. *Journal of Geophysical Research, Atmospheres*, **109**, D21201, doi:10.1029/2004JD004945.

**Jones**, P.D., A. Moberg, T. J. Osborn, and K. R. Briffa, 2003: Surface climate responses to explosive volcanic eruptions seen in long European temperature records and mid-to-high latitude tree-ring density around the Northern Hemisphere. In: *Volcanism and the Earth's Atmosphere*, [Robock, A. and C. Oppenheimer (eds.)], Geophysical Monograph Series no. **139**, American Geophysical Union, Washington D.C., 239-254.

**Jones**, P.D., T.J. Osborn, K.R. Briffa, C.K. Folland, E.B. Horton, L.V. Alexander, D.E. Parker, and N.A. Rayner, 2001: Adjusting for sampling density in grid box land and ocean surface temperature time series. *Journal of Geophysical Research, Atmospheres*, **106**, 3371-3380.

**Jones**, P.D., M. New, D.E Parker, S. Martin, and I.G. Rigor, 1999: Surface air temperature and its changes over the past 150 years. *Reviews of Geophysics*, **37**, 173-199.

**Jones**, P.D., 1994: Recent warming in global temperature series. *Geophysical Research Letters*, **21**, 1149-1152.

**Jones**, G.S., S.F.B. Tett, and P.A. Stott, 2003: Causes of atmospheric temperature change 1960-2000: a combined attribution analysis. *Geophysical Research Letters*, **30**, doi:10.1029/2002GL016377.

**Karoly**, D.J. and Q. Wu, 2005: Detection of regional surface temperature trends. *Journal of Climate*, **18**, 4337-4343.

**Karoly**, D.J., J.A. Cohen, G.A. Meehl, J.F.B. Mitchell, A.H. Oort, R.J. Stouffer, and R.T. Wetherald, 1994: An example of fingerprint detection of greenhouse climate change. *Climate Dynamics*, **10**, 97-105.

**Karoly**, D.J., K. Braganza, P.A. Stott, J.M. Arblaster, G.A. Meehl, A.J. Broccoli, and K.W. Dixon, 2003: Detection of a human influence on North American climate. *Science*, **302**, 1200-1203.

**Kiehl**, J.T., J.M. Caron, and J.J. Hack, 2005: On using global climate model simulations to assess the accuracy of MSU retrieval methods for tropospheric warming trends. *Journal of Climate*, **18**, 2533-2539.

**Klein Goldewijk**, K., 2001: Estimating global land use change over the past 300 years: The HYDE database. *Global Biogeochemical Cycles*, **15**, 417-433.

**Knight**, J.R., R.J. Allan, C.K. Folland, M. Vellinga, and M.E. Mann, 2005: A signature of persistent natural thermohaline circulation cycles in observed climate. *Geophysical Research Letters*, **32**. L20708, doi:10.1029/2005GL024233.

**Knutson**, T.R. and R.E. Tuleya, 2004: Impact of $CO_2$-induced warming on simulated hurricane intensity and precipitation: sensitivity to the choice of climate model and convective parameterization. *Journal of Climate*, **17**, 3477-3495.

**Knutson**, T.R., T.L. Delworth, K.W. Dixon, I.M. Held, J. Lu, V. Ramaswamy, D. Schwarzkopf, G. Stenchikov, and R.J. Stouffer, 2006: Assessment of twentieth-century regional surface temperature trends using the GFDL CM2 coupled models. *Journal of Climate* (in press).

**Koch**, D., 2001: Transport and direct radiative forcing of carbonaceous and sulfate aerosols in the GISS GCM. *Journal of Geophysical Research, Atmospheres*, **106**, 20311-20332.

**Koch**, D., D. Jacob, I. Tegen, D. Rind, and M. Chin, 1999: Tropospheric sulfur simulation and sulfate direct radiative forcing in the Goddard Institute for Space Studies general circulation model. *Journal of Geophysical Research, Atmospheres*, **104**, 23799-23822.

Krabill, W., E. Frederick, S. Manizade, C. Martin, J. Sonntag, R. Swift, R. Thomas, W. Wright, and J. Yungel, 1999: Rapid thinning of parts of the southern Greenland Ice Sheet. *Science*, **283**, 1522-1524.

Krishnan, R., and V. Ramanathan, 2002: Evidence of surface cooling from absorbing aerosols. *Geophysical Research Letters*, **29**, doi:10.1029/2002GL014687.

Lean, J., 2000: Evolution of the Sun's spectral irradiance since the Maunder Minimum. *Geophysical Research Letters*, **27**, 2425-2428.

Lean, J., J. Beer, and R. Bradley, 1995: Reconstruction of solar irradiance since 1610: Implications for climate change. *Geophysical Research Letters*, **22**, 3195-3198.

Leroy, S.S., 1998: Detecting climate signals: some Bayesian aspects. *Journal of Climate*, **11**, 640-651.

Levitus, S., J.I. Antonov, and T.P. Boyer, 2005: Warming of the world ocean, 1955-2003. *Geophysical Research Letters*, **32**, L02604, doi:10.1029/2004GL021592.

Levitus, S., J.I. Antonov, J.L. Wang, T.L. Delworth, K.W. Dixon, and A.J. Broccoli, 2001: Anthropogenic warming of Earth's climate system. *Science*, **292**, 267-270.

Levitus, S., J.I. Antonov, T.P. Boyer, and C. Stephens, 2000: Warming of the world ocean. *Science*, **287**, 2225-2229.

Lindzen, R.S., and C. Giannitsis, 1998: On the climatic implications of volcanic cooling. *Journal of Geophysical Research, Atmospheres*, **103**, 5929-5941.

Manabe, S. and R.J. Stouffer, 1980: Sensitivity of a global climate model to an increase of $CO_2$ concentration in the atmosphere. *Journal of Geophysical Research, Atmospheres*, **85**, 5529-5554.

Mao, J. and A. Robock, 1998: Surface air temperature simulations by AMIP general circulation models: volcanic and ENSO signals and systematic errors. *Journal of Climate*, **11**, 1538-1552.

Marshall, C.H., R.A. Pielke, L.T. Steyaert, and D.A. Willard, 2004: The impact of anthropogenic land-cover change on the Florida Peninsula sea breezes and warm season sensible weather. *Monthly Weather Review*, **132**, 28-52.

Matthews, H.D., A.J. Weaver, K.J. Meissner, N.P. Gillett, and M. Eby, 2004: Natural and anthropogenic climate change: incorporating historical land cover change, vegetation dynamics and the global carbon cycle. *Climate Dynamics*, **22**, 461-479.

Matthews, H.D., A.J. Weaver, M. Eby, and K.J. Meissner, 2003: Radiative forcing of climate by historical land cover change. *Geophysical Research Letters*, **30**, 1055, doi:10.1029/2002GL016098.

McAvaney, B.J., C. Covey, S. Joussaume, V. Kattsov, A. Kitoh, W. Ogana, A.J. Pitman, A.J. Weaver, R.A. Wood and Z.C. Zhao, 2001: Model evaluation. In: *Climate Change 2001 The Scientific Basis. Contribution of Working Group I to the Third Assessment Report of the Intergovernmental Panel on Climate Change* [Houghton, J.T., Y. Ding, D.J. Griggs, M. Noguer, P.J. van der Linden, X. Dai, K. Maskell, and C.A. Johnson, (eds.)]. Cambridge University Press, Cambridge, UK and New York, NY, 881 pp.

Mears, C.A. and F.W. Wentz, 2005: The effect of diurnal correction on satellite-derived lower tropospheric temperature. *Science*, **309**, 1548-1551.

Mears, C.A., M.C. Schabel, and F.W. Wentz, 2003: A reanalysis of the MSU channel 2 tropospheric temperature record. *Journal of Climate*, **16**, 3650-3664.

Meehl, G.A., 1984: Modeling the Earth's climate. *Climatic Change*, **6**, 259-286.

Meehl, G.A., G.J. Boer, C. Covey, M. Latif, and R.J. Stouffer, 2000: The Coupled Model Intercomparison Project (CMIP). *Bulletin of the American Meteorological Society*, **81**, 313-318.

Meehl, G.A., W.M. Washington, B.D. Santer, W.D. Collins, J.M. Arblaster, A. Hu, D.M. Lawrence, H. Teng, L.E. Buja, and W.G. Strand, 2006: Climate change in the 20th and 21st centuries and climate change commitment in the CCSM3. *Journal of Climate* (in press).

Menon, S., J. Hansen, L. Nazarenko, and Y.F. Luo, 2002: Climate effects of black carbon aerosols in China and India. *Science*, **297**, 2250-2253.

Michaels, P.J., and P.C. Knappenberger, 2000: Natural signals in the MSU lower tropospheric temperature record. *Geophysical Research Letters*, **27**, 2905-2908.

Michaels, P.J. and P.C. Knappenberger, 1996: Human effect on global climate? *Nature*, **384**, 523-524.

Min, S.-K., A. Hense, and W.-T. Kwon, 2005: Regional-scale climate change detection using a Bayesian detection method. *Geophysical Research Letters*, **32**, L03706, doi:10.1029/2004GL021028.

Minschwaner, K., R.W. Carver, B.P Briegleb, and A.E. Roche, 1998: Infrared radiative forcing and atmospheric lifetimes of trace species based on observations from UARS. *Journal of Geophysical Research, Atmospheres*, **103**, 23243-23253.

Mitchell, J.F.B., D.J. Karoly, G.C. Hegerl, F.W. Zwiers, M.R. Allen and J. Marengo, 2001: Detection of climate change and attribution of causes. In: *Climate Change 2001 The Scientific Basis. Contribution of Working Group I to the Third Assessment Report of the Intergovernmental Panel on Climate Change* [Houghton, J.T., Y. Ding, D.J. Griggs, M. Noguer, P.J. van der Linden, X. Dai, K. Maskell, and C.A. Johnson, (eds.)]. Cambridge University Press, Cambridge, UK, and New York, NY, 881 pp.

Murphy, J.M., D.M.H. Sexton, D.N. Barnett, G.S. Jones, M.J. Webb, M. Collins and D.A. Stainforth, 2004: Quantification of modeling uncertainties in a large ensemble of climate simulations. *Nature*, **430**, 768-772.

Myhre, G. and A. Myhre, 2003: Uncertainties in radiative forcing due to surface albedo changes caused by land use changes. *Journal of Climate*, **16**, 1511-1524.

NRC (National Research Council), 2005: *Radiative Forcing of Climate Change Expanding the Concept and Addressing Uncertainties*. National Academy Press, Washington DC, 168 pp.

**NRC** (National Research Council), 2000: *Reconciling Observations of Global Temperature Change.* National Academy Press, Washington DC, 85 pp.

**Nicholls**, N., G.V. Gruza, J. Jouzel, T.R. Karl, L.A. Ogallo, and D.E. Parker, 1996: Observed climate variability and change. In: *Climate Change 1995 The Science of Climate Change. Contribution of Working Group I to the Second Assessment Report of the Intergovernmental Panel on Climate Change* [Houghton, J.T., L.G. Meira Filho, B.A. Callander, N. Harris, A. Kattenberg, and K. Maskell, (eds.)]. Cambridge University Press, Cambridge, UK, and New York, NY, 572 pp.

**North**, G.R. and M.J. Stevens, 1998: Detecting climate signals in the surface temperature record. *Journal of Climate,* **11**, 563-577.

**North**, G.R., K.Y. Kim, S.S.P Shen, and J.W. Hardin, 1995: Detection of forced climate signals, part I, filter theory. *Journal of Climate,* **8**, 401-408.

**Oleson**, K.W., *G.B. Bonan,* S. Levis and M. Vertenstein, 2004: Effect of land use change on North American climate: impact of surface data sets and model biogeophysics. *Climate Dynamics,* **23**, 117-132.

**Pan**, Y.H. and A.H. Oort, 1983: Global climate variations connected with sea surface temperature anomalies in the eastern equatorial Pacific Ocean for the 1958-73 period. *Monthly Weather Review,* **111**, 1244-1258.

**Parker**, D.E., M. Gordon, D.P.N. Cullum, D.M.H. Sexton, C.K. Folland, and N. Rayner, 1997: A new global gridded radiosonde temperature data base and recent temperature trends. *Geophysical Research Letters,* **24**, 1499-1502.

**Paul**, F., A. Kaab, M. Maisch, T. Kellenberger, and W. Haeberli, 2004: Rapid disintegration of Alpine glaciers observed with satellite data. *Geophysical Research Letters,* **31**, L21402, doi:10.1029/2004GL020816.

**Pawson**, S., K. Labitzke, and S. Leder, 1998: Stepwise changes in stratospheric temperature. *Geophysical Research Letters,* **25**, 2157-2160.

**Penner**, J.E., M. Wang, A. Kumar, L. Rotstayn, and B.D. Santer, 2006: Effect of black carbon on mid-troposphere and surface temperature trends. In: *Integrated Assessment of Human Induced Climate Change.* [Schlesinger, M.E. (ed.)]. Cambridge University Press, Cambridge, UK, (in press).

**Penner**, J.E., S.Y. Zhang, and C.C. Chuang, 2003: Soot and smoke aerosol may not warm climate. *Journal of Geophysical Research, Atmospheres,* **108**, 4657, doi:10.1029/2003JD003409.

**Penner**, J.E., M. Andreae, H. Annegarn, L. Barrie, J. Feichter, D. Hegg, A. Jayaraman, R. Leaitch, D. Murphy, J. Nganga and G. Pitari, 2001: Aerosols, their direct and indirect effects. In: *Climate Change 2001 The Scientific Basis. Contribution of Working Group I to the Third Assessment Report of the Intergovernmental Panel on Climate Change* [Houghton, J.T., Y. Ding, D.J. Griggs, M. Noguer, P.J. van der Linden, X. Dai, K. Maskell and C.A. Johnson, (eds.)]. Cambridge University Press, Cambridge, UK and New York, NY, 881 pp.

**Pielke** Sr., R.A., 2003: Heat storage within the Earth system. *Bulletin of the American Meteorological Society,* **84**, 331-335.

**Pielke** Sr., R.A., 2004: Assessing global warming with surface heat content. *EOS,* **85**, 210-211.

**Pielke** Sr., R.A. and T.N. Chase, 2004: Comment on "Contributions of anthropogenic and natural forcing to recent tropopause height changes." *Science,* **303**, 1771c.

**Pierce**, D.W., T.P. Barnett, K.M. AchutaRao, P.J. Gleckler, J.M. Gregory, and W.M. Washington, 2006: Anthropogenic warming of the oceans: observations and model results. *Journal of Climate* (in press).

**Pitman**, A.J., G.T. Narisma, R.A. Pielke, and N.J. Holbrook, 2004: Impact of land cover change on the climate of southwest Western Australia. *Journal of Geophysical Research, Atmospheres,* **109**, doi:10.1029/2003JD004347.

**Ramachandran**, S., V. Ramaswamy, G.L. Stenchikov, and A. Robock, 2000: Radiative impact of the Mount Pinatubo volcanic eruption: lower stratospheric response. *Journal of Geophysical Research, Atmospheres,* **105**, 24409-24429.

**Ramanathan**, V., P.J. Crutzen, J. Lelieveld, A.P. Mitra, D. Althausen, J. Anderson, M.O. Andreae, W. Cantrell, G.R. Cass, C.E. Chung, A.D. Clarke, J.A. Coakley, W.D. Collins, W.C. Conant, F. Dulac, J. Heintzenberg, A.J. Heymsfield, B. Holben, S. Howell, J. Hudson, A. Jayaraman, J.T. Kiehl, T.N. Krishnamurti, D. Lubin, G. McFarquhar, T. Novakov, J A. Ogren, I.A. Podgorny, K. Prather, K. Priestley, J.M. Prospero, P.K. Quinn, K. Rajeev, P. Rasch, S. Rupert, R. Sadourny, S.K. Satheesh, G.E. Shaw, P. Sheridan, and F.P.J. Valero, 2001: The Indian Ocean experiment: An integrated analysis of the climate forcing and effects of the great Indo-Asian haze. *Journal of Geophysical Research, Atmospheres,* **106**, 28371-28398.

**Ramankutty**, N. and J.A. Foley, 1999: Estimating historical changes in global land cover: croplands from 1700 to 1992. *Global Biogeochemical Cycles,* **13**, 997-1027.

**Ramaswamy**, V., M.D. Schwarzkopf, W.J. Randel, B.D. Santer, B.J. Soden, and G.L. Stenchikov, 2006: Anthropogenic and natural influences in the evolution of lower stratospheric cooling. *Science,* **311**, 1138-1141.

**Ramaswamy**, V., M.L. Chanin, J. Angell, J. Barnett, D. Gaffen, M. Gelman, P. Keckhut, Y. Koshelkov, K. Labitzke, J.J.R. Lin, A. O'Neill, J. Nash, W. Randel, R. Rood, K. Shine, M. Shiotani, and R. Swinbank, 2001a: Stratospheric temperature trends: observations and model simulations. *Reviews of Geophysics,* **39**, 71-122.

**Ramaswamy**, V., V., O. Boucher, J. Haigh, D. Hauglustaine, J. Haywood, G. Myhre, T. Nakajima, G. Y. Shi, and S. Solomon, 2001b: Radiative forcing of climate change. In: *Climate Change 2001 The Scientific Basis. Contribution of Working Group I to the Third Assessment Report of the Intergovernmental Panel on Climate Change* [Houghton, J.T., Y. Ding, M. Noguer, P.J. van der Linden, X. Dai, K. Maskell, and C.A. Johnson, (eds.)]. Cambridge University Press, Cambridge, UK and New York, NY, 881 pp.

**Ramaswamy**, V., M.D. Schwarzkopf and W J. Randel, 1996: Fingerprint of ozone depletion in the spatial and temporal pattern of recent lower-stratospheric cooling. *Nature,* **382**, 616-618.

Randel, W.J. and F. Wu, 2006: Biases in stratospheric temperature trends derived from historical radiosonde data. *Journal of Climate* (in press)

Randel, W.J., F. Wu, and D.J. Gaffen, 2000: Interannual variability of the tropical tropopause derived from radiosonde data and NCEP reanalyses. *Journal of Geophysical Research, Atmospheres*, **105**, 15509-15523.

Randel, W.J. and F. Wu, 1999: Cooling of the Arctic and Antarctic polar stratosphere due to ozone depletion. *Journal of Climate*, **12**, 1467-1479.

Reichert, B.K., R. Schnur, and L. Bengtsson, 2002: Global ocean warming tied to anthropogenic forcing. *Geophysical Research Letters*, **29**, doi:10.1029/2001GL013954.

Rignot, E. and R.H. Thomas, 2002: Mass balance of polar ice sheets. *Science*, **297**, 1502-1506.

Robock, A., 2005: Comments on "Climate forcing by the volcanic eruption of Mount Pinatubo." *Geophysical Research Letters*, **32**, L20711, doi:10.1029/2005GL023287.

Robock, A. and C. Oppenheimer, 2003: *Volcanism and the Earth's Atmosphere*, Geophysical Monograph Series no. 139, American Geophysical Union, Washington DC, 360 pp.

Robock, A., 2000: Volcanic eruptions and climate. *Reviews of Geophysics*, **38**, 191-219.

Rodwell, M.J., D.P. Rowell, and C.K. Folland, 1999: Oceanic forcing of the wintertime North Atlantic Oscillation and European climate. *Nature*, **398**, 320-323.

Roeckner, E., E. Roeckner, L. Bengtsson, J. Feichter, J. Lelieveld, and H. Rodhe, 1999: Transient climate change simulations with a coupled atmosphere-ocean GCM including the tropospheric sulfur cycle. *Journal of Climate*, **12**, 3004-3032.

Santer, B.D., J.J. Hnilo, T.M.L. Wigley, J.S. Boyle, C. Doutriaux, M. Fiorino, D.E. Parker, and K.E. Taylor, 1999: Uncertainties in observationally-based estimates of temperature change in the free atmosphere. *Journal of Geophysical Research, Atmospheres*, **104**, 6305-6333.

Santer, B.D., K.E. Taylor, T.M.L. Wigley, T.C. Johns, P.D. Jones, D.J. Karoly, J.F.B. Mitchell, A.H. Oort, J.E. Penner, V. Ramaswamy, M.D. Schwarzkopf, R.J. Stouffer, and S. Tett, 1996a: A search for human influences on the thermal structure of the atmosphere. *Nature*, **382**, 39-46.

Santer, B.D., K.E. Taylor, T.M.L. Wigley, T.C. Johns, P.D. Jones, D.J. Karoly, J.F.B. Mitchell, A.H. Oort, J.E. Penner, V. Ramaswamy, M.D. Schwarzkopf, R.J. Stouffer, S. Tett, J.S. Boyle, and D.E. Parker, 1996b: Human effect on global climate? *Nature*, **384**, 522-524.

Santer, B.D., M.F. Wehner, T.M.L. Wigley, R. Sausen, G.A. Meehl, K.E. Taylor, C. Ammann, J. Arblaster, W.M. Washington, J.S. Boyle, and W. Bruggemann, 2003a: Contributions of anthropogenic and natural forcing to recent tropopause height changes. *Science*, **301**, 479-483.

Santer, B.D., T.M.L. Wigley, A.J. Simmons, P.W. Kållberg, G.A. Kelly, S.M. Uppala, C. Ammann, J.S. Boyle, W. Brüggemann, C. Doutriaux, M. Fiorino, C. Mears, G.A. Meehl, R. Sausen, K.E. Taylor, W.M. Washington, M F. Wehner, and F.J. Wentz, 2004: Identification of anthropogenic climate change using a second-generation reanalysis. *Journal of Geophysical Research, Atmospheres*, **109**, doi:10.1029/2004JD005075.

Santer, B.D., T.M.L. Wigley, C. Mears, F.J. Wentz, S.A. Klein, D.J. Seidel, K.E. Taylor, P.W. Thorne, M.F. Wehner, P.J. Gleckler, J.S. Boyle, W.D. Collins, K.W. Dixon, C. Doutriaux, M. Free, Q. Fu, J.E. Hansen, G.S. Jones, R. Ruedy, T.R. Karl, J.R. Lanzante, G.A. Meehl, V. Ramaswamy, G. Russell, and G.A. Schmidt, 2005: Amplification of surface temperature trends and variability in the tropical atmosphere. *Science*, **309**, 1551-1556.

Santer, B.D., T.M.L Wigley, C. Doutriaux, J.S. Boyle, J.E. Hansen, P.D. Jones, G.A. Meehl, E. Roeckner, S. Sengupta, and K.E. Taylor, 2001: Accounting for the effects of volcanoes and ENSO in comparisons of modeled and observed temperature trends. *Journal of Geophysical Research, Atmospheres*, **106**, 28033-28059.

Santer, B.D., T.M.L. Wigley, D.J. Gaffen, L. Bengtsson, C. Doutriaux, J.S. Boyle, M. Esch, J.J. Hnilo, P.D. Jones, G.A. Meehl, E. Roeckner, K.E. Taylor, and M.F. Wehner, 2000: Interpreting differential temperature trends at the surface and in the lower troposphere. *Science*, **287**, 1227-1232.

Santer, B.D., T.M.L. Wigley, G.A. Meehl, M.F. Wehner, C. Mears, M. Schabel, F.J. Wentz, C. Ammann, J. Arblaster, T. Bettge, W.M. Washington, K. E. Taylor, J. S. Boyle, W. Brüggemann, C. Doutriaux, 2003b: Influence of satellite data uncertainties on the detection of externally-forced climate change. *Science*, **300**, 1280-1284.

Satheesh, S.K. and V. Ramanathan, 2000: Large differences in tropical aerosol forcing at the top of the atmosphere and Earth's surface. *Nature*, **405**, 60-63.

Sato, M., J.E. Hansen, M.P. McCormick, and J.B. Pollack, 1993: Stratospheric aerosol optical depths, 1850-1990: *Journal of Geophysical Research, Atmospheres*, **98**, 22987-22994.

Seidel, D.J., and J.R. Lanzante, 2004: An assessment of three alternatives to linear trends for characterizing global atmospheric temperature changes. *Journal of Geophysical Research, Atmospheres*, **109**, doi:10.1029/2003JD004414.

Seidel, D.J., J.K. Angell, J. Christy, M. Free, S A. Klein, J.R. Lanzante, C. Mears, D. Parker, M. Schabel, R. Spencer, A. Sterin, P. Thorne, and F. Wentz, 2004: Uncertainty in signals of large-scale climate variations in radiosonde and satellite upper-air temperature data sets. *Journal of Climate*, **17**, 2225-2240.

Seidel, D.J., R.J. Ross, J.K. Angell, and G.C. Reid, 2001: Climatological characteristics of the tropical tropopause as revealed by radiosondes. *Journal of Geophysical Research, Atmospheres*, **106**, 7857-7878.

Sexton, D.M.H., D.P. Rowell, C.K. Folland, and D.J. Karoly, 2001: Detection of anthropogenic climate change using an atmospheric GCM. *Climate Dynamics*, **17**, 669-685.

**Sherwood**, S.C., J. Lanzante, and C. Meyer, 2005: Radiosonde daytime biases and late 20th century warming. *Science*, **309**, 1556-1559.

**Shindell**, D.T., G. Fulavegi, and N. Bell, 2003: Preindustrial-to-present-day radiative forcing by tropospheric ozone from improved simulations with the GISS chemistry-climate GCM. *Atmospheric Chemistry and Physics*, **3**, 1675-1702.

**Shine**, K.P., M.S. Bourqui, P.M.D. Forster, S.H.E. Hare, U. Langematz, P. Braesicke, V. Grewe, M. Ponater, C. Schnadt, C.A. Smiths, J.D. Haighs, J. Austin, N. Butchart, D.T. Shindell, W.J. Randel, T. Nagashima, R.W. Portmann, S. Solomon, D.J. Seidel, J. Lanzante, S. Klein, V. Ramaswamy, and M.D. Schwarzkopf, 2003: A comparison of model-simulated trends in stratospheric temperatures. *Quarterly Journal of the Royal Meteorological Society*, **129**, 1565-1588.

**Smith**, S.J., H. Pitcher, and T.M.L. Wigley, 2005: Future sulfur dioxide emissions. *Climatic Change*, **73**, 267-318.

**Smith**, S.J., H. Pitcher, and T.M.L. Wigley, 2001: Global and regional anthropogenic sulfur dioxide emissions. *Global and Planetary Change*, **29**, 99-119.

**Soden**, B.J., R.T. Wetherald, G.L. Stenchikov, and A. Robock, 2002: Global cooling after the eruption of Mt. Pinatubo: a test of climate feedback by water vapor. *Science*, **296**, 727-730.

**Soden**, B.J., 2000: The sensitivity of the tropical hydrological cycle to ENSO. *Journal of Climate*, **13**, 538-549.

**Stainforth**, D.A., T. Aina, C. Christensen, M. Collins, N. Faull, D.J. Frame, J.A. Kettleborough, S. Knight, A. Martin, J.M. Murphy, C. Piani, D. Sexton, L.A. Smith, R.A. Spicer, A.J. Thorpe and M.R. Allen, 2005: Uncertainties in predictions of the climate response to rising levels of greenhouse gases. *Nature*, **433**, 403-406.

**Stott**, P.A., 2003: Attribution of regional-scale temperature changes to anthropogenic and natural causes. *Geophysical Research Letters*, **30**, doi:10.1029/2003GL017324.

**Stott**, P.A. and S.F.B. Tett, 1998: Scale-dependent detection of climate change. *Journal of Climate*, **11**, 3282-3294.

**Stott**, P.A., D A. Stone, and M.R. Allen, 2004: Human contribution to the European heatwave of 2003. *Nature*, **423**, 61-614.

**Stott,** P.A., J.F.B. Mitchell, M.R. Allen, T.L. Delworth, J.M. Gregory, G.A. Meehl, and B.D. Santer, 2006: Observational constraints on past attributable warming and predictions of future global warming. *Journal of Climate* (in press).

**Stott**, P.A., S.F.B. Tett, G.S. Jones, M.R. Allen, J.F.B. Mitchell, and G.J. Jenkins, 2000: External control of 20th century temperature by natural and anthropogenic forcings. *Science*, **290**, 2133-2137.

**Sun**, S. and J.E. Hansen, 2003: Climate simulations for 1951-2050 with a coupled atmosphere-ocean model. *Journal of Geophysical Research, Atmospheres*, **16**, 2807-2826.

**Tett**, S.F.B. and P.W. Thorne, 2004: Comment on "Tropospheric temperature series from satellites." *Nature*, **432**, doi:10.1038/nature03208.

**Tett**, S.F.B., G.S. Jones, P.A. Stott, D.C. Hill, J.F.B. Mitchell, M.R. Allen, W.J. Ingram, T.C. Johns, C.E. Johnson, A. Jones, D.L. Roberts, D.M.H. Sexton, and M.J. Woodage, 2002: Estimation of natural and anthropogenic contributions to twentieth century temperature change. *Journal of Geophysical Research, Atmospheres*, **107**, doi:10.1029/2000JD000028.

**Tett**, S.F.B., P.A. Stott, M.R. Allen, W.J. Ingram, and J.F.B. Mitchell, 1999: Causes of twentieth-century temperature change near the Earth's surface. *Nature*, **399**, 569-572.

**Tett**, S.F.B., J.F.B. Mitchell, D.E. Parker, and M.R. Allen, 1996: Human influence on the atmospheric vertical temperature structure: detection and observations. *Science*, **274**, 1170-1173.

**Thorne**, P.W., D.E. Parker, S.F.B. Tett, P.D. Jones, M. McCarthy, H. Coleman, and P. Brohan, 2005: Revisiting radiosonde upper-air temperatures from 1958 to 2002. *Journal of Geophysical Research*, **110**, D18105, doi:10.1029/2004JD00575.

**Thorne**, P.W., P.D. Jones, S.F.B. Tett, M.R. Allen, D.E. Parker, P.A. Stott, G.S. Jones, T.J. Osborn, and T.D. Davies, 2003: Probable causes of late twentieth century tropospheric temperature trends. *Climate Dynamics*, **21**, 573-591.

**Thorne**, P.W., P.D. Jones, T.J. Osborn, T.D. Davies, S.F.B. Tett, D.E. Parker, P.A. Stott, G.S. Jones, and M.R. Allen, 2002: Assessing the robustness of zonal mean climate change detection. *Geophysical Research Letters*, **29**, doi: 10.1029/2002GL015717.

**Tie**, X.X., S. Madronich, S. Walters, D.P. Edwards, P. Ginoux, N. Mahowald, R.Y. Zhang, C. Lou, and G. Brasseur, 2005: Assessment of the global impact of aerosols on tropospheric oxidants. *Journal of Geophysical Research, Atmospheres*, **110**, 03204, doi:10.1029/2004JD005359.

**Trenberth**, K.E., 1992: *Climate System Modeling*. Cambridge University Press, Cambridge, UK, 788 pp.

**Trenberth**, K.E. and L. Smith, 2005: The mass of the atmosphere: a constraint on global analyses. *Journal of Climate*, **18**, 860-875.

**Trenberth**, K.E. and T.J. Hoar, 1996: The 1990-1995 El Niño-southern oscillation event: longest on record. *Geophysical Research Letters*, **23**, 57-60.

**Trenberth**, K.E., J. Fasullo, and L. Smith, 2005: Trends and variability in column-integrated atmospheric water vapor. *Climate Dynamics*, **24**, 7-8, doi:10.1007/s00382-005-0017-4.

**Wallace**, J.M., Y. Zhang, and J.A. Renwick, 1995: Dynamic contribution to hemispheric mean temperature trends. *Science*, **270**, 780-783.

**Washington**, W M , J.W. Weatherly, G.A. Meehl, A.J. Semtner, T.W. Bettge, A.P. Craig, W.G. Strand, J. Arblaster, V.B. Wayland, R. James, and Y. Zhang, 2000: Parallel Climate Model (PCM) control and transient simulations. *Climate Dynamics*, **16**, 755-774.

**Weber**, G.R., 1996: Human effect on global climate? *Nature*, **384**, 524-525.

**Wehner**, M.F , 2000: A method to aid in the determination of the sampling size of AGCM ensemble simulations. *Climate Dynamics*, **16**, 321-331.

Wentz, F.J., and M. Schabel, 2000: Precise climate monitoring using complementary satellite data sets. *Nature*, **403**, 414-416.

Wielicki, B.A., T.M. Wong, R.P. Allan, A. Slingo, J.T. Kiehl, B.J. Soden, C.T. Gordon, A.J. Miller, S.K. Yang, D.A. Randall, F. Robertson, J. Susskind, and H. Jacobowitz, 2002: Evidence for large decadal variability in the tropical mean radiative energy budget. *Science*, **295**, 841-844.

Wigley, T.M.L., 2000: ENSO, volcanoes, and record-breaking temperatures. *Geophysical Research Letters*, **27**, 4101-4104.

Wigley, T.M.L., C.M. Ammann, B.D. Santer, and K.E. Taylor, 2005b: Using the Mount Pinatubo volcanic eruption to determine climate sensitivity: comments on "Climate forcing by the volcanic eruption of Mount Pinatubo." *Geophysical Research Letters*, **32**, L20709, doi:10.1029/2005GL023312.

Wigley, T.M.L., C.M. Ammann, B.D. Santer, and S.C.B. Raper, 2005a: The effect of climate sensitivity on the response to volcanic forcing. *Journal of Geophysical Research, Atmospheres*, **110**, D09107, doi:10.1029/2004JD005557.

Willis, J.K., D. Roemmich, and B. Cornuelle, 2004: Interannual variability in upper-ocean heat content, temperature, and thermosteric expansion on global scales. *Journal of Geophysical Research*, **109**, C12036, doi:10.1029/2003JC002260.

Yulaeva, E. and J.M. Wallace, 1994: The signature of ENSO in global temperature and precipitation fields derived from the Microwave Sounding Unit. *Journal of Climate*, **7**, 1719-1736.

Zwiers, F.W. and X. Zhang, 2003: Towards regional-scale climate change detection. *Journal of Climate*, **16**, 793-797.

## CHAPTER 6 REFERENCES

Christy, J.R., and W.B. Norris, 2004: What may we conclude about tropospheric temperature trends? *Geophysical Research Letters*, **31**, L06211

CCSP, 2004: *Strategic Plan for the Climate Change Science Program Final Report, July 2003*. Available at http://www.climatescience.gov/Library/stratplan2003/final/default.htm

Davey, C.A. and R.A. Pielke, 2005: Microclimatic exposures of surface-based weather stations. *Bulletin of the American Meteorological Society*, **86**, 497-504.

Donlon, C J., 2005: *Proceedings of the Sixth Global Ocean Data Assimilation Experiment (GODAE) [and] High Resolution Sea Surface Temperature Pilot Project (GHRSST-PP) Science Team[s] Meeting, Meteorological Office, Exeter, May 16-20, 2005*, International GHRSST-PP Project Office, Hadley Centre, Exeter, UK, 212 pp.

Folland, C.K., N. Rayner, P. Frich, T. Basnett, D. Parker, and E.B. Horton, 2000: Uncertainties in climate data sets - a challenge for WMO. *WMO Bulletin*, **49**, 59-68.

GCOS, 2004: *Global Climate Observing System Implementation Plan for the Global Observing System for Climate in support of the UNFCCC*. GCOS-92, WMO-TD 1219, WMO, Geneva, 136 pp. Available at: http://www.wmo.int/web/gcos/gcoshome.html and on CD-ROM.

GEOSS, 2005: *Global Earth Observation System of Systems, GEOSS; 10-Year Implementation Plan Reference Document*. GEO1000R/ESA SP-1284, ESA Publications, Noordwijk, Netherlands. 209 pp. (Available as CD-ROM)

Houghton, J.T., Y. Ding, D. J. Griggs, M. Noguer, P.J. van der Linden, X. Dai, K. Maskell, and C.A. Johnson, (eds.), 2001: *Climate Change 2001 The Scientific Basis. Contribution of Working Group I to the Third Assessment Report of the Intergovernmental Panel on Climate Change*. Cambridge University Press, Cambridge, UK, and New York, NY, 881 pp.

NRC (National Research Council), 2000a: *Reconciling observations of global temperature change*. National Academy Press, 85 pp.

NRC (National Research Council), 2000b: *Improving atmospheric temperature monitoring capabilities*. (Letter Report from the Panel on Reconciling Temperature Observations), 18 pp.

NRC (National Research Council), 2003: *Understanding climate change feedbacks*. National Academy Press, Washington, DC, 166 pp.

NRC (National Research Council), 2004: *Climate data records from environmental satellites*. National Academy Press, Washington, DC, 150 pp.

NRC (National Research Council), 2005: *Radiative forcing of climate change expanding the concept and addressing uncertainties*. National Academy Press, Washington, DC. 224 pp.

Nash, J., R. Smout, T. Oakley, B. Pathack, and S. Kurnosenko, 2005: WMO Intercomparison of High Quality Radiosonde Systems, Vacoas, Mauritius, 2-25 February 2005, Final Report, WMO Commission on Instruments and Methods of Observation, 118 pp., www.wmo.ch/web/www/IMOP/reports/2003-2007/RSO-IC-2005_Final_Report.pdf

Rayner, N.A., P. Brohan, D.E. Parker, C.K. Folland, J.J. Kennedy, M. Vanicek, T. Ansell, and S.F.B. Tett, 2006: Improved analyses of changes and uncertainties in sea surface temperature measured *in situ* since the mid-nineteenth century: the HadSST2 data set. *Journal of Climate*, **19**, 446-469.

Sherwood, S.C., J. Lanzante, and C. Meyer, 2005: Radiosonde daytime biases and late 20th century warming. *Science*, **309**, 1556-1559.

Simmons, A.J., P.D. Jones, V. da Costa Bechtold, A.C.M. Beljaars, P.W. Kållberg, S. Saarinen, S.M. Uppala, P. Viterbo and N. Wedi, 2004: Comparison of trends and low-frequency variability in CRU, ERA-40 and NCEP/NCAR analyses of surface air temperature. *Journal of Geophysical Research*, **109**, D24115, doi:10.1029/2004JD006306

Thorne, P.W., D.E. Parker, J.R. Christy, and C.A. Mears, 2005: Uncertainties in climate trends: lessons from upper-air temperature records. *Bulletin of the American Meteorological Society*, 86, 1437-1442.

## APPENDIX A  REFERENCES

**Santer**, B.D., T.M.L. Wigley, J.S. Boyle, D.J. Gaffen, J.J. Hnilo, D. Nychka, D.E. Parker, and K.E. Taylor, 2000: Statistical significance of trends and trend differences in layer-average temperature time series. *Journal of Geophysical Research*, **105**, 7337–7356.

**Thorne**, P.W., D.E. Parker, J.R. Christy, and C.A. Mears, 2005: Uncertainties in climate trends: lessons from upper-air temperature records. *Bulletin of the American Meteorological Society*, **86**, 1437–1442.

**Lanzante**, J.R., 2005: A cautionary note on the use of error bars. *Journal of Climate*, **18**, 3699–3703.

## PHOTOGRAPHY CREDITS

**Cover/Title Page/Table of Contents:**
Image 2 (man in storm), also on all pages relating to Chapter 1, pages 15-28, WeatherStock Volume 1. Copyright 1993, Warren Faidley/WeatherStock
Images 3 through 7: Grant Goodge, STG, Asheville, NC
**Page 2:**
Image 1 (weather observation), Neal Sanders, Asheville, NC
**Page 28:**
Images 1 (sunset) and 3 (volcano), WeatherStock Volume 1B, Copyright 1993, Warren Faidley/WeatherStock
**Pages 29-46:**
Image 1 (Craggy Mountains of Western North Carolina), Grant Goodge, STG, Asheville, NC
**Pages 47-70:**
Image 1 (cumulus in top of haze layer - lower troposphere), Grant Goodge, STG, Asheville, NC
**Pages 71-88:**
Image 1 (Bryce Canyon, Utah), Grant Goodge, STG, Asheville, NC
**Pages 89-118:**
Image 1 (tropical cumulus over Honolulu, HI), Grant Goodge, STG, Asheville, NC
**Pages 119-128:**
Image 1 (sunset over Mt. Pisgah, NC), Grant Goodge, STG, Asheville, N.C.
**Pages 91, 103, 118:**
All images (glaciers): Benjamin Santer, DOE LLNL

# Contact Information

Global Change Research Information Office
c/o Climate Change Science Program Office
1717 Pennsylvania Avenue, NW
Suite 250
Washington, DC 20006
202-223-6262 (voice)
202-223-3065 (fax)

The Climate Change Science Program incorporates the U.S. Global Change Research Program and the Climate Change Research Initiative.

To obtain a copy of this document, place an order at the Global Change Research Information Office (GCRIO) web site: http://www.gcrio.org/orders.

---

# Climate Change Science Program and the Subcommittee on Global Change Research

**James R. Mahoney**
Department of Commerce
National Oceanic and Atmospheric Administration
Director, Climate Change Science Program
Chair, Subcommittee on Global Change Research

**Jack Kaye,** Vice Chair
National Aeronautics and Space Administration

**Margaret S. Leinen**, Vice Chair
National Science Foundation

**James Andrews**
Department of Defense

**Allen Dearry**
Department of Health and Human Services

**Jerry Elwood**
Department of Energy

**Mary Glackin**
National Oceanic and Atmospheric Administration

**William Hohenstein**
Department of Agriculture

**Linda Lawson**
Department of Transportation

**Patrick Leahy**
U.S. Geological Survey

**Patrick Neale**
Smithsonian Institution

**Jacqueline Schafer**
U.S. Agency for International Development

**Joel Scheraga**
Environmental Protection Agency

**Harlan Watson**
Department of State

## EXECUTIVE OFFICE AND OTHER LIAISONS

**Khary Cauthen**
Council on Environmental Quality

**Stephen Eule**
Department of Energy
Director, Climate Change Technology Program

**Teresa Fryberger**
Office of Science and Technology Policy

**Margaret R. McCalla**
Office of the Federal Coordinator for Meteorology

**Andrea Petro**
Office of Management and Budget

www.ingramcontent.com/pod-product-compliance
Lightning Source LLC
Chambersburg PA
CBHW081448170526
45166CB00008B/2353